Deep Learning-Based Action Recognition

Deep Learning-Based Action Recognition

Editor

Hyo Jong Lee

MDPI • Basel • Beijing • Wuhan • Barcelona • Belgrade • Manchester • Tokyo • Cluj • Tianjin

Editor
Hyo Jong Lee
Jeonbuk National University
Korea

Editorial Office
MDPI
St. Alban-Anlage 66
4052 Basel, Switzerland

This is a reprint of articles from the Special Issue published online in the open access journal *Applied Sciences* (ISSN 2076-3417) (available at: https://www.mdpi.com/journal/applsci/special_issues/deep_learning_recognition).

For citation purposes, cite each article independently as indicated on the article page online and as indicated below:

LastName, A.A.; LastName, B.B.; LastName, C.C. Article Title. *Journal Name* **Year**, *Volume Number*, Page Range.

ISBN 978-3-0365-5199-9 (Hbk)
ISBN 978-3-0365-5200-2 (PDF)

Contents

About the Editor

Hyo Jong Lee

Hyo Jong Lee received his B.S., M.S., and Ph.D. degrees in computer science from the University of Utah, USA, where he was involved in computer graphics, image processing, and parallel processing. He is currently Professor of the Division of Computer Science, Jeonbuk National University, Jeonju, South Korea. He is also the president of AI Tech Co., Ltd. and member of IEEE Computing Society and the Association for Computing Machinery. He shares authorship of over 120 papers. His research interests include image processing, artificial intelligence, medical imaging, parallel algorithms, and brain science.

applied sciences

MDPI

Editorial

Special Issue on Deep Learning-Based Action Recognition

Hyo Jong Lee

Division of Computer Science and Engineering, CAIIT, Jeonbuk National University, Jeonju 54896, Korea; hlee@jbnu.ac.kr

1. Introduction

Human action recognition (HAR) has gained popularity because of its various applications, such as human–object interaction [1], intelligent surveillance [2], virtual reality [3], and autonomous driving [4]. The demand for HAR applications as well as gesture and pose estimation is growing rapidly. In response to this growing demand, various methods to apply human action recognition have been introduced. Features from images or videos can be extracted by multiple descriptors, such as local binary pattern, scale-invariant feature transformation, histogram of oriented gradient, and histogram of optic flow identifying action types. Recently, deep learning networks have been deployed in many challenging areas, such as image classification and object detection. Action recognition is also an ideal area for the application of deep learning networks. One of the primary advantages of deep learning is its ability to automatically learn representative features from large-scale data. As long as sufficient data are available, action recognition coupled with a deep learning network can perform more efficiently than traditional image processing methods.

2. Scope of Action Recognition

Based on the above understanding, the research results of deep learning-based HAR were primarily interpreted. However, given the challenging nature of HAR, further research is needed to study it from various aspects.

The recognition of an object's posture must precede the action recognition. The pose estimation is usually based on a skeleton model, which consists of joint points and their connections. It is possible to predict specific action by estimating the pose of a person using the joints and skeletal information.

The common network of action recognition may be either a regular convolution neural network (CNN) or a graph CNN. Unlike the recognition of a pose estimation from a fixed time point, it is possible to increase the efficiency of action recognition by adding temporal information along with the spatial information of an object. In some cases, the subject of action recognition is one person, but when multiple people are apparent in the same scene, it is important to process the action recognition of all the people in the scene. Including temporal information of an object's movement is of great help in recognizing specific actions because it can detect movements each minute that cumulatively constitute specific actions. This technique can be used to target action recognition for when multiple people are present in a scene. If the static action recognition is provided with sufficient temporal data, it will be possible to use static action to analyze action captured from videos.

Gestures can convey intentions through various local movements of the arms or fingers in a confined space with a limited range of motions. Therefore, gesture recognition can be used as an important component of action recognition. Thus, this special issue has published research papers focused on gesture recognition.

3. Deep Learning-Based Action Recognition

Many researchers are interested in and conducting deep learning-based action recognition research. Approximately 25 papers have been submitted to this special issue, and

Citation: Lee, H.J. Special Issue on Deep Learning-Based Action Recognition. *Appl. Sci.* **2022**, *12*, 7834. https://doi.org/10.3390/app12157834

Received: 1 August 2022
Accepted: 3 August 2022
Published: 4 August 2022

Publisher's Note: MDPI stays neutral with regard to jurisdictional claims in published maps and institutional affiliations.

12 of them were accepted (i.e., 34.2% acceptance rate). This special issue mainly consists of training data, pose estimation of objects, action recognition and gesture recognition. Rey et al. [5] present an approach on how to solve a data shortage problem in deep learning, by extracting synthesized acceleration and gyro norms data from video for human activity recognition scenarios.

There are two papers focused on pose estimation—the first one by S. Kim and H. Lee introduces the Lightweight Stacked Hourglass Network [6], which expands the convolutional receptive field while also reducing the computational load and providing scale invariance. The second paper, authored by J. Wu and H. Lee [7], proposes a Partition Pose Representation, which integrates the instance of a person and their body joints based on joint offset. They also propose a Partitioned Center Pose Network, which can detect people and their body joints simultaneously, then group all body joints.

Four papers deal with action recognition directly by using convolutional networks. The first paper, authored by Dong et al. [8], introduces high-order spatial and temporal features of skeleton data, such as velocity, acceleration, and relative distance, to construct graph convolutional networks. The other three papers adapt the spatio-temporal concept to extract better features. Tasnim et al. [9] suggest a spatio-temporal image formation technique of 3D skeleton joints by capturing spatial information and temporal changes for action discrimination. J. Kim and J. Cho [10] proposes a low-cost embedded model to extract spatial feature maps by applying CNN to the images that develop the video and using the frame change rate of sequential images as temporal information. The low complexity was achieved by transforming the weighted spatial feature maps into spatio-temporal features, and then inputting the spatio-temporal features into multilayer perceptrons. K. Hu et al. [11] propose an improved Long Short-Term Memory (LSTM) network, which is able to extract time information. They enhanced the input differential feature module and spatial memory state differential module to enhance features of actions. A. Stergiou et al. [12] introduce the concept of class regularization, which regularizes feature map activations based on the classes of the examples used. The proposed method essentially amplifies or suppresses activations based on an educated guess of the given class.

There are four papers focused on gesture recognition. Gesture recognition generally consists of a series of continuous actions, so it is necessary to memorize past actions. Four papers independently propose a unique method for gesture recognition. N. Nguyen et al. [13] present a dynamic gesture recognition approach using multi-features extracted from RGB frame and 3D skeleton joint information. N. Do et al. [14] exploit depth and skeletal data for the dynamic hand gesture recognition problem. The paper also explores a multi-level feature LSTM with pyramid and the LSTM block, which deal with the diversity of hand features. Y. Chu et al. [15] present a neural network for sensor-based hand gesture recognition, which is extended from the PairNet. N. Nguyen et al. [16] present another dynamic hand gesture recognition approach with two modules: gesture spotting and gesture classification, which uses bidirectional LSTM and a single LSTM, respectively.

4. Future Action Recognition

Traditionally, action recognition has been performed directly from videos or images in a single layered manner. The spatio-temporal features are extracted as 2D feature descriptors. Classes of action recognition are rather simple, such as walking, jumping or raising a hand. However, as computing power improves and deep learning techniques are naturally applied to action recognition, many researchers are optimistic about the potential of action recognition. Every day, new data on human actions are being accumulated and learning skills are improving. The need for various applications related to action recognition is also rapidly increasing. Therefore, recognition is attempted by extracting 3D feature values for each intrinsic action. Various modifications of deep networks to reduce the complexity of computations are also being attempted. Ultimately, a deep learning method that can recognize complex actions occurring in the real world is expected to be developed in the future.

Funding: This research was supported by Basic Science Research Program through the National Research Foundation of Korea (NRF), funded by the Ministry of Education (GR2019R1D1A3A03103736) and in part by project for Joint Demand Technology R&D of Regional SMEs funded by Korean Ministry of SMEs and Startups in 2021. (S3035805).

Institutional Review Board Statement: Not applicable.

Informed Consent Statement: Not applicable.

Data Availability Statement: Not applicable.

Conflicts of Interest: The author declares no conflict of interest.

References

1. Fortes Rey, V.; Garewal, K.K.; Lukowicz, P. Translating Videos into Synthetic Training Data for Wearable Sensor-Based Activity Recognition Systems Using Residual Deep Convolutional Networks. *Appl. Sci.* **2021**, *11*, 3094. [CrossRef]
2. Dawar, N.; Kehtarnavaz, N. Continuous detection and recognition of actions of interest among actions of non-interest using a depth camera. In Proceedings of the IEEE International Conference on Image Processing, Beijing, China, 17–20 September 2017.
3. Hu, K.; Zheng, F.; Weng, L.; Ding, Y.; Jin, J. Action Recognition Algorithm of Spatio–Temporal Differential LSTM Based on Feature Enhancement. *Appl. Sci.* **2021**, *11*, 7876. [CrossRef]
4. Wei, H.; Laszewski, M.; Kehtarnavaz, N. Deep Learning-Based Person Detection and Classification for Far Field Video Surveillance. In Proceedings of the 13th IEEE Dallas Circuits and Systems Conference, Dallas, TX, USA, 2–12 November 2018.
5. Chu, Y.-C.; Jhang, Y.-J.; Tai, T.-M.; Hwang, W.-J. Recognition of Hand Gesture Sequences by Accelerometers and Gyroscopes. *Appl. Sci.* **2020**, *10*, 6507. [CrossRef]
6. Fangbemi, A.; Liu, B.; Yu, N.; Zhang, Y. Efficient Human Action Recognition Interface for Augmented and Virtual Realty Applications Based on Binary Descriptor. In Proceedings of the 5th International Conference, AVR 2018, Ontranto, Italy, 24–27 June 2018.
7. Wu, J.; Lee, H.-J. A New Multi-Person Pose Estimation Method Using the Partitioned CenterPose Network. *Appl. Sci.* **2021**, *11*, 4241. [CrossRef]
8. Kim, S.-T.; Lee, H.J. Lightweight Stacked Hourglass Network for Human Pose Estimation. *Appl. Sci.* **2020**, *10*, 6497. [CrossRef]
9. Tasnim, N.; Islam, M.K.; Baek, J.-H. Deep Learning Based Human Activity Recognition Using Spatio-Temporal Image Formation of Skeleton Joints. *Appl. Sci.* **2021**, *11*, 2675. [CrossRef]
10. Chen, L.; Ma, N.; Wang, P.; Li, J.; Wang, P.; Pang, G.; Shi, X. Survey of pedestrian action recognition techniques for autonomous driving. *Tsinghua Sci. Technol.* **2020**, *25*, 458–470. [CrossRef]
11. Nguyen, N.-H.; Phan, T.-D.-T.; Kim, S.-H.; Yang, H.-J.; Lee, G.-S. 3D Skeletal Joints-Based Hand Gesture Spotting and Classification. *Appl. Sci.* **2021**, *11*, 4689. [CrossRef]
12. Kim, J.; Cho, J. Low-Cost Embedded System Using Convolutional Neural Networks-Based Spatiotemporal Feature Map for Real-Time Human Action Recognition. *Appl. Sci.* **2021**, *11*, 4940. [CrossRef]
13. Dong, J.; Gao, Y.; Lee, H.J.; Zhou, H.; Yao, Y.; Fang, Z.; Huang, B. Action Recognition Based on the Fusion of Graph Convolutional Networks with High Order Features. *Appl. Sci.* **2020**, *10*, 1482. [CrossRef]
14. Nguyen, N.-H.; Phan, T.-D.-T.; Lee, G.-S.; Kim, S.-H.; Yang, H.-J. Gesture Recognition Based on 3D Human Pose Estimation and Body Part Segmentation for RGB Data Input. *Appl. Sci.* **2020**, *10*, 6188. [CrossRef]
15. Stergiou, A.; Poppe, R.; Veltkamp, R.C. Learning Class-Specific Features with Class Regularization for Videos. *Appl. Sci.* **2020**, *10*, 6241. [CrossRef]
16. Do, N.-T.; Kim, S.-H.; Yang, H.-J.; Lee, G.-S. Robust Hand Shape Features for Dynamic Hand Gesture Recognition Using Multi-Level Feature LSTM. *Appl. Sci.* **2020**, *10*, 6293. [CrossRef]

applied sciences

MDPI

Article

A New Multi-Person Pose Estimation Method Using the Partitioned CenterPose Network

Jiahua Wu and Hyo-Jong Lee *

Division of Computer Science and Engineering, Jeonbuk National University, Jeonju 54896, Korea; salmon2wu@gmail.com
* Correspondence: hlee@jbnu.ac.kr

Abstract: In bottom-up multi-person pose estimation, grouping joint candidates into the appropriately structured corresponding instance of a person is challenging. In this paper, a new bottom-up method, the Partitioned CenterPose (PCP) Network, is proposed to better cluster the detected joints. To achieve this goal, we propose a novel approach called Partition Pose Representation (PPR) which integrates the instance of a person and its body joints based on joint offset. PPR leverages information about the center of the human body and the offsets between that center point and the positions of the body's joints to encode human poses accurately. To enhance the relationships between body joints, we divide the human body into five parts, and then, we generate a sub-PPR for each part. Based on this PPR, the PCP Network can detect people and their body joints simultaneously, then group all body joints according to joint offset. Moreover, an improved l_1 loss is designed to more accurately measure joint offset. Using the COCO keypoints and CrowdPose datasets for testing, it was found that the performance of the proposed method is on par with that of existing state-of-the-art bottom-up methods in terms of accuracy and speed.

Keywords: multi-person pose estimation; partitioned centerpose network; partition pose representation

Citation: Wu, J.; Lee, H.J. A New Multi-Person Pose Estimation Method Using the Partitioned CenterPose Network. *Appl. Sci.* **2021**, *11*, 4241. https://doi.org/10.3390/app11094241

Academic Editors: Antonio Fernández-Caballero

Received: 8 March 2021
Accepted: 5 May 2021
Published: 7 May 2021

1. Introduction

Driven by extensive research efforts, significant progress has been made in human pose estimation. The goal of human pose estimation is to obtain the posture of a human body from monocular images or videos. Pose estimation is a fundamental computer vision task providing vital information for many applications such as action detection and recognition [1], human tracking [2], and medical assistance among others [3].

With the rapid progress in deep learning technology, human pose estimation performance has improved greatly over recent years. However, finding a balance between efficiency and accuracy remains challenging. Multi-person pose estimation methods are generally classified based on their starting point for prediction as either top-down or bottom-up [4]. Top-down methods [5–11] first identify and localize instances of people using an existing person detector system and then conduct pose estimation for each person individually. Generally, top-down methods are effective since these methods profit from advances in person detectors. However, the computational cost of such methods linearly increases with the number of people in an image because single-person pose estimation must be carried out repeatably, in sequence, for each person in the image, as such, such methods are usually too slow to achieve real-time detection.

In contrast, bottom-up strategies [12–16] first identify all the body joints in the entire image, then these joints are grouped into corresponding instances of people. Unlike top-down methods, bottom-up methods avoid higher joint detection and are more robust as the number of people in an image increase. In many cases, performance when clustering the joint candidates determines the final accuracy of detection. Cao et al. [12] proposed the use of Part Affinity Fields (PAFs) to encode the coordinates and angles of limbs to assist in grouping joints into different people; this approach ignores the relationship between each body joint and instance of a person. Newell et al. [13] constructed associative embedding maps to tag each joint on the corresponding person pose. This method adds a link between each body joint and the corresponding instance of a person, however, it neglects information relevant to adjacent body joints. Consequently, it is difficult to simultaneously maintain relationships between different joints in a single limb and link each joint from the corresponding instance of a person.

To overcome this issue, we first propose a novel pose representation technique, termed Partition Pose Representation (PPR), which combines the position information from instances of people and their body joints. Inspired by [17], we first represent each instance of a person with a single point at the center of their bounding box. Then, the positions of body joints are encoded by their offset from the center point, as shown in Figure 1b. In this way, the relationship between adjacent body joints is severed. To maintain some correlation between adjacent body joints, we further divide the human body into five parts: the head, left arm, right arm, left leg, and right leg, we then extend PPR to sub-PPR for each part. The respective center points of each part are the nose, left elbow, right elbow, left knee, and right knee. With the addition of sub-PPR, human poses generate stable connections with their instance of a person, as shown in Figure 1c.

(**a**) (**b**) (**c**)

Figure 1. Different pose representations captured from image (**a**). (**b**) Traditional pose representation, in which each joint is represented by absolute coordinates. (**c**) Proposed partition pose representation.

To exploit the advantages of PPR, we introduce a new bottom-up model, the Partitioned CenterPose (PCP) Network, to identify the poses of multiple people. The PCP Network can simultaneously locate the position of an instance of a person and identify all joint candidates. Meanwhile, a parallel prediction branch in the PCP Network, called the offset prediction head, builds an associative embedding map to predict the offset for each body center. Here we introduce an improved l_1 loss to obtain more accurate joint offset values. Supported by PPR, the joint candidates can be assigned to the corresponding body center using the offset as a guide.

Experiments on the MS COCO and CrowdPose datasets demonstrate the efficiency and effectiveness of the proposed method. It achieves competitive performance and superior speed versus state-of-the-art methods. Our work makes three main contributions.

(1) We propose a novel partition pose representation method to construct a relationship between body joints and the body center, while preserving correlations between adjacent body joints.
(2) We propose a new bottom-up model with an improved l_1 loss to efficiently and robustly predict and partition body joints to multiple people.
(3) In experiments, our PCP Network is competitive with state-of-the-art methods using the MS COCO and CrowdPose datasets while achieving a higher inference speed.

2. Related Work

2.1. Multi-Person Pose Estimation

Multi-person pose estimation is a comprehensive task that combines the challenges of person detection and keypoint estimation. With the incredible advancements over recent years in object detection and single-person pose estimation methods [4,5,8,18–24], the performance of multi-person pose estimation has also improved, getting good results even on some complex datasets. Based on how calculations for a particular method are started, multi-person pose estimation methods are often divided into top-down methods and bottom-up methods.

Top-down methods. Top-down approaches typically first use an object detector to obtain an instance of a person and then independently estimate the pose for each person identified. G-RMI [6] produces a heatmap and offset map for each joint before combining this information using an aggregation procedure. RMPE [11] introduced using a parametric pose NMS for refining pose candidates. He et al. [5] proposed an extension of the Mask R-CNN framework that synchronously predicts keypoints and human masks. In these top-down methods, predicting keypoint heatmaps is made easier by restricting the search to the detected person's bounding box. However, the top-down strategy incurs extra computational costs while initially detecting each person's bounding box.

Bottom-up methods. Bottom-up approaches first detect body joints and then assign these joints to individuals. With the increasing demands to carry out image processing tasks on mobile devices, finding appropriate lightweight methods has become a new research hotspot. Motivated by bottom-up approaches being faster and more capable of achieving real-time estimation, our approach is based on previous bottom-up approaches and aims to obtain better performance while maintaining high computational efficiency.

Existing bottom-up methods mainly focus on how to associate detected keypoints with the corresponding instance of a person. The PersonLab approach [14] introduced a greedy decoding scheme together with Hough voting to determine grouping. CMU-Pose [12] proposed Part Affinity Fields (PAFs) to encode the location and orientation of limbs, this work was further developed in the PifPaf technique [15]. However, the computational efficiency of these two-stage methods is limited by the quality of the greedy algorithm. Newell et al. [13] propose a one-stage method to detect joints and group them in one pipeline. Based on this one-stage strategy and HRNet [8], Cheng et al. [16] presented a Scale-Aware High-Resolution Network (HigherHRNet) to solve the scale variation challenge. However, existing research only focuses on the features of joints (like in PifPaf), or only uses the connection between joints and an instance of a person to cluster (like in AssocEmbedding and HigherHRNet). The novelty of our method is to use Partition Pose Representation (PPR) to combine position information from instances of a person with structure information about body joints. In PPR, we utilize tailored semantic information and information on the offset of joints from the body center to replace information from tags in associative embedding maps. Moreover, we divide the human body into five parts, define the pivot joint in these parts as the part's center. Assisted by these part centers, the relationships between different joints in a single limb become enhanced by the offset of the body joint to the part center.

2.2. Backbone Network

The backbone networks of multi-person pose estimation methods are designed to extract keypoint features and instances of people; the accuracy with which they do so largely determines the quality of the prediction results. To ensure the effectiveness of the proposed method, three different backbones architectures, Hourglass [4], Deep Layer Aggregation (DLA) [25], and HRNet [8], are comprehensively considered.

Hourglass: The stacked Hourglass Network [4] consists of overlapping residual blocks [26], each of which is linked by a skip connection to effectively process and consolidate multi-scale features. With an encoder–decoder architecture and an intermediate supervision process, the Hourglass network shows robust performance in some complex environments, such as in cases with occlusion or cases where similar parts from nearby people are present [27]. The size of this network is quite large, which results in graceful keypoint estimation performance. The structure of an hourglass module is illustrated in Figure 2a.

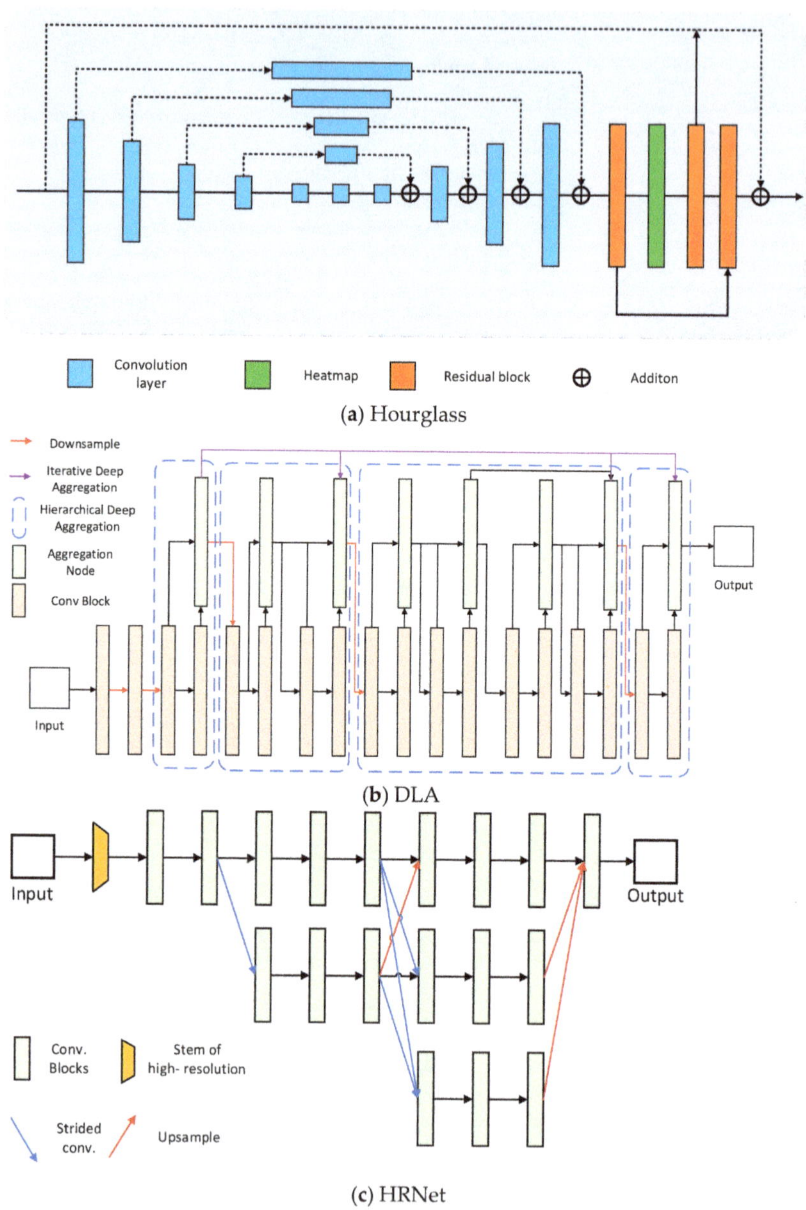

Figure 2. Structures of three state-of-the-art backbone networks for human pose estimation.

DLA: DLA [25] is an image classification network with hierarchical skip connections, in which aggregation is defined as the combination of different layers throughout a network. DLA uses iterative deep aggregation to symmetrically increase feature map resolution, preventing loss of information in dense predictions. Moreover, DLA hierarchically merges features to create networks with better accuracy and fewer parameters. The structure of a DLA network is illustrated in Figure 2b.

HRNet: HRNet [8] aims to maintain high-resolution features throughout the entire network. This network can be divided into parallel multi-resolution convolutions and repeated multi-resolution fusions. High- to low-resolution convolution streams generate multi-scale feature maps in parallel. The goal of the fusion module is to merge information across multi-resolution representations. The structure of HRNet with three parallel branches is illustrated in Figure 2c.

3. Partition Pose Representation

In this section, we describe the proposed PPR in detail. Unlike traditional grouping methods, PPR is committed to generating connections between each body joint and instance of a person while simultaneously strengthening the correlations between different body joints. Let $I \in R^{W \times H \times 3}$ denote an input image of width W and height H and $p^k = \left\{p_1^k, p_2^k, \ldots, p_N^k\right\}$ denote N joint candidates from the kth persons in I. $\left(x_n^k, y_n^k\right)$ is the spatial coordinate of p^k, and $\left(x_{lt}^k, y_{lt}^k, x_{rb}^k, y_{rb}^k\right)$ is the bounding box of the kth instance of a person. Inspired by CenterNet [17], the body center is denoted by $\left(\hat{x}_0^k, \hat{y}_0^k\right) = \left(x_{lt}^k + x_{rb}^k, y_{lt}^k + y_{rb}^k\right)/2$.

PPR aims to aggregate the instance of a person and body pose with an offset to the body center. So, the coordinates of the nth joint of person k can be defined as:

$$\left(x_n^k, y_n^k\right) = \left(\hat{x}_0^k + \delta x_n^k, \hat{y}_0^k + \delta y_n^k\right) \tag{1}$$

where $\left(\delta x_n^k, \delta y_n^k\right)$ is the offset of the nth joint to the body center.

However, Equation (1) only considers unification of an instance of a person and body pose; it ignores the relationship between adjacent joints. Using additional information from correlated joints, the offset vector can be more accurately mapped to the position of the pose by the prediction model. Naturally, PPR divides the human body into five parts: (1) head, including nose, left eye, right eye, left ear, and right ear; (2) left arm, including left shoulder, left elbow, and left wrist; (3) right arm, including right shoulder, right elbow, and right wrist; (4) left leg, including left hip, left knee, and left ankle; and (5) right leg, including right hip, right knee, and right ankle. Then, we use the same approach as used in Equation (1) to represent the joints in each part. Here, the center points of each part p_c^k are no longer the body center, but the nose, left elbow, right elbow, left knee, and right knee are taken as the centers of the five respective body parts. Some complex environments may mean a part center is not visible; this will affect encoding by PPR. In this situation, we calculate the center of the remaining joints in this part to replace the part center; we call this point the illusion center. Thus, the complete PPR can be formulated as:

$$\left(x_n^k, y_n^k\right) = \begin{cases} \left(\hat{x}_0^k + \delta x_n^k, \hat{y}_0^k + \delta y_n^k\right) \text{if} p_n^k \in p_c^k \\ \left(\hat{x}_m^k + \delta \hat{x}_n^k, \hat{y}_m^k + \delta \hat{y}_n^k\right) \text{otherwise} \end{cases} \tag{2}$$

when the part center is visible, $\left(\hat{x}_m^k, \hat{y}_m^k\right)$ is the coordinates of the center point of the mth part and $\left(\delta \hat{x}_n^k, \delta \hat{y}_n^k\right)$ is the offset of the nth joint from the corresponding part center. When the part center is not visible, $\left(\hat{x}_m^k, \hat{y}_m^k\right)$ is the coordinate of the illusion center of the mth part and $\left(\delta \hat{x}_n^k, \delta \hat{y}_n^k\right)$ is the offset of the nth joint from the corresponding illusion center.

Using the offset from the part center to the body center, PPR establishes the connection between a body pose and the instance of a person. At the same time, PPR retains global information related to the limbs and generates correlations between body joints in one part through the offset of other joints to the part center.

4. Partitioned CenterPose Network

In conjunction with PPR, we propose the box-free bottom-up PCP Network to detect body joints of multiple people. Motivated by the recent success of keypoint-based object detection approaches [17,28], we implement the PCP Network with a simple one-stage model. Below, we will describe the network architecture, training, and inference details of the PCP Network. The overall pipeline for the proposed network is shown in Figure 3.

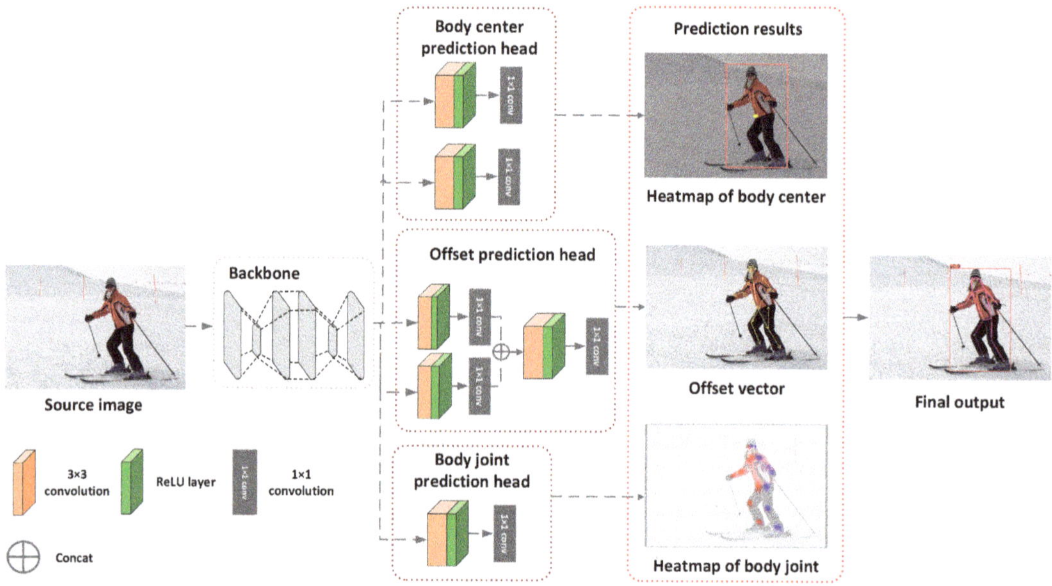

Figure 3. The architecture of the Partitioned CenterPose Network. A convolutional backbone network applies three sets of prediction heads to predict instance location, joint offset, and joint heatmap. The final output is generated by combining these three prediction results.

4.1. Network Architecture

In the PCP Network, a convolutional backbone network is applied for feature extraction. Then, we use three sets of prediction heads (body center prediction head, offset prediction head, and body joint prediction head) to process the output features. First, we will discuss the structure of the offset prediction head. In PPR, the offset vector is the key to connecting an instance of a person with their body joints; as such, it is very important to obtain an accurate offset vector. Directly regressing the value of an offset vector is inefficient as it is a highly non-linear task and difficult to learn the mapping [3]. Inspired by [13], we use two associative embedding maps to record the vector value of each offset. As shown in Figure 3, the output of the backbone is passed through two parallel branches. The output channel of the first branch is twice the number of part centers, which focus on the 2D vector value of the offset from the body center to the part centers. The second branch looks at the offset of the remaining joints to the part center. Then, we concatenate the output of these two branches and pass it through a simple convolutional module to acquire the final embedding maps. When the coordinates of the body center or part center are obtained, the feature value of the embedding map at this position can be regarded as the corresponding offset vector value. In the body center prediction head, follow the approach used by CenterNet [17], we use a simple convolutional module, which contains only a separate 3×3 convolution, ReLU, and a 1×1 convolution, to predict the body center and the bounding box using two parallel branches. The body joint prediction head estimates a heatmap of each body joint $\left(x_n^k, y_n^k\right)$ using the same structure as used for the body center prediction head to reduce computational complexity.

4.2. Training and Inference

Training. An improved l_1 loss was designed for the PCP Network to better train the system to identify the offset between the joint and part center. As shown in Figure 4, the lengths of the offset vectors in the head part are short, but the structures of the offset vectors in different people are relatively similar. Thus, enhancing the weight of offset length in the loss function allows the network to understand small differences in head structure more accurately. Conversely, the offset vectors of the limbs of different people differ more in terms of angle while the lengths tend to be quite similar. Accordingly, based on the l_1 loss, we designed two different loss functions for the offset vector in the head and in the limbs:

$$L_{off}^{head} = \frac{1}{N}\sum_{i}^{N}\left(||\vec{O}_i - \vec{O'}_i||_1 + 0.5||\vec{O}_i - \vec{O'}_i||_2\right) \tag{3}$$

$$()()L_{off}^{limb} = \frac{1}{N}\sum_{i}^{N}\left(||\vec{O}_i - \vec{O'}_i||_1 + \left|arctan\frac{\vec{O}_i}{||\vec{O}_i||_2} - arctan\frac{\vec{O'}_i}{||\vec{O'}_i||_2}\right|\right) \tag{4}$$

where $\vec{O}$ is the predicted offset vector and $\vec{O'}$ is the corresponding ground truth. N is the number of body joints in the body part. $|\cdot|$ is the absolute value, and $||\cdot||_1$ and $||\cdot||_2$ are the l_1-norm and l_2-norm, respectively. In Section 5.4, we discuss an ablation experiment to demonstrate the effect of the improved l_1 loss.

The total loss of the improved l_1 loss is shown below:

$$L = L_{bct} + \alpha L_{bsize} + L_{off}^{pct} + L_{off}^{pj} + L_{bj} \tag{5}$$

$$L_{off}^{pct} = L_{off}^{limb} \tag{6}$$

$$L_{off}^{pj} = \left(L_{off}^{head} + 4 * L_{off}^{limb}\right)/5 \tag{7}$$

where L_{bct} and L_{bj} denote the focal losses [29], which are used to train the network to detect the body center and body joint heatmaps, respectively. The focal loss is defined as:

$$L_{focal} = \frac{-1}{N}\sum_{n}\begin{cases}(1-\hat{H}_p)^{\beta}\log(\hat{H}_p) & \text{if } H_p = 1\\ (1-H_p)^{\gamma}(\hat{H}_p)^{\beta}\log(1-\hat{H}_p) & \text{otherwise}\end{cases} \tag{8}$$

where β and γ are hyper-parameters used to reduce the imbalance between an easy example and a hard example. H_p is the ground truth heatmap and $\hat{H}_p$ is the heatmap of p^k. Following [28], β is set to 2 and γ is set to 4. L_{bsize} is the l_1 loss [30] used to regress the size of the bounding box. L_{off}^{pct} is the loss function used to train the offset between the part center and body center, while L_{off}^{pj} is the loss function used to train the offset between the joint and part center. α is a constant weight parameter that is set to 0.1.

Inference. Following PPR, we group the detected keypoints by offset vector. Given a test image of width W and height H, the outputs of the PCP Network include a body center heatmap $H_{bc} \in R^{W\times H\times 1}$, bounding box maps $H_{bb} \in R^{W\times H\times 2}$, offset maps $H_{off} \in R^{W\times H\times 34}$, joint heatmaps $H_{bj} \in R^{W\times H\times 17}$. We first choose the top N_η high-confidence instances of people (100 was used in our implementation) and extract their body centers $\left(\hat{x}_0^k, \hat{y}_0^k\right)$ from H_{bc}. With the coordinates of body centers, the size of the bounding box $\left(w^k, h^k\right)$ and the offset of the part centers to the body center $\vec{O_n^k}$ can be extracted from $H_{bb}\left(\hat{x}_0^k, \hat{y}_0^k\right)$ and $H_{off}^{n,n+1}\left(\hat{x}_0^k, \hat{y}_0^k\right)$, respectively. For the kth body center, we extract the coordinates of part center candidates $\left(x_n^k, y_n^k\right)$ from the joint heatmaps H_{bj}, where the candidates are selected from inside the bounding box of the kth body center. Then, the offset $\vec{O}_{ht}$ from the part center candidates to the body center are calculated by:

$$\vec{O}_{ht} = \left(x_n^k - \hat{x}_0^k, y_n^k - \hat{y}_0^k\right) \tag{9}$$

Figure 4. PPR of the head (magnified area) and PPR of the limbs in different people. (**a**,**b**) are two samples from COCO dataset [31]. The lengths of the offset vectors in the head part are much shorter than those in the limb parts.

Next, each part center candidate is assigned by argmin $i \in N_c\left(\vec{O}_{ht} - \vec{O_n^k}\right)$ to identify the closest predicted offset vector $\vec{O_n^k}$. Here, N_c is the total number of part centers. After grouping the part centers, we can extract the offset of the remaining joints to the part center $\vec{O_m^k}$ from $H_{off}^{m,m+1}\left(x_n^k, y_n^k\right)$. If the part center is not visible, $\left(x_n^k, y_n^k\right)$ will be replaced by $\left(\hat{x}_0^k, \hat{y}_0^k\right) + \vec{O_n^k}$. Using the same strategy, we can group the remaining body joints to corresponding instances of a person. Finally, the complete human skeletons of multiple people are formed using the default connections between the predicted body joints.

The network structure of the prediction heads is simple and lightweight, the body centers are obtained directly from keypoint estimation without the need for IoU-based non-maxima suppression or other greedy algorithms. In the inference post-processing, due to the constraints of the bounding box, the number of joint candidates can be reduced greatly to only in the candidates in small areas of the image, this not only improves accuracy it also reduces computing time. Therefore, in our method, post-processing does not take too long while the computational efficiency is similar to one-stage methods.

5. Experiments

5.1. Dataset

The experiments were performed using the MS-COCO dataset [31]. This dataset contains more than 250,000 instances of people with 17 body joints, the dataset is divided into *train*, *val* and *test-dev* sets with 57 k, 5 k, and 20 k images, respectively. We use the *train* set for training and test the results on the *test-dev* set. The *val* set is used to perform ablation studies and visualization experiments.

The MS-COCO dataset uses Object Keypoint Similarity (OKS)-based AP (average precision) and AR (average recall) metrics to evaluate the performance of a detector. OKS is inspired by the IoU index in object detection, this calculates the distance between predicted body joints and the ground truth, normalized to the scale of the person [32]. OKS can be defined as:

$$OKS_p = \frac{\sum_i exp\left\{-d_{pi}^2/2S_p^2\sigma_i^2\right\}\delta(v_{pi}=1)}{\sum_i \delta(v_{pi}=1)} \tag{10}$$

where p denotes the pth person in an image and i is the ith keypoint of this person. d_{pi} is the Euclidian distance between the ground truth keypoint and predicted keypoint. S_p is the scale factor of the person, which is equal to the square root of the object segment area. σ_i is the normalization factor of the ith keypoint, which reflects the difficulty of labeling this keypoint. $v_{pi} = 1$ indicates that the ith keypoint of the pth person is visible.

In this section, we mainly use AP (mean AP score in OKS = 0.5, 0.55, ..., 0.90, 0.95), $AP^{0.5}$, $AP^{0.75}$, AP^M, AP^L, and AR as metrics, where 0.5 and 0.75 are the threshold values for OKS, M and L represent medium objects (32^2 < area < 96^2) and large objects (area > 96^2), respectively [33,34].

5.2. Experimental Setup

We experimented on using four backbones in our method: DLA-34 [25], ResNet-101 [26], Hourglass-104 [4], and HRNet-w32 [8]. All these models were written using PyTorch software [35]. The resolution of the input image was 512 × 512, leading to heatmaps with a size of 128 × 128. The ground-truth heatmap was constructed by applying a Gaussian kernel with the same parameters as used in [36] to filter all body joints. Each sample was augmented by rotating, scaling, and flipping. We utilized Adam [37] as the optimizer and trained the PCP Network on a RTX2080ti GPU. For the DLA-34 backbone, we trained with a batch size of 48 and a learning rate of 3×10^{-4} for 300 epochs; the learning rate was decreased by 0.1 in epochs 250 and 280. For the ResNet-101 backbone, we trained with a batch size of 48 and a learning rate of 1×10^{-3} for 300 epochs; the learning rate was decreased by 0.1 in epochs 250 and 280. For the Hourglass-104 backbone, we trained with a batch size of 24 and a learning rate of 2.5×10^{-4} for 150 epochs; the learning rate was decreased by 0.1 in epochs 110 and 130. For the HRNet-w32 backbone, we trained with a batch size of 32 and a learning rate of 2×10^{-4} for 320 epochs; the learning rate decreased by 0.1 in epochs 270 and 300.

5.3. Experimental Results

To assess the performance of our PCP Network, we compared the results of our method with those of six current mainstream bottom-up pose estimation methods, including CMU-Pose [12], Mask-RCNN [5], G-RMI [6], AssocEmbedding [13], PifPaf [15], PersonLab [14], and HigherHRNet [16]. Table 1 summarizes the experimental results on the test-dev dataset. The differences between HigherHRNet-1 and HigherHRNet-2 are the backbone and input size. As shown in Table 1, our method is slightly inferior to PersonLab and HigherHRNet-2, which both use a more powerful backbone and larger training images. However, when using the same backbone and same input size, the performance of our method is better than Mask-RCNN, G-RMI, AssocEmbedding, PifPaf, and HigherHRNet-1. In addition to performance, we also consider the inference time of each method.

As shown in Table 1, the speed of our PCP Network is outstanding, especially when DLA is used as the backbone. Even with the HRNet backbone, the inference speed of our PCP Network was 5× faster than that of PersonLab. These results verify that our method has superior efficiency due to its excellent inference speed while maintaining very competitive performance for multi-person pose estimation tasks.

Table 1. Comparisons of our model to other state-of-the-art models on the MSCOCO test-dev daTable 2080. ti GPU. Superscripts M, L of AP stand for medium and large objects. The highest values are indicated in bold.

Method	Backbone	Input Size	AP	$AP^{0.5}$	$AP^{0.75}$	AP^{M}	AP^{L}	AR	Time [s]
CMU-Pose [12]	-	-	0.618	0.849	0.675	0.571	0.682	0.665	0.5
Mask-RCNN [5]	ResNet-101	-	0.631	0.873	0.687	0.578	0.714	-	0.2
G-RMI [6]	ResNet-101	353	0.649	0.855	0.713	0.623	0.700	0.697	-
AssocEmbedding [13]	Hourglass	512	0.655	0.868	0.723	0.606	0.726	0.710	0.19
PifPaf [15]	-	-	0.667	-	-	0.624	0.729	-	-
PersonLab [14]	ResNet-152	1401	0.687	0.890	0.754	0.641	0.755	**0.754**	0.381
HigherHRNet-1 [16]	HRNet-W32	512	0.664	0.875	0.728	0.612	0.742	-	0.052
HigherHRNet-2 [16]	HRNet-W48	640	**0.705**	**0.893**	**0.772**	**0.666**	**0.758**	0.749	0.142
Ours (DLA)	DLA-34	512	0.634	0.864	0.693	0.575	0.739	0.698	**0.039**
Ours (ResNet)	ResNet-101	512	0.651	0.868	0.703	0.642	0.737	0.721	0.073
Ours (Hourglass)	Hourglass	512	0.663	0.881	0.731	0.662	0.747	0.748	0.132
Ours (HRNet)	HRNet-W32	512	0.668	0.883	0.740	0.665	0.748	0.751	0.078

To further prove that the performance of the proposed method is satisfactory, we also show some results from the proposed method that show intuitively that our approach is able to identify joints on a human skeleton accurately. Figure 5 shows qualitative examples from the MSCOCO dataset, including the intermediate body joint heatmaps and final predicted human poses. It is clear that our method performs well even on scenes with some challenging attributes such as sub-optimal scale, appearance variation, occlusion, or crowding.

Figure 5. Qualitative results on the MSCOCO dataset. For each pair, we show the predicted human pose (**left**) and intermediate heatmap (**right**). In the predicted human pose, each color corresponds to a particular human instance.

5.4. Ablation Analysis

We perform several ablation experiments on the COCO *val* set to better understand the gain of the proposed PPR and improved l_1 loss. Here, HRNet is used as the backbone of our network.

First, to demonstrate the effect of the proposed PPR, we trained the PCP Network with traditional pose representations (Figure 1b). Here, the body center prediction head was removed. As shown in Table 2, this network achieved an AP of 0.648. Using the proposed PPR, our PCP Network outperformed the above network by +0.12 AP (AP = 0.660). Table 3 shows the performance results from using the original l_1 loss and the improved l_1 loss. When the improved l_1 loss was used, the performance of our model increased from AP = 0.657 to 0.660. These results verify the effectiveness of the proposed PPR and improved l_1 loss. Table 3 also shows that the increase in AP for poses of large people is significantly higher than for other methods. This indicates that the improved l_1 loss works better on instances of large people.

Table 2. Ablation study results: traditional pose representation (TPR) vs. proposed PPR on the COCO2017 val dataset. Superscripts M, L of AP stand for medium and large objects. The highest values are indicated in bold.

Method	AP	$AP^{0.5}$	$AP^{0.75}$	AP^{M}	AP^{L}	AR
PCP Network (TPR)	0.648	0.854	0.715	0.603	0.700	0.697
PCP Network (PPR)	**0.660**	**0.869**	**0.725**	**0.608**	**0.742**	**0.704**

Table 3. Ablation study results: original l_1 loss vs. improved l_1 loss on the COCO2017 val dataset. Superscripts M, L of AP stand for medium and large objects. The highest values are indicated in bold.

Method	AP	$AP^{0.5}$	$AP^{0.75}$	AP^{M}	AP^{L}	AR
PCP Network (original loss)	0.657	0.867	0.722	0.607	0.728	0.701
PCP Network (improved loss)	**0.660**	**0.869**	**0.725**	**0.608**	**0.742**	**0.704**

5.5. CrowdPose

We demonstrated the proposed method has a state-of-the-art human pose estimation performance on the CrowdPose [38] dataset, which contained crowd scenes to make it more challenging. The training, validation, and testing subset contained 10K, 2K, and 8K images, respectively. The CrowdPose dataset also used the AP from the COCO dataset as an evaluation metric and split it into three crowding levels: easy, medium, hard. In this section, for metrics, we mainly use AP, $AP^{0.5}$, $AP^{0.75}$, AP^{E} (for easy images), AP^{M} (for medium images), and AP^{H} (for hard images). We trained the models on the training and validation subsets and reported the results achieved on the testing subset. The experimental setup follows that of COCO exactly.

The experimental results are shown in Table 4. Our method outperforms traditional top-down methods (Mask-RCNN and AlphaPose) and bottom-up method (CMU-Pose) by a large margin in terms of AP. SPPE is an efficient crowded scene pose estimation method which is a global refinement of AlphaPose; the performance of our method is comparable to AlphaPose without additional optimization. Multi-scale testing can improve the precision of predictions for small people, especially in crowd scenes. After multi-scale testing, HigherHRNet achieves the best performance on the CrowdPose dataset. While, without the optimization of multi-scale testing, the performance of our method is on par with HigherHRNet even the latter significant advantages in terms of the backbone used and the input size. The experimental results in Table 4 show the great potential of our method in complex environments and challenging scenes.

Table 4. Comparisons of our model to other state-of-the-art models on the CrowdPose test dataset. Superscripts E, M, H of AP stand for easy, medium and hard. * indicates multi-scale testing. The highest values are indicated in bold.

Method	Backbone	Input Size	AP	$AP^{0.5}$	$AP^{0.75}$	AP^{E}	AP^{M}	AP^{H}
				Top-down methods				
Mask-RCNN [5]	ResNet-101	-	0.572	0.835	0.603	0.694	0.579	0.458
AlphaPose [11]	-	-	0.610	0.813	0.660	0.712	0.614	0.511
SPPE [38]	ResNet-101	-	0.660	0.842	0.715	0.755	0.663	0.574
				Bottom-up methods				
CMU-Pose [12]	-	-	-	-	-	0.627	0.487	0.323
HigherHRNet [16]	HRNet-W48	640	0.659	0.864	0.706	0.733	0.665	0.579
HigherHRNet * [16]	HRNet-W48	640	**0.676**	**0.874**	**0.726**	**0.758**	**0.681**	**0.589**
Ours (HRNet)	HRNet-W32	512	0.657	0.855	0.705	0.742	0.668	0.574

6. Conclusions

In this paper, we proposed a new bottom-up multi-person pose estimation method which strikes a balance between efficiency and accuracy. The grouping of candidate joints into a corresponding pose in a limited amount of time is the main challenge in bottom-up multi-person pose estimation. To solve this problem, we first introduced Partition Pose Representation (PPR) for multi-person pose estimation. PPR builds relationships between each joint and the corresponding instance of a person using the offset between the joint and the body center. Moreover, PPR further divides the human body into five constituent parts and utilizes another offset to the center of these parts to rebuild relationships between adjacent joints. With PPR, it is possible to group candidate joints simply and quickly without the need for any additional complex algorithms.

To leverage the advantages of PPR, we proposed the Partitioned CenterPose (PCP) Network to estimate instances of people and their body joints, PCP then groups all body joints by joint offset. By considering the different characteristics of the offsets of joints on different parts of the human body, we proposed an improved l_1 loss to enhance the accuracy of the predicted joint offsets. Extensive experiments and subjective evaluation of predictions on the COCO and CrowdPose datasets demonstrate that our method performs well both in terms of efficiency and prediction accuracy. A future study that extends PPR to 3D human pose estimation is planned. Considering the complexity of human poses in 3D space, we must reconsider how we define the center of the human body and design different loss functions to obtain more accurate offsets.

Author Contributions: Conception and design of the proposed method: H.J.L. and J.W.; performance of the experiments: J.W.; writing of the paper: J.W.; paper review and editing: H.J.L. All authors have read and agreed to the published version of the manuscript.

Funding: This research was supported by the Basic Science Research Program through the NRF of Korea funded by the Ministry of Education (GR 2019R1D1A3A03103736). This research was also conducted by the 'Leaders in Industry-university Cooperation +', supported by the Ministry of Education and National Research Foundation of Korea

Institutional Review Board Statement: Not applicable.

Informed Consent Statement: Informed consent was obtained from all subjects involved in the study.

Data Availability Statement: Not applicable.

Conflicts of Interest: The authors declare no conflict of interest.

References

1. Dong, J.; Gao, Y.; Lee, H.J.; Zhou, H.; Yao, Y.; Fang, Z.; Huang, B. Action Recognition Based on the Fusion of Graph Convolutional Networks with High Order Features. *Appl. Sci.* **2020**, *10*, 1482. [CrossRef]
2. Insafutdinov, E.; Andriluka, M.; Pishchulin, L.; Tang, S.; Levinkov, E.; Andres, B.; Schiele, B. Arttrack: Articulated mul-ti-person tracking in the wild. In Proceedings of the 2017 IEEE Conference on Computer Vision and Pattern Recognition (CVPR), Honolulu, HI, USA, 21–26 July 2017; Institute of Electrical and Electronics Engineers (IEEE): Piscataway, NJ, USA, 2017; pp. 6457–6465.
3. Chen, Y.; Tian, Y.; He, M. Monocular human pose estimation: A survey of deep learning-based methods. *Comput. Vis. Image Underst.* **2020**, *192*, 102897. [CrossRef]
4. Newell, A.; Yang, K.; Deng, J. Stacked hourglass networks for human pose estimation. In Proceedings of the European Conference on Computer Vision, Amsterdam, The Netherlands, 11–14 October 2016; pp. 483–499.
5. He, K.; Gkioxari, G.; Dollar, P.; Girshick, R. Mask R-CNN. In Proceedings of the 2017 IEEE International Conference on Computer Vision (ICCV), Venice, Italy, 22–29 October 2017; pp. 2980–2988.
6. Papandreou, G.; Zhu, T.; Kanazawa, N.; Toshev, A.; Tompson, J.; Bregler, C.; Murphy, K. Towards Accurate Multi-Person Pose Estimation in the Wild. In Proceedings of the 2017 IEEE Conference on Computer Vision and Pattern Recognition (CVPR), Honolulu, HI, USA, 21–26 July 2017; pp. 3711–3719.
7. Xiao, B.; Wu, H.; Wei, Y. Simple Baselines for Human Pose Estimation and Tracking. In Proceedings of the European Conference on Computer Vision, GASTEIG Cultural Center, Munich, Germany, 10–13 September 2018; pp. 472–487.
8. Sun, K.; Xiao, B.; Liu, D.; Wang, J. Deep high-resolution representation learning for human pose estimation. In Proceedings of the IEEE Computer Society Conference on Computer Vision and Pattern Recognition, Long Beach, CA, USA, 16–20 June 2019; pp. 5686–5696.
9. Chen, Y.; Wang, Z.; Peng, Y.; Zhang, Z.; Yu, G.; Sun, J. Cascaded Pyramid Network for Multi-person Pose Estimation. In Proceedings of the 2018 IEEE Conference on Computer Vision and Pattern Recognition (CVPR), Salt Lake City, UT, USA, 18–22 June 2018; pp. 7103–7112.
10. Sun, X.; Xiao, B.; Wei, F.; Liang, S.; Wei, Y. Integral Human Pose Regression. In Proceedings of the European Conference on Computer Vision, GASTEIG Cultural Center, Munich, Germany, 10–13 September 2018; pp. 536–553.
11. Fang, H.-S.; Xie, S.; Tai, Y.-W.; Lu, C. RMPE: Regional Multi-person Pose Estimation. In Proceedings of the 2017 IEEE International Conference on Computer Vision (ICCV), Venice, Italy, 22–29 October 2017; pp. 2353–2362.
12. Cao, Z.; Šimon, T.; Wei, S.-E.; Sheikh, Y. Realtime multi-person 2d pose estimation using part affinity fields. In Proceedings of the 2017 IEEE Conference on Computer Vision and Pattern Recognition (CVPR), Honolulu, HI, USA, 21–26 July 2017; pp. 1302–1310.
13. Newell, A.; Huang, Z.; Deng, J. Associative embedding: End-to-end learning for joint detection and grouping. In Proceedings of the 31st International Conference on Neural Information Processing Systems (NIPS'17), Red Hook, NY, USA, 4–9 December 2017; pp. 2274–2284.
14. Papandreou, G.; Zhu, T.; Chen, L.-C.; Gidaris, S.; Tompson, J.; Murphy, K. PersonLab: Person Pose Estimation and Instance Segmentation with a Bottom-Up, Part-Based, Geometric Embedding Model. In Proceedings of the European Conference on Computer Vision, GASTEIG Cultural Center, Munich, Germany, 10–13 September 2018; pp. 282–299.
15. Kreiss, S.; Bertoni, L.; Alahi, A. PifPaf: Composite Fields for Human Pose Estimation. In Proceedings of the 2019 IEEE Conference on Computer Vision and Pattern Recognition (CVPR), Long Beach, CA, USA, 15–20 June 2019; pp. 11969–11978.
16. Cheng, B.; Xiao, B.; Wang, J.; Shi, H.; Huang, T.S.; Zhang, L. Higherhrnet: Scale-aware representation learning for bot-tom-up human pose estimation. In Proceedings of the International Conference on Computer Vision and Pattern Recogni-tion (CVPR), Seattle, WA, USA, 16–28 June 2020; pp. 5386–5395.
17. Zhou, X.; Wang, D.; Krähenbühl, P. Objects as points. *arXiv* **2019**, arXiv:1904.07850.
18. Tompson, J.; Goroshin, R.; Jain, A.; LeCun, Y.; Bregler, C. Efficient object localization using Convolutional Networks. In Proceedings of the 2015 IEEE Conference on Computer Vision and Pattern Recognition (CVPR), Boston, MA, USA, 7–12 June 2015; pp. 648–656.
19. Wei, S.-E.; Ramakrishna, V.; Kanade, T.; Sheikh, Y. Convolutional Pose Machines. In Proceedings of the 2016 IEEE Conference on Computer Vision and Pattern Recognition (CVPR), Las Vegas, NV, USA, 27–30 June 2016; pp. 4724–4732.
20. Bulat, A.; Tzimiropoulos, G. Human Pose Estimation via Convolutional Part Heatmap Regression. In Proceedings of the Haptics: Science, Technology, Applications, London, UK, 4–7 July 2016; Springer Science and Business Media LLC: Berlin/Heidelberg, Germany, 2016; Volume 9911, pp. 717–732.
21. Chu, X.; Yang, W.; Ouyang, W.; Ma, C.; Yuille, A.L.; Wang, X. Multi-context Attention for Human Pose Estimation. In Proceedings of the 2017 IEEE Conference on Computer Vision and Pattern Recognition (CVPR), Honolulu, HI, USA, 21–26 July 2017; pp. 5669–5678.
22. Lifshitz, I.; Fetaya, E.; Ullman, S. Human Pose Estimation Using Deep Consensus Voting. In *European Conference on Computer Vision*; Springer Science and Business Media LLC: Berlin/Heidelberg, Germany, 2016; Volume 9906, pp. 246–260.
23. Carreira, J.; Agrawal, P.; Fragkiadaki, K.; Malik, J. Human Pose Estimation with Iterative Error Feedback. In Proceedings of the 2016 IEEE Conference on Computer Vision and Pattern Recognition (CVPR), Las Vegas, NV, USA, 27–30 June 2016; pp. 4733–4742.
24. Hu, P.; Ramanan, D. Bottom-Up and Top-Down Reasoning with Hierarchical Rectified Gaussians. In Proceedings of the 2016 IEEE Conference on Computer Vision and Pattern Recognition (CVPR), Las Vegas, NV, USA, 27–30 June 2016; pp. 5600–5609.

25. Yu, F.; Wang, D.; Shelhamer, E.; Darrell, T. Deep Layer Aggregation. In Proceedings of the 2018 IEEE/CVF Conference on Computer Vision and Pattern Recognition, Salt Lake City, UT, USA, 18–22 June 2018; pp. 2403–2412.
26. He, K.; Zhang, X.; Ren, S.; Sun, J. Deep Residual Learning for Image Recognition. *arXiv* **2015**, arXiv:1512.03385.
27. Kim, S.-T.; Lee, H.J. Lightweight Stacked Hourglass Network for Human Pose Estimation. *Appl. Sci.* **2020**, *10*, 6497. [CrossRef]
28. Law, H.; Deng, J. Cornernet: Detecting objects as paired keypoints. In Proceedings of the European Conference on Computer Vision, GASTEIG Cultural Center, Munich, Germany, 10–13 September 2018; pp. 734–750.
29. Lin, T.-Y.; Goyal, P.; Girshick, R.B.; He, K.; Dollar, P. Focal Loss for Dense Object Detection. *IEEE Trans. Pattern Anal. Mach. Intell.* **2020**, *42*, 318–327. [CrossRef] [PubMed]
30. Girshick, R. Fast R-CNN. In Proceedings of the 2017 IEEE International Conference on Computer Vision (ICCV), Santiago, Chile, 13–16 December 2015; pp. 1440–1448.
31. Lin, T.Y.; Maire, M.; Belongie, S.; Hays, J.; Perona, P.; Ramanan, D.; Dollár, P.; Zitnick, C.L. Microsoft COCO: Common objects in context. In *Computer Vision—ECCV ECCV Lecture Notes in Computer Science*; Springer: Berlin/Heidelberg, Germany, 6–12 September 2014; pp. 740–755.
32. Li, Y.; Wang, X.; Liu, W.; Feng, B. Pose Anchor: A Single-stage Hand Keypoint Detection Network. *IEEE Trans. Circuits Syst. Video Technol.* **2019**, *30*, 1. [CrossRef]
33. Xia, H.; Zhang, T. Self-Attention Network for Human Pose Estimation. *Appl. Sci.* **2021**, *11*, 1826. [CrossRef]
34. Duan, K.; Bai, S.; Xie, L.; Qi, H.; Huang, Q.; Tian, Q. CenterNet: Keypoint Triplets for Object Detection. In Proceedings of the 2019 IEEE/CVF International Conference on Computer Vision (ICCV), Seoul, Korea, 27 October–2 November 2019; pp. 6568–6577.
35. Paszke, A.; Gross, S.; Chintala, S.; Chanan, G.; Yang, E.; DeVito, Z.; Lin, Z.; Desmaison, A.; Antiga, L.; Lerer, A. Automatic differentiation in pytorch. In Proceedings of the NeurIPS Workshop, Long Beach, CA, USA, 4–9 December 2017.
36. Tompson, J.J.; Jain, A.; LeCun, Y.; Bregler, C. Joint training of a convolutional network and a graphical model for human pose estimation. In *Advances in Neural Information Processing Systems*; MIT Press: Cambridge, MA, USA, 2014.
37. Kingma, D.P.; Ba, J. Adam: A method for stochastic optimization. In Proceedings of the International Conference Learn, Represent, (ICLR), San Diego, CA, USA, 5–8 May 2015.
38. Li, J.; Wang, C.; Zhu, H.; Mao, Y.; Fang, H.-S.; Lu, C. CrowdPose: Efficient Crowded Scenes Pose Estimation and a New Benchmark. In Proceedings of the 2019 IEEE/CVF Conference on Computer Vision and Pattern Recognition (CVPR), Long Beach, CA, USA, 2019, 16–20 June; pp. 10855–10864.

applied sciences

MDPI

Article

Lightweight Stacked Hourglass Network for Human Pose Estimation

Seung-Taek Kim and Hyo Jong Lee *

Division of Computer Science and Engineering, Jeonbuk National University, Jeonju 54896, Korea; kis7279@jbnu.ac.kr

* Correspondence: hlee@jbnu.ac.kr

Received: 17 August 2020; Accepted: 14 September 2020; Published: 17 September 2020

Featured Application: The proposed lightweight hourglass network can be applied as an alternative to existing methods that use the hourglass model as a backbone network.

Abstract: Human pose estimation is a problem that continues to be one of the greatest challenges in the field of computer vision. While the stacked structure of an hourglass network has enabled substantial progress in human pose estimation and key-point detection areas, it is largely used as a backbone network. However, it also requires a relatively large number of parameters and high computational capacity due to the characteristics of its stacked structure. Accordingly, the present work proposes a more lightweight version of the hourglass network, which also improves the human pose estimation performance. The new hourglass network architecture utilizes several additional skip connections, which improve performance with minimal modifications while still maintaining the number of parameters in the network. Additionally, the size of the convolutional receptive field has a decisive effect in learning to detect features of the full human body. Therefore, we propose a multidilated light residual block, which expands the convolutional receptive field while also reducing the computational load. The proposed residual block is also invariant in scale when using multiple dilations. The well-known MPII and LSP human pose datasets were used to evaluate the performance using the proposed method. A variety of experiments were conducted that confirm that our method is more efficient compared to current state-of-the-art hourglass weight-reduction methods.

Keywords: pose estimation; stacked hourglass network; deep learning; convolutional receptive field

1. Introduction

Human pose estimation is a fundamental method for detecting human behavior, and it is applied in virtual cinematography using computer graphics, human behavior recognition, and building security systems. Joint position varies greatly depending on a variety of factors, such as camera angle, clothing, and context. The traditional method estimates or tracks the human pose using additional equipment, such as depth sensors. However, with the advent of convolutional neural networks (CNNs), it is possible to efficiently infer the entire spatial feature from a single image without the need for additional equipment. Accordingly, many studies on human pose estimation using convolutional neural networks are currently underway, and these have achieved great results in terms of their accuracy [1–3].

The stacked hourglass network [2] is one of the best-known methods for resolving performance problems in human pose estimation. It has a stacked structure of hourglass modules composed of residual blocks [4]. Since the hourglass network performs promisingly in resolving the human pose estimation problem, a number of studies have used it as a backbone or modified the original hourglass network to improve performance [5–10]. Ning et al. [11] developed a stacked hourglass design and inception-resnet module with encoded external features for human pose estimation.

Ke et al. [12] introduced the multiscale supervision network with a multiscale regression network using the stacked hourglass module to increase robustness in keypoint localization for complex background and occlusions. Zhang et al. [13] suggested a method to overcome information loss in a repetitive cycle of down-sampling and up-sampling in a feature map. They used a dilated convolution and skip connections applied to a stacked hourglass network to optimize performance while adding extra subnetworks and large parameter sizes.

However, the residual block used in the hourglass network consists of a relatively small kernel with a fixed size. This may not be conducive to extracting the relationship between joints in the entire human body, and performance may significantly deteriorate depending on the size of the person in the input image. Additionally, since the network is stacked, very large memory capacity and computational powers are required. In this paper, we focus on how to reduce the parameter size and achieve the best performance, while others [5–14] deploy an extra subnetwork or layers to the stacked hourglass network. The purpose of this paper is to illustrate an optimized hourglass network with minimized parameter size without sacrificing the quality of the network.

Previous studies investigating human pose estimation [2,6,15] have confirmed that the size of the convolutional receptive field is a major factor in understanding the whole human body. If the receptive field is too small, it is difficult for the network to understand the relationship between each joint. Conversely, if it is too large, information that is not relevant to pose estimation is used for calculation, leading to impaired performance. In addition, if the convolution operation using a large kernel size is used several times to increase the receptive field, the size of the network may be too large.

The goal of this study is to improve the efficiency of the stacked hourglass network [2] in human pose estimation. We propose a method for designing an efficient hourglass network that is lightweight and greatly reduces the number of network parameters. We also propose a residual block that, by expanding the convolutional receptive field of the residual block, enables performance to be maintained on a multiscale basis through multidilation. To verify the performance, we used the MPII dataset [16] and Leeds Sports Poses (LSP) dataset [17], which are widely used human pose estimation datasets, and demonstrated the effectiveness of our approach through various experiments. In summary, our main contributions are threefold:

- A proposed stacked hourglass network structure improves performance in human pose estimation with fewer parameters (Sections 3.1, 3.2.1 and 3.2.2).
- A new structured residual block, known as a multidilated light residual block, which expands the receptive field of convolution operation, effectively represents the relationship of the full human body, and supports multiscale performance through multiple dilations (Section 3.2).
- An additional skip connection in an encoder of the hourglass module that reflects the low-level features of previous stacks on a subsequent stack. This additional skip connection improves performance in pose estimation without increasing the number of parameters. (Section 3.1).

2. Related Work

2.1. Network Architecture

The structure of the original hourglass network is shown in Figure 1. The network consists of stacked hourglass modules. An hourglass module is composed of residual blocks [4], and there is a skip connection between each stack. Each module has an encoder–decoder architecture. Loss can be calculated using heat maps obtained from each stack, and the network can perform more stable learning by adjusting to repeated predictions; this process is known as intermediate supervision. The hourglass network has been used as a backbone network for many other studies that have since been shown to perform efficiently to overcome human pose estimation problems. We designed a new residual block to reduce the parameters and improve the performance of the network. In addition, we propose a new network architecture based on an additional skip connection.

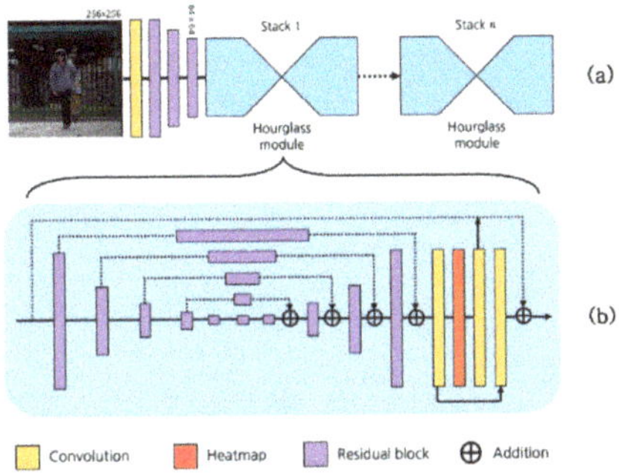

Figure 1. (**a**) Original hourglass network architecture and (**b**) hourglass module. The hourglass architecture is composed of hourglass modules stacked n times.

2.2. Residual Block

An hourglass module basically consists of residual blocks [4]. In [8,18], a hierarchical structure of residual blocks was proposed, and the convolution operation was binarized to increase the efficiency of the network. Compared to problems such as object detection, a key problem in human pose estimation is analyzing the full human body by extracting global spatial information. In [9], performance was improved by expanding the receptive field of convolution in residual blocks. Researchers in [5,9] also proposed a multiscale residual block model.

2.3. Human Pose Estimation

It has been well established that convolutional neural networks (CNNs) represent a significant step forward in recognizing spatial features from images. Accordingly, numerous studies have been conducted that use CNNs in human pose estimation where spatial information recognition is critical. The researchers in [19] made the first attempt to use CNNs for human pose estimation problems and showed a dramatic improvement in performance compared to traditional computer vision methods, such as [20,21]. Initial pose estimation methods using CNNs [19,22,23] predicted the coordinates of joints using CNNs and fully connected layers (FCLs). In [3], a method using a heat map generated by a CNN was proposed; this is currently used in most human pose estimation research [1,2,8,15,24–27].

3. Proposed Method

3.1. Network Architecture

Several studies have confirmed that the encoder–decoder architecture makes the network lighter and improves performance [28–30]. The hourglass network [2] used in our study is a productive approach to solving problems in human pose estimation because the network is able to learn more complex features by stacking modules. An encoder first extracts features by reducing the image resolution, while a decoder increases the image resolution and reassembles features. In an hourglass network, the encoder function is connected to the decoder using a skip connection so that the decoder can restore features well. According to recent work [31,32], extracting features is a more important process than simply restoring them. An hourglass network is structured so that input from a previous stack ($n-1$) is reflected in the current stack (n), in addition to the output of the previous stack ($n-1$) via the skip connection, as shown in Figure 1. In this structure, only relatively high-level features reconstructed by the decoder are reiterated in the next stack. Since our goal is to resolve this problem while also making the network lighter, we propose a method that enhances feature extraction performance while requiring minimal modification.

Figure 2 shows the proposed hourglass module. A feature extracted by the encoder is transferred to the next stack by the simple addition of parallel skip connections, as shown in Figure 2; this does not require a significant increase in computation. The proposed structure improves the encoder's extraction performance by transferring features to subsequent stack encoders. This structure produces better performance than the original hourglass network. An architecture with additional skip connections improves performance by increasing the performance of the encoder, even though the size of the network itself remains substantially unchanged.

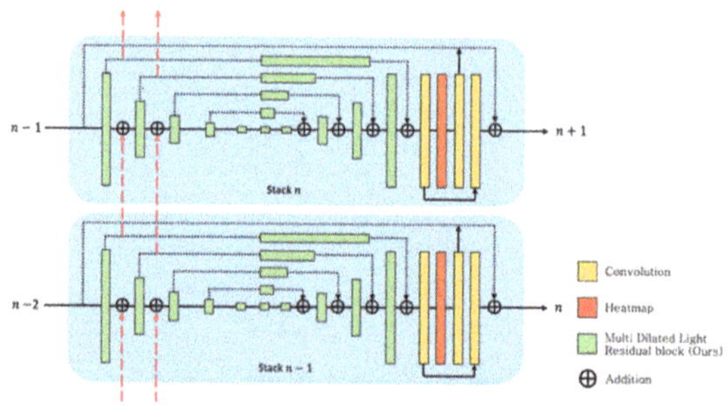

Figure 2. Proposed hourglass architecture, which utilizes an additional skip connection (red dashed arrow in the figure) from the previous stack ($n-1$) to the current stack (n).

3.2. Residual Block Design

3.2.1. Dilated Convolution

In human pose estimation, it is important to increase the receptive field so that the network can learn to recognize the features of the full human body. However, if the kernel size is increased to widen the receptive field, the computational cost also increases. Since our goal was to design an efficient hourglass network, we constructed a residual block using dilated convolution [33]. The number of parameters in the standard convolution is shown in Equation (1), where the kernel size is K, the input channel size is C, and the output channel size is M. If the size of the input and output is the same as $H \times W$, the required computational cost is shown by Equation (2):

$$\# \, param = K^2CM \quad (1)$$

$$Computational\ Cost = K^2CMHW \quad (2)$$

In the case of the dilated convolution, if the kernel size remains the same, the number of parameters and the computational cost remain the same as the standard convolution; however, the receptive field is wider, depending on the dilation size D. For dilated convolutions with a 3×3 kernel size when $D_1 = 2$, the kernel size and computational cost are the same as for the 3×3 standard convolution, but the receptive field is the same as for the 5×5 standard convolution. Additionally, as shown when $D_1 = 2$ and $D_2 = 3$ in Figure 3, dilated convolution has zero padding inside the kernel; thus, the computational cost is slightly lower than it is for standard convolution with the same kernel size.

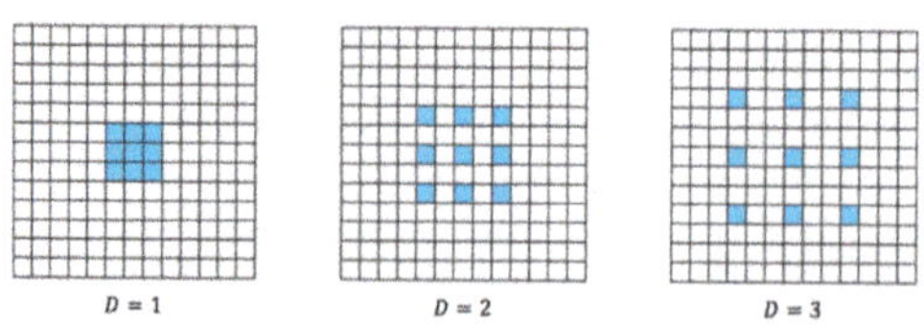

Figure 3. Standard convolution with 3×3 kernel size and dilated convolution. $D = 1$ is the same as the standard convolution. $D = 2$, 3 are calculated by placing zero padding inside the kernel, as shown in the figure.

3.2.2. Depthwise Separable Convolution

We used depthwise separable convolution, as proposed in [34], with dilated convolution for our residual block. Depthwise separable convolution performed pointwise (1 × 1) convolution after depthwise convolution, which was performed using an independent kernel for each channel. Figure 4 shows the concept of the depthwise separable convolution. This method shows lower performance than it does with standard convolution, but the number of parameters was significantly reduced, and the computation speed was faster. We interpolated the reduced performance due to depth-size separable convolution into a new residual block with dilated convolution.

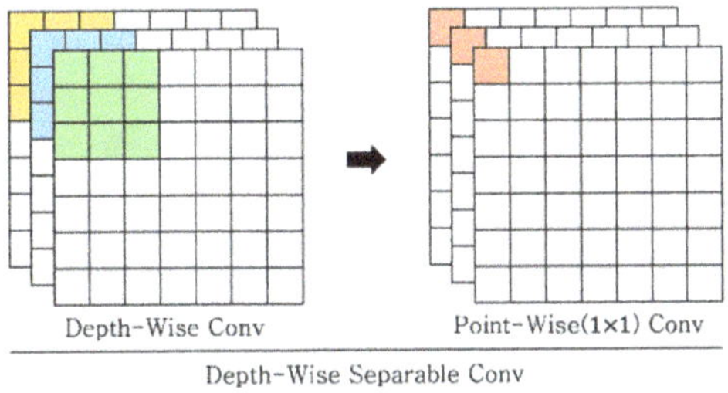

Figure 4. Depthwise separable convolution. Each channel performs convolution using an independent kernel, which is referred to as depthwise convolution. Pointwise convolution is then performed with a 1 × 1 kernel.

3.2.3. Proposed Multidilated Light Residual Block

The original stacked hourglass network constructed an hourglass module using a preactivated residual block, as shown in Figure 5a. The structure of the preactivated residual block is [ReLU→Batch Normalization→Convolution]; this is unlike that of a conventional residual block, which is designed as [Convolution→Batch Normalization→ReLU]. This structure is advantageous in building a deep network and improves the training speed [35]. However, residual blocks were originally designed for image classification or object detection problems (in which it is important to learn local features) where the convolutional receptive field is relatively small. Moreover, in the deep network architecture, although a bottleneck structure is used to reduce the number of parameters and computation cost, the residual block with a bottleneck structure is still large when designing a stacked multistage network, such as an hourglass network. Therefore, there is a need for a residual block with a new structure that can improve performance in human pose estimation by reducing the size of the network and expanding the receptive field.

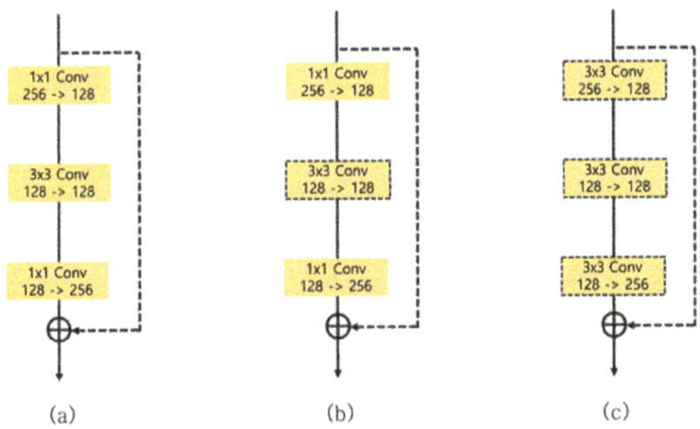

Figure 5. *Cont.*

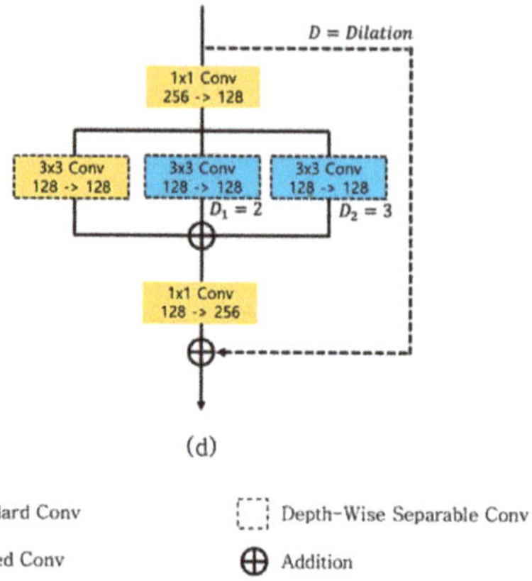

Figure 5. (**a**) Preactivation residual block used in the vanilla hourglass network. (**b**) Structure where the 3×3 convolution layer of (**a**) is changed to a depthwise separable convolution. (**c**) Structure in which the 1×1 layer of (**b**) is changed to 3×3. (**d**) Our proposed multidilated light residual block.

In this study, to design a residual block with a new structure capable of solving the aforementioned problems, experiments were performed on residual blocks with various structures. Figure 5b shows a structure in which the middle layer of the preactivated residual block shown in Figure 5a has been changed to a depthwise separable convolution. Using this residual block, we carried out experiments to observe the effect of the depthwise separable convolution on the network size and performance in human pose estimation. Since preactivated residual blocks are bottlenecks with the first and last layers having 1×1 convolutions, it did not make sense to reduce the number of parameters by changing the layer to a depthwise separable convolution. The researchers in [34] declared that if a nonlinear function is used between a depthwise convolution and a 1×1 convolution, the performance is significantly reduced. Therefore, all of the depthwise separable convolutions used in this paper consist of a structure that does not use an activation function between the depthwise convolution and 1×1 convolution, such as [ReLU→Batch Normalization→Depthwise Convolution→1×1 Convolution].

To evaluate the effect of the bottleneck structure of residual blocks while using a depthwise separable convolution, we designed the new module shown in Figure 5c. Figure 5c is a modified structure of Figure 5b, where we changed the standard convolutions of the first layer [256→128, 1×1] and the last layer [128→256, 1×1] into depthwise separable convolutions of [256→128, 3×3] and [128→256, 3×3], respectively. Figure 5d is our proposed multidilated light residual block, where a residual block of a new structure was used for improving the performance and reducing the number of parameters. Table 1 below shows the detailed structure of the proposed residual block.

Table 1. Proposed multidilated light residual block structure.

Type		Input Size (C, H, W)	Output Size (C, H, W)	Kernel Size	Dilation Scale (D)
I	INPUT	(256, H, W)	-	-	-
		Layer 1			
	ReLU	(256, H, W)	(256, H, W)	-	-
	Batch Norm	(256, H, W)	(256, H, W)	-	-
	Conv	(256, H, W)	(128, H, W)	1×1	1
		Layer 2			
	ReLU	(128, H, W)	(128, H, W)	-	-
B_1	Batch Norm	(128, H, W)	(128, H, W)	-	-
	Conv	(128, H, W)	(128, H, W)	3×3	1

Table 1. *Cont.*

Type		Input Size (C, H, W)	Output Size (C, H, W)	Kernel Size	Dilation Scale (D)
B_2	ReLU	(128, H, W)	(128, H, W)	-	-
	Batch Norm	(128, H, W)	(128, H, W)	-	-
	Depthwise Conv	(128, H, W)	(128, H, W)	3×3	2
	Conv	(128, H, W)	(128, H, W)	3×3	1
B_3	ReLU	(128, H, W)	(128, H, W)	-	-
	Batch Norm	(128, H, W)	(128, H, W)	-	-
	Depthwise Conv	(128, H, W)	(128, H, W)	3×3	3
	Conv	(128, H, W)	(128, H, W)	3×3	1
ADD ($B_1 + B_2 + B_3$)		(128, H, W)	(128, H, W)	-	-
Layer 3					
B_4	ReLU	(128, H, W)	(128, H, W)	-	-
	Batch Norm	(128, H, W)	(128, H, W)	-	-
	Conv	(128, H, W)	(256, H, W)	1×1	1
ADD ($B_4 + I$)		(256, H, W)	(256, H, W)	-	-
OUTPUT		-	(256, H, W)	-	-

In this study, the multidilated light residual block was used to greatly lighten the stack hourglass network, while multidilated convolution was used to expand the receptive field to increase the immutability of scale, resulting in improved performance in human pose estimation.

4. Experiments and Results

4.1. Dataset and Evaluation Matrix

The well-known human pose estimation datasets MPII and LSP were used to evaluate the performance of the proposed additional interstack skip connection and multidilated light residual blocks. The MPII dataset contains over 40,000 images of people with joint information, of which around 25,000 images were collected in real-world contexts. For human pose estimation, 16 coordinates for each joint were labeled for each person. In addition, we conducted experiments using the LSP and its extended training datasets [36] for objective evaluation. The LSP dataset contains 12,000 images with challenging athletic poses. In this dataset, each full body is annotated with a total of 14 joints.

To evaluate the performance of our method, we compared the performance with the state-of-the-art lightweight method for the stacked hourglass network [8] with various experiments. As an evaluation method, we used the percentage of correct key-points (PCK) on the LSP datasets and the modified PCK measure, which is the percentage of correct key-points on the head (PCKh) with the MPII dataset, as used in [32]. PCKh@0.5 uses 50% of the ground-truth head segment's length as a threshold. If the error rate is lower than the threshold value when comparing the predicted value with the ground truth, it is determined to be the correct answer.

4.2. Training Details

We followed the same training process as used for the original stacked hourglass network [2] with an input image size of 256 × 256. For the data augmentation required for training, rotation (±30°), scaling (±0.25), and flipping were performed. The model used in all experiments was written using PyTorch software [37]. We used the Adam optimizer [38] for training with a batch size of eight. The number of training epochs was 300, and the initial learning rate was 2.5×10^{-4}, which was reduced to 2.5×10^{-5} and 2.5×10^{-6} in the 150th and 220th epochs, respectively. The network was initialized by a normal distribution $\mathcal{N}(m, \sigma^2)$ with mean $m = 0$ and standard deviation $\sigma = 0.001$.

$$\mathcal{L} = \frac{1}{N}\sum_{n=1}^{N}\sum_{ij}\|H_n(i,\ j) - \hat{H}_n(i,\ j)\|^2 \tag{3}$$

The ground-truth heat map $H = \{H_k\}_{k=1}^{K}$ was generated by applying a Gaussian around k body joints, as shown in [3]. The loss $\mathcal{L}$ between the heat map $\hat{H} = \{\hat{H}_k\}_{k=1}^{K}$ and H predicted by the network used the mean squared error (MSE). Losses were calculated using the predicted heat maps from each stack and summed up by intermediate supervision. Figure 6a visualizes the loss in the training process, and Figure 6b visualizes the accuracy of PCKh@0.5 in the MPII validation set.

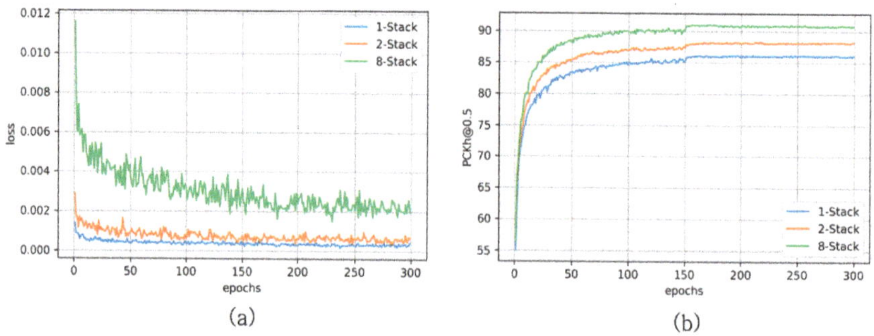

Figure 6. (**a**) Loss during training and (**b**) PCKh@0.5 on the MPII validation set.

4.3. Lightweight and Bottleneck Structure

To evaluate the network weight-reduction performance, we used depthwise separable convolution and looked at the effect of the bottleneck structure. Table 2 shows the experimental result obtained by constructing a single-stack hourglass network with each residual block in Figure 5; the number of parameters in the table represents the total number of parameters in a single hourglass network.

Table 2. Results for MPII validation sets in a single-stack hourglass network, depending on the type of residual block. (# Params is total number of parameters in a single-stack hourglass network.).

Residual Block	# Params	PCKh@0.5 (Mean)
Bottleneck (Original) (Figure 5a)	3.6M	86.21
Bottleneck (Original) + Depthwise Separable (Figure 5b)	1.4M	85.41
Bottleneck (3 × 3) + Depthwise Separable (Figure 5c)	1.6M	85.67
Multidilated Light Residual Block (Ours) (Figure 5d)	2.0M	86.12

In general, in a problem involving localization, such as human pose estimation, performance reduction occurs when using residual blocks in a bottleneck structure that uses 1 × 1 convolution to reduce the size of the network [18]. However, using 1 × 1 convolution was inevitable in this study because the network was made lighter by using depthwise separable convolution. Therefore, in order to confirm the effect of the bottleneck structure using 1 × 1 convolution in this experiment, the 1 × 1 convolutions of the first and last layers of the original residual block (Figure 5a) used 3 × 3 kernels. We experimented with a residual block (Figure 5c) that increased the kernel size to 3 × 3 and applied the depthwise separable convolution. Although the accuracy and parameters increased slightly, our result confirmed that the 1 × 1 convolution had no significant effect on our experiment.

Through this experiment, we confirmed that the multidilated light residual blocks proposed in this paper showed improved performance in terms of achieving a more lightweight network through a 56% reduction in the number of parameters. In addition, it was confirmed that the PCKh@0.5 performance was reduced by approximately 0.09, despite the large reduction in the number of parameters, as compared to the original residual block (Figure 5a). This slight reduction in accuracy was overcome by using the additional skip connection structure described in Section 3.1.

4.4. Additional Skip Connection

To confirm the effect of the additional skip connection (Section 3.1) on network performance, experiments were conducted on a dual-stack hourglass network (Table 3). When we applied only the proposed method, without using a modified residual block, we observed that the number of parameters remained the same, but the accuracy was greatly increased. The number of parameters in the dual-stack network, using both the proposed network structure (Section 3.1) and the residual block (Figure 5d), was reduced to the level of the original single-stack hourglass network. However, it was confirmed that the accuracy was similar to that of the original dual-stack hourglass network. From this experiment, it was confirmed that the proposed hourglass network using an additional skip connection showed significant results.

Table 3. Comparison of different network architectures with the MPII validation dataset. (# Params is the total number of parameters in a dual-stack hourglass network).

Network Architecture	Residual Block	# Params	PCKh@0.5 (Mean)
Hourglass (Original)	Bottleneck (Original) (Figure 5a)	6.7M	87.72
Ours	Bottleneck (Original) (Figure 5a)	6.7M	88.68
Ours	Ours (Figure 5d)	3.9M	87.60

4.5. Effect of the Dilation Scale

The dilated convolution in our residual block used zero padding equal to the dilation value D to fit the size of the input and output. Therefore, to check the effect of zero padding on the pose estimation problem according to dilation size, the dilation sizes of $D_1 = 2$ and $D_2 = 3$ and the increased sizes of $D_1 = 3$ and $D_2 = 5$ (proposed in this work) were compared (Table 4).

Table 4. Comparison of different dilation scales with the MPII validation dataset.

Method [×Stack]	PCKh@0.5 (Mean)
Ours ($D_1 = 2$, $D_2 = 3$) [×1]	86.12
Ours ($D_1 = 3$, $D_2 = 5$) [×1]	85.67
Ours ($D_1 = 2$, $D_2 = 3$) [×2]	87.89
Ours ($D_1 = 3$, $D_2 = 5$) [×2]	87.19

The receptive field of the 3×3 dilated convolution extended by $D = 2$, 3, and 5 was the same as the standard convolution using kernels of 5×5, 7×7, and 11×11, respectively. In this experiment, optimum dilation enhanced the performance of pose estimation; however, when the dilation size became too large, too much zero padding caused the network to fail to learn the spatial features, resulting in a loss of the ability to localize joints. In the human pose estimation problem, it was confirmed that zero padding due to the size of the receptive field and dilation had a significant effect on performance, while the optimum dilation size was determined to be $D_1 = 2$ and $D_2 = 3$.

4.6. Results and Analysis

To evaluate our method, we compared it with current state-of-the-art lightweight hourglass network methods. The authors of [8] proposed a new hourglass architecture using hierarchical residual blocks and evaluated the performance of network binarization in human pose estimation. Single-stack and an eight-stack networks are implemented for comparison.

As shown in Table 5, our method enhanced performance in pose estimation, despite an approximately 40% reduction in the number of parameters, as compared to state-of-the-art lightweight hourglass networks. It can be seen that the human pose estimation performance using our method was superior.

Table 5. Comparison with state-of-the-art lightweight hourglass methods with the MPII validation dataset.

Method [×Stack]	[8]-Real [×1]	[8]-Real [×8]	Ours [×1]	Ours [×8]
Head	96.8	97.4	96.3	98.1
Shoulder	93.8	96.0	94.1	96.2
Elbow	86.4	90.7	85.7	90.9
Wrist	80.3	86.2	80.4	87.2
Hip	87.0	89.6	85.6	89.8
Knee	80.4	86.1	80.3	87.3
Ankle	75.7	83.2	76.0	83.5
# Params	6M	25M	2M	14.8M
PCKh@0.5 (Mean)	85.5	89.8	86.1	90.8

We compared our results with those of existing methods on the MPII and LSP datasets. Table 6 presents the PCKh scores from different methods with the MPII dataset. Table 7 shows the PCK scores with the LSP dataset. The results confirmed that the proposed method shows improved human pose performance compared to the existing methods.

Table 6. Accuracy comparison with existing methods using the MPII validation dataset (PCKh@0.5).

Method	Head	Sho.	Elb.	Wri.	Hip	Knee	Ank.	Mean
Pishchulin et al. [39]	74.3	49.0	40.8	34.1	36.5	34.4	35.2	44.1
Tompson et al. [3]	95.8	90.3	80.5	74.3	77.6	69.7	62.8	79.6
Carreira et al. [40]	95.7	91.7	81.7	72.4	82.8	73.2	66.4	81.3
Tompson et al. [41]	96.1	91.9	83.9	77.8	80.9	72.3	64.8	82.0
Hu et al. [42]	95.0	91.6	83.0	76.6	81.9	74.5	69.5	82.4
Pishchulin et al. [43]	94.1	90.2	83.4	77.3	82.6	75.7	68.6	82.4
Lifshitz et al. [44]	97.8	93.3	85.7	80.4	85.3	76.6	70.2	85.0
Gkioxary et al. [45]	96.2	93.1	86.7	82.1	85.2	81.4	74.1	86.1
Rafi et al. [46]	97.2	93.9	86.4	81.3	86.8	80.6	73.4	86.3
Belagiannis et al. [24]	97.7	95.0	88.2	83.0	87.9	82.6	78.4	88.1
Insafutdinov et al. [47]	96.8	95.2	89.3	84.4	88.4	83.4	78.0	88.5
Wei et al. [15]	97.8	95.0	88.7	84.0	88.4	82.8	79.4	88.5
Bulat et al. [8]	97.9	95.1	89.9	85.3	89.4	85.7	81.7	89.7
Ours	98.1	96.2	90.9	87.2	89.8	87.3	83.5	90.8

Table 7. Accuracy comparison with existing methods using the LSP validation dataset (PCK@0.2).

Method	Head	Shoulder	Elbow	Wrist	Hip	Knee	Ankle	Mean
Lifshitz et al. [44]	96.8	89.0	82.7	79.1	90.9	86.0	82.5	86.7
Pishchulin et al. [43]	97.0	91.0	83.8	78.1	91.0	86.7	82.0	87.1
Insafutdinov et al. [47]	97.4	92.7	87.5	84.4	91.5	89.9	87.2	90.1
Wei et al. [15]	97.8	92.5	87.0	93.9	91.5	90.8	89.9	90.5
Bulat et al. [8]	97.2	82.1	88.1	85.2	92.2	91.4	88.7	90.7
Ours	98.1	93.1	89.2	86.1	92.7	92.8	90.4	91.7

Table 8 shows results from the experiment comparing the number of parameters for the MPII dataset. Figure 7 presents a schematic diagram of Table 8. This experiment confirmed that the method proposed in this paper significantly reduced the number of parameters while enhancing pose estimation performance.

Table 8. Parameter and accuracy comparison with existing methods using the MPII validation dataset (PCKh@0.5).

Method	# Params	PCKh@0.5 (Mean)
Bulat et al. [1]	58.1M	89.7
Insafutdinov et al. [47]	42.6M	88.5
Bulat et al. [8]	25.0M	89.8
Newell et al. [2]	25.1M	90.9
Ours	14.8M	90.8

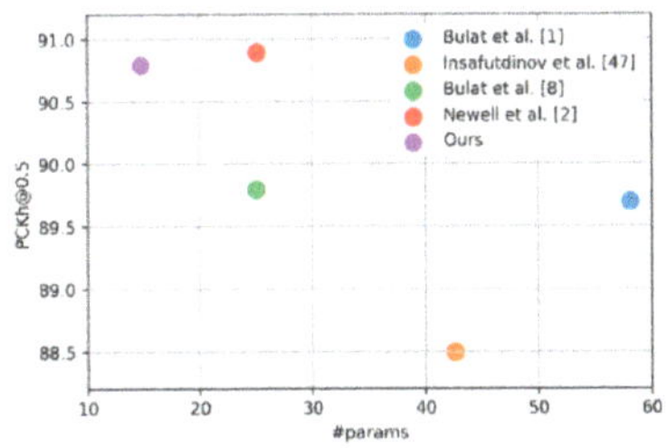

Figure 7. Visualization of performance versus number of parameters among studies using the MPII dataset.

The number of parameters and accuracy according to the number of stacks used are summarized in Table 9. Figure 8 visualizes the pose estimation results for the MPII dataset in the eight-stack network, in which the joints in the areas covered or crossed by the body are correctly estimated. These experiments also confirmed that the proposed method represents an improvement over existing methods in terms of efficiency and performance.

Table 9. Results for MPII validation datasets by number of stacks.

Method	# Params	PCKh@0.5 (Mean)
[×1]	2.0M	86.12
[×2]	3.9M	87.89
[×8]	14.8M	90.78

Figure 8. Prediction results of the proposed method for the MPII dataset.

5. Conclusions

In this paper, we proposed a lightweight stacked hourglass network for human pose estimation. The problem with existing stacked hourglass networks is that they continuously transmit only relatively high-level features from one stack to the next. To solve this problem, we proposed a new hourglass network structure utilizing additional interstack skip connections at the front end of the encoder. These improve the performance by reflecting relatively low-level features extracted by the encoder in the next stack to allow the gradient to flow smoothly during the learning process, even in the case of a deep stack. Moreover, since the skip connection involves a simple elementwise sum operation, performance can be improved without increasing the number of parameters, which assists in constructing a lightweight network.

To maintain accuracy, a multidilated light residual block was also proposed to reduce the number of parameters in the network by about 40% compared to an existing hourglass network. Using a multidilated light residual block improves performance by expanding the receptive field using dilated convolution, significantly reducing both the number of parameters and the computational load by applying depthwise separable convolution. In this paper, a variety of experiments was conducted for objective performance evaluation of the proposed methods, and the results confirmed that our proposed methods demonstrate an effective step forward in meeting the challenges of human pose estimation.

Author Contributions: Conception and design of the proposed method: H.J.L. and S.-T.K.; performance of the experiments: S.-T.K.; writing of the paper: S.-T.K.; paper review and editing: H.J.L. All authors have read and agreed to the published version of the manuscript.

Funding: This research was supported by the Basic Science Research Program through the National Research Foundation of Korea funded by the Ministry of Education (GR2019R1D1A3A03103736).

Conflicts of Interest: The authors declare no conflict of interests.

References

1. Bulat, A.; Tzimiropoulos, G. Human pose estimation via convolutional part heatmap regression. In Proceedings of the Haptics: Science, Technology, Applications, London, UK, 4–7 July 2016; Springer Science and Business Media LLC: Berlin/Heidelberg, Germany, 2016; Volume 9911, pp. 717–732.
2. Newell, A.; Yang, K.; Deng, J. Stacked hourglass networks for human pose estimation. In Proceedings of the European Conference on Computer Vision, Amsterdam, The Netherlands, 11–14 October 2016; Springer: Berlin/Heidelberg, Germany, 2016.
3. Tompson, J.J.; Jain, A.; LeCun, Y.; Bregler, C. Joint training of a convolutional network and a graphical model for human pose estimation. In *Advances in Neural Information Processing Systems*; MIT Press: Cambridge, MA, USA, 2014.
4. He, K.; Zhang, X.; Ren, S.; Sun, J. Deep residual learning for image recognition. In Proceedings of the 2016 IEEE Conference on Computer Vision and Pattern Recognition (CVPR), Las Vegas, NV, USA, 27–30 June 2016; Institute of Electrical and Electronics Engineers (IEEE): Piscataway, NJ, USA, 2016; pp. 770–778.
5. Wang, R.; Cao, Z.; Wang, X.; Liu, Z.; Zhu, X.; Wanga, X. Human pose estimation with deeply learned Multi-scale compositional models. *IEEE Access* **2019**, *7*, 71158–71166. [CrossRef]
6. Chu, X.; Yang, W.; Ouyang, W.; Ma, C.; Yuille, A.L.; Wang, X. Multi-context Attention for Human Pose Estimation. In Proceedings of the 2017 IEEE Conference on Computer Vision and Pattern Recognition (CVPR), Honolulu, HI, USA, 21–26 July 2017; Institute of Electrical and Electronics Engineers (IEEE): Piscataway, NJ, USA, 2017; pp. 5669–5678.
7. Peng, X.; Tang, Z.; Yang, F.; Feris, R.S.; Metaxas, D. Jointly optimize data augmentation and network training: Adversarial data augmentation in human pose estimation. In Proceedings of the 2018 IEEE/CVF Conference on Computer Vision and Pattern Recognition, Salt Lake City, UT, USA, 18–22 June 2018; Institute of Electrical and Electronics Engineers (IEEE): Piscataway, NJ, USA, 2018; pp. 2226–2234.
8. Bulat, A.; Tzimiropoulos, G. Hierarchical binary CNNs for landmark localization with limited resources. *IEEE Trans. Pattern Anal. Mach. Intell.* **2020**, *42*, 343–356. [CrossRef] [PubMed]

9. Yang, W.; Li, S.; Ouyang, W.; Li, H.; Wang, X. Learning feature pyramids for human pose estimation. In Proceedings of the 2017 IEEE International Conference on Computer Vision (ICCV), Venice, Italy, 22–29 October 2017; Institute of Electrical and Electronics Engineers (IEEE): Piscataway, NJ, USA, 2017; pp. 1290–1299.
10. Tang, W.; Wu, Y. Does learning specific features for related parts help human pose estimation? In Proceedings of the 2019 IEEE/CVF Conference on Computer Vision and Pattern Recognition (CVPR), Long Beach, CA, USA, 15–21 June 2019; Institute of Electrical and Electronics Engineers (IEEE): Piscataway, NJ, USA, 2019; pp. 1107–1116.
11. Ning, G.; Zhang, Z.; He, Z. Knowledge-guided deep fractal neural networks for human pose estimation. *IEEE Trans. Multimed.* **2018**, *20*, 1246–1259. [CrossRef]
12. Ke, L.; Chang, M.-C.; Qi, H.; Lyu, S. Multi-scale structure-aware network for human pose estimation. In Proceedings of the Haptics: Science, Technology, Applications, Pisa, Italy, 13–16 June 2018; Springer Science and Business Media LLC: Berlin/Heidelberg, Germany, 2018; pp. 731–746.
13. Zhang, Y.; Liu, J.; Huang, K. Dilated hourglass networks for human pose estimation. *Chin. Autom. Congr.* **2018**, 2597–2602. [CrossRef]
14. Artacho, B.; Savakis, A. UniPose: Unified human pose estimation in single images and videos. In Proceedings of the 2020 IEEE/CVF Conference on Computer Vision and Pattern Recognition (CVPR), Seattle, WA, USA, 13–19 June 2020; Institute of Electrical and Electronics Engineers (IEEE): Piscataway, NJ, USA, 2020; pp. 7033–7042.
15. Wei, S.-E.; Ramakrishna, V.; Kanade, T.; Sheikh, Y. Convolutional pose machines. In Proceedings of the 2016 IEEE Conference on Computer Vision and Pattern Recognition (CVPR), Las Vegas, NV, USA, 27–30 June 2016; Institute of Electrical and Electronics Engineers (IEEE): Piscataway, NJ, USA, 2016; pp. 4724–4732.
16. Andriluka, M.; Pishchulin, L.; Gehler, P.; Schiele, B. 2D human pose estimation: New benchmark and state of the art analysis. In Proceedings of the 2014 IEEE Conference on Computer Vision and Pattern Recognition, Columbus, OH, USA, 23–28 June 2014; Institute of Electrical and Electronics Engineers (IEEE): Piscataway, NJ, USA, 2014; pp. 3686–3693.
17. Johnson, S.; Everingham, M. Clustered pose and nonlinear appearance models for human pose estimation. In Proceedings of the British Machine Vision Conference 2010, Aberystwyth, UK, 31 August–3 September 2010; British Machine Vision Association and Society for Pattern Recognition: Durham, UK, 2010.
18. Bulat, A.; Tzimiropoulos, G. Binarized convolutional landmark localizers for human pose estimation and face alignment with limited resources. In Proceedings of the 2017 IEEE International Conference on Computer Vision (ICCV), Venice, Italy, 22–29 October 2017; Institute of Electrical and Electronics Engineers (IEEE): Piscataway, NJ, USA, 2017; pp. 3726–3734.
19. Toshev, A.; Szegedy, D. DeepPose: Human pose estimation via deep neural networks. In Proceedings of the 2014 IEEE Conference on Computer Vision and Pattern Recognition, Columbus, OH, USA, 23–28 June 2014; Institute of Electrical and Electronics Engineers (IEEE): Piscataway, NJ, USA, 2014; pp. 1653–1660.
20. Yang, Y.; Ramanan, D. Articulated pose estimation with flexible mixtures-of-parts. In Proceedings of the CVPR 2011, Providence, RI, USA, 20–25 June 2011; Institute of Electrical and Electronics Engineers (IEEE): Piscataway, NJ, USA, 2011; pp. 1385–1392.
21. Ferrari, V.; Marín-Jiménez, M.J.; Zisserman, A. Progressive search space reduction for human pose estimation. In Proceedings of the 2008 IEEE Conference on Computer Vision and Pattern Recognition, Anchorage, Alaska, USA, 24–26 June 2008; Institute of Electrical and Electronics Engineers (IEEE): Piscataway, NJ, USA, 2008; pp. 1–8.
22. Li, S.; Liu, Z.; Chan, A.B. Heterogeneous multi-task learning for human pose estimation with deep convolutional neural network. *Int. J. Comput. Vis.* **2014**, *113*, 19–36. [CrossRef]
23. Jain, A.; Tompson, J.; Andriluka, M.; Taylor, G.W.; Bregler, C. Learning human pose estimation features with convolutional networks. *arXiv* **2013**, arXiv:1312.7302.
24. Belagiannis, V.; Zisserman, A. Recurrent human pose estimation. In Proceedings of the 2017 12th IEEE International Conference on Automatic Face & Gesture Recognition (FG 2017), Washington, DC, USA, 30 May–3 June 2017; Institute of Electrical and Electronics Engineers (IEEE): Piscataway, NJ, USA, 2017; pp. 468–475.

25. Chu, X.; Ouyang, W.; Li, H.; Wang, X. Structured feature learning for pose estimation. In Proceedings of the 2016 IEEE Conference on Computer Vision and Pattern Recognition (CVPR), Las Vegas, NV, USA, 27–30 June 2016; Institute of Electrical and Electronics Engineers (IEEE): Piscataway, NJ, USA, 2016; pp. 4715–4723.
26. Yang, W.; Ouyang, W.; Li, H.; Wang, X. End-to-end learning of deformable mixture of parts and deep convolutional neural networks for human pose estimation. In Proceedings of the 2016 IEEE Conference on Computer Vision and Pattern Recognition (CVPR), Las Vegas, NV, USA, 27–30 June 2016; Institute of Electrical and Electronics Engineers (IEEE): Piscataway, NJ, USA, 2016; pp. 3073–3082.
27. Cao, Z.; Šimon, T.; Wei, S.-E.; Sheikh, Y. Realtime multi-person 2d pose estimation using part affinity fields. In Proceedings of the 2017 IEEE Conference on Computer Vision and Pattern Recognition (CVPR), Honolulu, HI, USA, 21–26 July 2017; Institute of Electrical and Electronics Engineers (IEEE): Piscataway, NJ, USA, 2017; pp. 1302–1310.
28. Noh, H.; Hong, S.; Han, B. Learning deconvolution network for semantic segmentation. In Proceedings of the IEEE International Conference on Computer Vision, Santiago, Chile, 7–13 December 2015; pp. 1520–1528.
29. Badrinarayanan, V.; Badrinarayanan, V.; Cipolla, R. SegNet: A Deep Convolutional Encoder-Decoder Architecture for Image Segmentation. *IEEE Trans. Pattern Anal. Mach. Intell.* **2017**, *39*, 2481–2495. [CrossRef] [PubMed]
30. Ronneberger, O.; Fischer, P.; Brox, T. U-net: Convolutional networks for biomedical image segmentation. In *International Conference on Medical Image Computing and Computer-Assisted Intervention*; Springer: New York, NY, USA, 2015; pp. 234–241.
31. Sun, K.; Xiao, B.; Liu, D.; Wang, J. Deep High-Resolution Representation Learning for Human Pose Estimation. In Proceedings of the 2019 IEEE/CVF Conference on Computer Vision and Pattern Recognition (CVPR), Long Beach, CA, USA, 15–21 June 2019; pp. 5686–5696.
32. Xiao, B.; Wu, H.; Wei, Y. Simple Baselines for Human Pose Estimation and Tracking. In Proceedings of the Haptics: Science, Technology, Applications, Munich, Germany, 8–14 September 2018; Springer Science and Business Media LLC: Berlin/Heidelberg, Germany, 2018; pp. 472–487.
33. Yu, F.; Koltun, V. Multi-scale context aggregation by dilated convolutions. *arXiv* **2015**, arXiv:1511.07122.
34. Chollet, F. Xception: Deep Learning with Depthwise Separable Convolutions. In Proceedings of the 2017 IEEE Conference on Computer Vision and Pattern Recognition (CVPR), Honolulu, HI, USA, 21–26 July 2017; pp. 1800–1807.
35. He, K.; Zhang, X.; Ren, S.; Sun, J. Identity Mappings in Deep Residual Networks. In Proceedings of the Haptics: Science, Technology, Applications, London, UK, 4–7 July 2016; Springer Science and Business Media LLC: Berlin/Heidelberg, Germany, 2016; Volume 9908, pp. 630–645.
36. Johnson, S.; Everingham, M. Learning effective human pose estimation from inaccurate annotation. In Proceedings of the CVPR 2011, Providence, RI, USA, 20–25 June 2011; Institute of Electrical and Electronics Engineers (IEEE): Piscataway, NJ, USA, 2011; pp. 1465–1472.
37. Paszke, A.; Gross, S.; Chintala, S.; Chanan, G.; Yang, E.; DeVito, Z.; Lin, Z.; Desmaison, A.; Antiga, L.; Lerer, A. Automatic differentiation in pytorch. In Proceedings of the NeurIPS Workshop, Long Beach, CA, USA, 4–9 December 2017.
38. Kingma, D.P.; Ba, J. Adam: A method for stochastic optimization. *arXiv* **2014**, arXiv:1412.6980.
39. Pishchulin, L.; Andriluka, M.; Gehler, P.; Schiele, B. Strong appearance and expressive spatial models for human pose estimation. In Proceedings of the 2013 IEEE International Conference on Computer Vision, Sydney, Australia, 1–8 December 2013; pp. 3487–3494.
40. Carreira, J.; Agrawal, P.; Fragkiadaki, K.; Malik, J. Human pose estimation with iterative error feedback. In Proceedings of the 2016 IEEE Conference on Computer Vision and Pattern Recognition (CVPR), Las Vegas, NV, USA, 27–30 June 2016; pp. 4733–4742.
41. Tompson, J.; Goroshin, R.; Jain, A.; LeCun, Y.; Bregler, C. Efficient object localization using convolutional NETWORKS. In Proceedings of the 2015 IEEE Conference on Computer Vision and Pattern Recognition (CVPR), Boston, MA, USA, 7–12 June 2015; pp. 648–656.
42. Hu, P.; Ramanan, D. Bottom-Up and Top-Down Reasoning with Hierarchical Rectified Gaussians. In Proceedings of the 2016 IEEE Conference on Computer Vision and Pattern Recognition (CVPR), Las Vegas, NV, USA, 27–30 June 2016; pp. 5600–5609.

43. Pishchulin, L.; Insafutdinov, E.; Tang, S.; Andres, B.; Andriluka, M.; Gehler, P.; Schiele, B. Deepcut: Joint subset partition and labeling for multi person pose estimation. In Proceedings of the 2016 IEEE Conference on Computer Vision and Pattern Recognition (CVPR), Las Vegas, NV, USA, 27–30 June 2016; pp. 4929–4937.
44. Lifshitz, I.; Fetaya, E.; Ullman, S. Human pose estimation using deep consensus voting. In *European Conference on Computer Vision*; Springer Science and Business Media LLC: Berlin/Heidelberg, Germany, 2016; Volume 9906, pp. 246–260.
45. Gkioxari, G.; Toshev, A.; Jaitly, N. Chained Predictions Using Convolutional Neural Networks. In Proceedings of the Haptics: Science, Technology, Applications, London, UK, 4–7 July 2016; Springer Science and Business Media LLC: Berlin/Heidelberg, Germany, 2016; Volume 9908, pp. 728–743.
46. Rafi, U.; Leibe, B.; Gall, J.; Kostrikov, I.; Wilson, R.C.; Hancock, E.R.; Smith, W.A.P.; Pears, N.E.; Bors, A.G. An efficient convolutional network for human pose estimation. In Proceedings of the British Machine Vision Conference 2016, York, UK, 19–22 September 2016; British Machine Vision Association and Society for Pattern Recognition: Durham, UK, 2016.
47. Insafutdinov, E.; Pishchulin, L.; Andres, B.; Andriluka, M.; Schiele, B. Deepercut: A deeper, stronger, and faster multi-person pose estimation model. In Proceedings of the Haptics: Science, Technology, Applications, London, UK, 4–7 July 2016; Springer Science and Business Media LLC: Berlin/Heidelberg, Germany, 2016; Volume 9910, pp. 34–50.

applied sciences

MDPI

Article

Deep Learning Based Human Activity Recognition Using Spatio-Temporal Image Formation of Skeleton Joints

Nusrat Tasnim [1], Mohammad Khairul Islam [2] and Joong-Hwan Baek [1,*]

1 School of Electronics and Information Engineering, Korea Aerospace University, Goyang 10540, Korea; tasnim.nishu70@kau.kr

2 Department of Computer Science and Engineering, University of Chittagong, Chittagong 4331, Bangladesh; mkislam@cu.ac.bd

* Correspondence: jhbaek@kau.ac.kr; Tel.: +82-2-300-0125

Abstract: Human activity recognition has become a significant research trend in the fields of computer vision, image processing, and human–machine or human–object interaction due to cost-effectiveness, time management, rehabilitation, and the pandemic of diseases. Over the past years, several methods published for human action recognition using RGB (red, green, and blue), depth, and skeleton datasets. Most of the methods introduced for action classification using skeleton datasets are constrained in some perspectives including features representation, complexity, and performance. However, there is still a challenging problem of providing an effective and efficient method for human action discrimination using a 3D skeleton dataset. There is a lot of room to map the 3D skeleton joint coordinates into spatio-temporal formats to reduce the complexity of the system, to provide a more accurate system to recognize human behaviors, and to improve the overall performance. In this paper, we suggest a spatio-temporal image formation (STIF) technique of 3D skeleton joints by capturing spatial information and temporal changes for action discrimination. We conduct transfer learning (pretrained models- MobileNetV2, DenseNet121, and ResNet18 trained with ImageNet dataset) to extract discriminative features and evaluate the proposed method with several fusion techniques. We mainly investigate the effect of three fusion methods such as element-wise average, multiplication, and maximization on the performance variation to human action recognition. Our deep learning-based method outperforms prior works using UTD-MHAD (University of Texas at Dallas multi-modal human action dataset) and MSR-Action3D (Microsoft action 3D), publicly available benchmark 3D skeleton datasets with STIF representation. We attain accuracies of approximately 98.93%, 99.65%, and 98.80% for UTD-MHAD and 96.00%, 98.75%, and 97.08% for MSR-Action3D skeleton datasets using MobileNetV2, DenseNet121, and ResNet18, respectively.

Keywords: spatio-temporal image formation; human activity recognition; deep learning; fusion strategies; transfer learning

Citation: Tasnim, N.; Islam, M.K.; Baek, J.-H. Deep Learning Based Human Activity Recognition Using Spatio-Temporal Image Formation of Skeleton Joints. *Appl. Sci.* **2021**, *11*, 2675. https://doi.org/10.3390/app11062675

Academic Editor: Hyo Jong Lee

Received: 22 February 2021
Accepted: 15 March 2021
Published: 17 March 2021

1. Introduction

Human action recognition has been grabbing more attention among researchers due to its multitude of real-world applications, for example, human–machine or human–object interaction [1–3], smart and intelligent surveillance system [4–6], content-based data retrieval [7], virtual reality/augmented reality [8,9], health care system [10,11], autonomous driving [12], and games [13]. The demand for human action recognition as well as pose estimation [14] is also expanding significantly, as a result, to manage the time, avoid intimate contact during the pandemic of diseases, and provide comfortable interaction for the rehabilitees. The human action recognition system focuses on identifying the activity accurately about what type of behavior is undertaken in a sequence of frames of so-called video.

With the progress of modern technology, various sensor devices such as Microsoft Kinect, Z-Depth, Leap Motion Controller, and Intel RealSense [15] have been invented to capture human activity data. Different modalities, for instance, RGB, depth, and skeleton data [16] are mainly used by the researchers for video-based human action recognition. Even though notable success has been appeared in the human action recognition system in the last few years, it is still challenging to accurately predict human activity for various restrictions such as camera orientation, occlusions, background, the variation of length, and speed of action [17]. Several methods offered a variety of ideas about human action recognition using hand-crafted features as well as using deep-learning features.

Hand-crafted features are usually referred to as the extraction of meaningful information (such as edges and corners) present in the images or videos using various descriptors e.g., local binary pattern (LBP) [18], space-time interest points (STIP) [19], scale-invariant feature transform (SIFT) [20], speeded up robust features (SURF) [21], histograms of oriented gradient (HOG) [22], and histograms of optic flow (HOF) [23]. These descriptors generate discriminative features by extracting information from locally important patches to represent action sequences. Abdul et al. [19] proposed a method for action recognition using trajectory-based feature representation where meaningful spatial information was preserved by detecting STIPs with SIFT and temporal change between the consecutive frames in an RGB video was computed by matching the interest points. Another hand-crafted feature-based method was introduced by Mahjoub et al. [24] with spatio-temporal interest point to detect the interest points in the video. They took the help of HOG and HOF descriptors to extract appearance and motion features to perform classification using support vector machine (SVM). In [25], Akam et al. combined local and global features extracted from RGB sequences and performed classification with SVM. They designed the shape descriptor as a local descriptor by integrating 3D trajectory and motion boundary histogram and extracted global features with gist descriptor for classification.

The engagement of modern researchers in the field of deep learning helps to design several learning models such as convolutional neural network (CNN) [26], recurrent neural network (RNN) [27], and long short-term memory (LSTM) [27]. These well-known models are widely being used to extract deep features for human action recognition. A two-stream CNN model was designed by Simonyan et al. [28] for human action recognition in RGB videos. The proposed CNN models consisting of spatial and temporal networks covered both local and global changes in the sequences for action discrimination. Zhang et al. [29] considered the time complexity to calculate the optical flow and suggested a deeply transferred motion vector CNN model to take the scope of optical flow. In [30], an effective deep 3D CNN model was introduced by Tran et al. to dig up the spatio-temporal features and recognized the action classes. Donahue et al. [31] proposed a long-term recurrent convolutional network with doubly deep compared to a fixed simple spatio-temporal receptive field for sequential processing. By using a long-term recurrent network, they captured complex dynamic to classify action groups.

Dataset captured in RGB format suffers from view dependency, background and illumination sensitivity, and computational complexity. While acquired an image or video of action in RGB format, it generates a lot of pixel values that makes it more complicated to differentiate from the background. The afore-mentioned obstacles hinder the performance of RGB video-based human action recognition and persuade to adopt the capturing devices to generate the depth and other formats of images or videos. Cheng et al. [32] proposed an efficient method for action recognition by extracting LBP features from depth motion map representation of three different views in depth sequences. They performed the classification by combining two fusion methods (feature-level fusion and decision-level). The integration of local and global features generated from depth sequences was introduced by Wu et al. [33]. They took the advantages of the correlation among the poses in neighboring frames of action to get the modified bag-of-words model called the bag of correlated poses. The dimensionality of the feature map was reduced with the help of principal components analysis and classified using SVM. Trelinski et al. [34] presented a CNN features-based method for human action recognition using depth sequences. The features were extracted by training a CNN model with multi-channel inputs such as two consecutive frames and projected view on the Cartesian plane. Finally, a LSTM was trained to determine the classification results. In [35], Wang et al. encoded the depth maps into weighted hierarchical depth motion maps (WHDMM) in three orthogonal planes. Then a three-branch CNN was conducted to pick out the discriminative features from WHDMM for the classification of human action.

Even though the depth images or videos require very-little storage compared to RGB, it also sustains from color, texture, and shape information. Meanwhile, both RGB and depth videos are captured using traditional cameras in 2D space. Thus, human action recognition based on RGB and depth sequences lack more deep information due to its 2D orientation and cannot capture the 3D structure of the action. By considering the following limitation, many sensor devices such as Microsoft Kinect provides more optimized information about the human body in terms of twenty skeleton joints. The skeleton information is represented in a deep 3D structure that is view-independent. Thus, the acquisition of an image or video in 3D skeleton format is much faster, lightweight in storage, and easy to use in the fields of human action recognition. Human action recognition based on skeleton data requires the exploitation of spatial and temporal changes of 3D skeleton joints in the sequences of action. There are several methods that suggest skeleton-based human action discrimination ranging from hand-crafted features based on human action classification using traditional machine learning algorithms to deep features based on human action recognition using deep learning. To provide more discriminative local and temporal features to improve recognition performance, skeleton joints information is represented in spatial format by capturing motion. Thus, it is very important to map the 3D skeleton joints in such a way that can cumulate both spatial and temporal information.

In this paper, we propose a new 3D skeleton-based human action recognition method by mapping the skeleton joints into spatio-temporal image by joining line between the same joints in two adjacent frames. Initially, we draw the position of joints by putting pixels with different colors in the jet color map (a color generated from the jet color map is the array of red, green, and blue intensities ranging from 0 to 1) in each frame. Then we draw lines between the same joints in two consecutive frames with different colors in the jet color map and combine them to get the final spatio-temporal image that helps to maintain both intra-frame and inter-frames information. To overcome the view dependency, we map the 3D skeleton joint coordinates along XY, YZ, and ZX planes (front, side, and top views). As the popularity of deep learning is increasing along with the recognition performance, we conduct the pretrained deep learning models to perform the classification of human action. First, we use the pretrained model to extract the discriminative features from the spatio-temporal images obtained from the front, side, and top views and then fuse the feature maps to get the final outputs. We also fine-tune the pretrained model to reduce the complexity and improve the recognition performance.

The remaining sections of this paper are illustrated as follows: The primitive knowledge including transfer learning and dataset are described in Section 2. In Section 3, we try to discuss the major ideas and limitations of the state-of-the-art studies about skeleton-based human action recognition. Section 4 elaborately explains the methodology of the proposed system in a step-by-step manner. The recognition results and performance comparisons are included in Section 5. We provide an analysis and discussion in Section 6. Finally, we add the conclusive words about the proposed method in Section 7.

2. Background Study

In this section, we present the elementary knowledge required for the proposed system. First, we interpret deep learning, more specifically transfer learning. Then we clarify the dataset used in overall illustrations and experiments.

2.1. Transfer Learning

With the tremendous improvement of modern technology over the past few years, we have seen much success of deep neural networks in different fields, particularly in recognition, classification, and machine translation tasks. These significant changes are also accelerated to the network architectural design such as AlexNet, ResNet, SuffleNet, GoogleNet, DenseNet, and MobileNet [36]. Even though the achievement of the deep neural network design has significant contributions in this domain, it still requires advanced knowledge and huge time to design an effective and efficient model. Due to the design and time complexity of the deep learning, we use the pretrained model trained with the ImageNet dataset for classification tasks known as transfer learning for human action recognition. We fine-tune the pretrained model to get better results for our proposed method. We consider three well-known pretrained deep learning models: MobileNetV2 [37], DenseNet121 [38], and ResNet18 [39] as the backbone of CNN models which are commonly applied in detection, recognition, and classification problems. To emphasize on the strength of STIF representation from the 3D skeleton data, we consider both less parameterized model (MobileNetV2) and heavy parameterized models (DenseNet121 and ResNet18).

MobileNetV2 integrates inverted residual blocks [37] in which the shortcut connection is established between the thin bottleneck layers. In MobileNetV2, an inverted residual block with a linear bottleneck first increases the dimensionality of the feature maps from low-dimensional inputs to high-dimensional outputs. Then a light-weight depth-wise separable convolution is introduced to filter the outputs. The feature maps are processed from low-dimension to high-dimension and again back into low-dimension. The main scenario is similar to narrow-wide-narrow concepts that reduce the number of parameters significantly. Figure 1 shows an inverted residual block of MobileNetV2 in which there are three blocks. The first block performs 1×1 convolution along with batch normalization and rectified linear unit (ReLU) operations to generate wide-dimensional feature maps. This wide-dimensional feature maps are then passed through the second block that accomplishes the depth-wise 3×3 separable convolution to reduce the computation. Finally, the third block conducts 1×1 convolution and batch normalization and reduces the dimensionality of the feature maps.

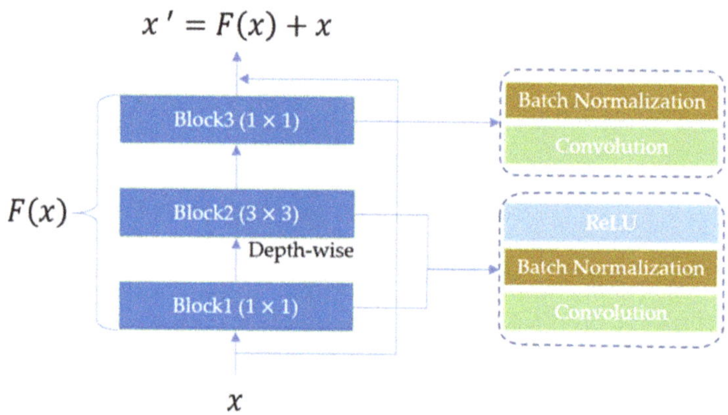

Figure 1. Inverted residual block in MobileNetV2.

DenseNet (dense convolutional network) focuses on the extraction of deeper features and tries to make it more efficient for training. DenseNet consists of two basic blocks: (i) dense block and (ii) transition block [38]. Each layer in DenseNet is connected to all other deeper layers in the network. The first layer is connected to the second, third, fourth, and so on. The second layer is joined with the third, fourth, fifth, and so on. Figure 2a shows a dense block in DenseNet in which each layer is connected with all other deeper layers. The transition layer is inserted to reduce the dimensionality of the output features in which a batch normalization, a convolution with 1×1 kernel, and an average pooling with 2×2 kernel operations are performed. Figure 2b depicts the graphical representation of the transition block in the DenseNet. In our experiments, we consider DenseNet121 for human behaviors classification.

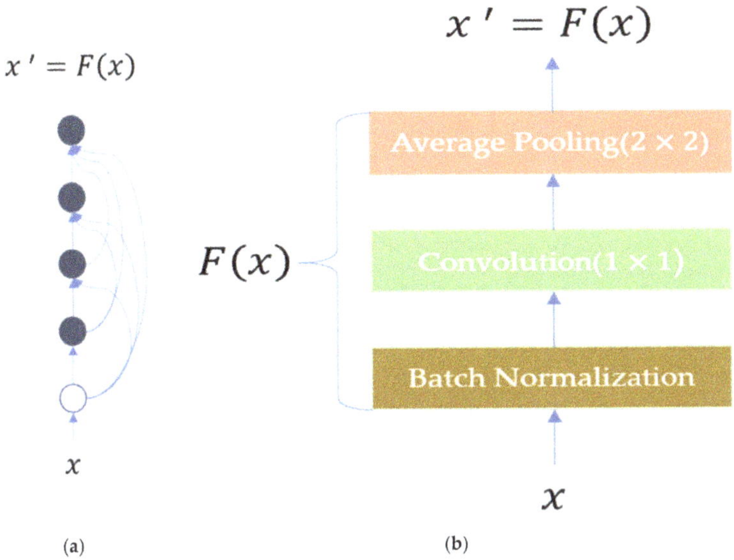

Figure 2. DenseNet (**a**) dense and (**b**) transition blocks.

ResNet (residual network) is designed by adding skip or shortcut connections from the previous layer to the current layer known as residual information. Compared with other well-known deep learning models, ResNet helps us to train up to hundreds or thousands of layers to get deeper features and improve the performance, particularly in recognition and classification tasks. The main component of ResNet is the residual blocks [39]. Figure 3 visualizes two kinds of residual blocks. The first residual block integrates skip connection directly to the output features, as shown in Figure 3a. However, the second residual block in Figure 3b adjusts the channels and resolution by conducting a 1×1 convolution before joining the skip connection. We adopt ResNet18 for our experiments.

2.2. Dataset

In this section, we provide a detail description of the datasets used in the proposed system. We use two publicly available datasets: UTD multi-modal human action dataset (UTD-MHAD) [40] and MSR-Action3D dataset [41].

2.2.1. UTD-MHAD

The members at embedded systems and signal processing (ESSP) Laboratory of the University of Texas at Dallas captured the UTD-MHAD skeleton dataset by using Microsoft Kinect camera. They extracted twenty skeleton joint coordinates of human body along X, Y, and Z axes as shown in Figure 4. Figure 4a shows the representation of the twenty skeleton joints with the corresponding names: *'Head'*, *'Shoulder Center'*, *'Spine'*, *'Hip Center'*, *'Shoulder Left'*, *'Elbow Left'*, *'Wrist Left'*, *'Hand Left'*, *'Shoulder Right'*, *'Elbow Right'*, *'Wrist Right'*, *'Hand Right'*, *'Hip Left'*, *'Knee Left'*, *'Ankle Left'*, *'Foot Left'*, *'Hip Right'*, *'Knee Right'*, *'Ankle Right'*, *'Foot Right'*. For better understanding and analysis, we order the joints from (x_1, y_1, z_1) to (x_{20}, y_{20}, z_{20}) as shown in Figure 4b. UTD-MHAD contains a total of 27 actions dataset which is accomplished by 8 persons (4 females and 4 males). Each person performs each action 4 times. There are three corrupted data removed and a total of 861 sequences are kept for the experiments. The names of the action classes are as follows: *'SwipeLeft'*, *'SwipeRight'*, *'Wave'*, *'Clap'*, *'Throw'*, *'ArmCross'*, *'BasketballShoot'*, *'DrawX'*, *'DrawCircle(CLW)'*,

'DrawCircle(counter CLW)', *'DrawTriangle'*, *'Bowling'*, *'Boxing'*, *'BaseballSwing'*, *'TennisSwing'*, *'ArmCurl'*, *'TennisServe'*, *'Push'*, *'Knock'*, *'Catch'*, *'PickUpandThrow'*, *'Jog'*, *'Walk'*, *'SitToStand'*, *'StandToSit'*, *'Lunge'*, *'Squat'*. Most of actions in this dataset are captured by hands. Three actions such as *'Jog'*, *'Walk'*, *'Lunge'* is performed by the leg, and only two actions *'SitToStand'* and *'StandToSit'* are done by the full body. We consider the dataset obtained by the first 5 subjects for training and 3 subjects for testing.

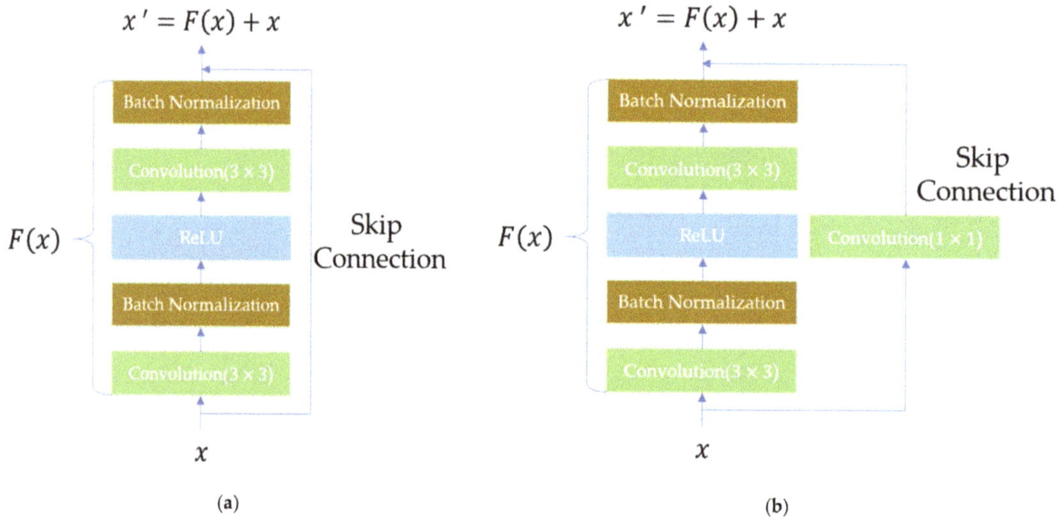

Figure 3. ResNet residual blocks (**a**) without and (**b**) with 1 × 1 convolution block.

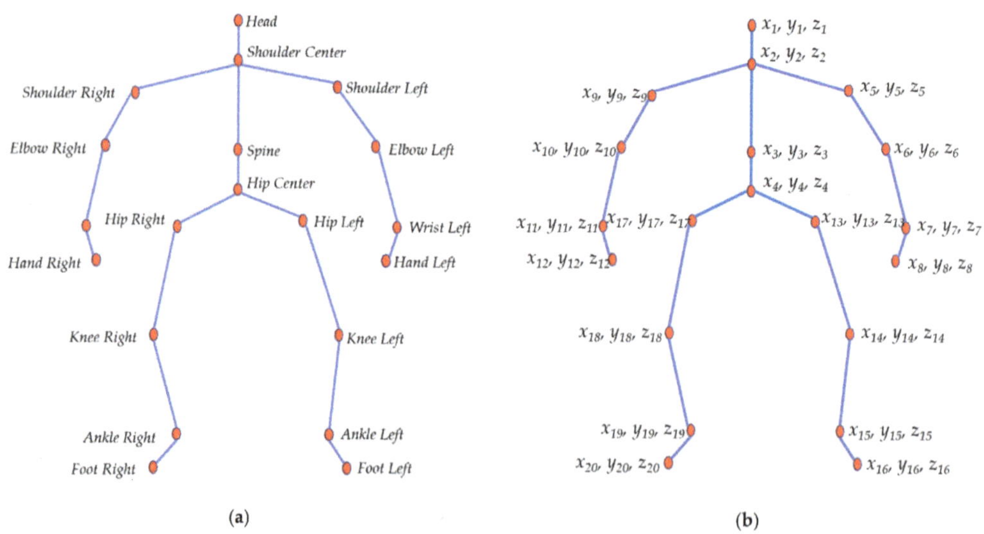

Figure 4. Human skeleton joints views; (**a**) twenty joints names and (**b**) corresponding joints numbering.

2.2.2. MSR-Action3D

Wanqing Li captured the MSR-Action3D dataset with the help of the Communication and Collaboration Systems Group at Microsoft Research Red-mond. This dataset contains 20 classes of actions which are done by 10 persons. Each person repeated each action three times. There are several corrupted sequences that are discarded and a total of 567 action sequences are maintained. As described in the previous Section 2.2.1 for UTD-MHAD dataset, the MSR-Action3D dataset also has twenty skeleton joint coordinates of human body along X, Y, and Z axes. The names and order are the same as the UTD-MHAD dataset as depicted in Figure 4. The action classes are as follows: *'HighArmWave'*, *'HoizontalArmWave'*, *'Hammer'*, *'HandCatch'*, *'ForwardPunch'*, *'HighThrow'*, *'DrawCross'*, *'DrawTick'*, *'DrawCircle'*, *'HandClap'*, *'TwoHandWave'*, *'SideBoxing'*, *'Band'*, *'ForwardKick'*, *'SideKick'*, *'Jogging'*, *'TennisSwing'*, *'TennisServe'*, *'GolfSwing'*, *'PickUpandThrow'*. The dataset from the first 6 subjects is used for training and the rest of the 4 subjects for testing.

3. Related Works

The shortcoming [17] of human behaviors capturing devices leads to the optimization of information about the human body. The 3D human skeleton dataset provides optimal and meaningful joint coordinates to recognize the genre of human attitudes. Several state-of-the-art review letters have been published on human action recognition particularly, 3D skeleton-based human action recognition [42–68]. The interpretations about the categories of 3D skeleton-based action recognition are separated into broad groups: trajectory and pose-based classification [42], joint and part-based classification [16,43], hand-crafted and deep learning-based classification [16,17,42,44], joints, mined joints, and dynamic based action classification [45], and spatio-temporal representation-based classification [42]. The spatio-temporal representation with CNN, RNN, LSTM, and the integration of CNN with RNN are also conducted for human action recognition using skeleton data [42]. All the referred categories fall into two major groups: 3D skeleton-based human action recognition with traditional classifiers using hand-crafted features and with deep learning features. The principal ideas of the prior works are briefly composed in the later sections.

3.1. *Traditional Features-Based Classifiers with Skeleton Dataset for Human Action Recognition*

The commonly used classifiers including SVM, k-nearest neighbors, hidden markov model (HMM), dynamic time warping (DTW), extreme learning machine (ELM), and Bayesian classifier are considered for skeleton-based human activity recognition. The above-mentioned discrimination methods require meaningful hand-crafted features for the classification of human behaviors. Video-based human action recognition depends on the spatial information of each frame and temporal changes of all frames. To perform hand-crafted features-based human action discrimination, the spatial information of each frame and temporal information between neighboring frames must have to be captured and combined.

Xia et al. [46] introduced new features called the histogram-based representation of 3D human posture from the projected views of depth sequences using linear discriminant analysis. The temporal information was kept by using HMM. In [47], Yang et al. presented a novel approach where they extracted features based on EigenJoints of joint positions differences and conducted Naïve Bayes Nearest Neighbor classifier for action recognition. Zhu et al. [48] demonstrated a new method for human action recognition by fusing spatio-temporal motion information and frame difference with the pairwise distance of skeleton joint coordinates. The spatio-temporal information determined by 3D interest point detection and local feature descriptor using Harris3D detector. The geometric relationship between various human body parts was exposed by Vemulapalli et al. [43] using rotations and transformation in 3D space. The performance was evaluated with DTW, Fourier temporal pyramid, and linear SVM by the representation of the lie group. Evangelidis et al. [49] illustrated a local descriptor from skeleton joints that maintained view-invariant features. They conducted Fisher kernel to explain the skeleton quads in an action and performed the classification using traditional SVM. A fast and powerful method for human action recognition using the 3D skeleton dataset was published by Zanfir et al. [50] in which they considered a moving pose descriptor containing pose information, speed, and acceleration of joints. Agahian et al. [51] assumed the skeleton sequences of action as a set of spatio-temporal poses. First, they normalized the joint coordinate and computed temporal variations. Then SVM was used to differentiate among the bag of poses (BOP) and finally histogram features were mined for the purposes of action classification using ELM. An effective multi-instance learning idea was brought out by Yang et al. [52] to find the discriminative multi-instance multi-task learning (MIMTL) features to expose the relationship among the skeleton joints. They also conducted multi-task learning model to recognize human action with MIMTL.

The previous hand-crafted features-based methods require an active engagement along with a lot of efforts to extract the feature of spatial and temporal information from the skeleton sequences. Sometimes, it becomes more complicated to design discriminative features from the 3D skeleton videos that degrades the performance of the system.

3.2. Deep Convolutional Neural Network for Skeleton-Based Human Action Recognition

Deep learning such as CNN, RNN, and LSTM naturalizes the process of human action classification by assembling the automated feature extraction and discrimination stages. Diverse methods revealed different techniques to recognize human behaviors from skeleton sequences using deep learning [53–68]. The most prominent issues in the prior methods based on deep learning can be divided into two categories: the use of raw 3D skeleton joint coordinates and the spatial representation of 3D skeleton joints.

Various skeleton-based human action recognition methods focused on the design of more powerful deep learning models to dig up the significant features from the skeleton joints. Du et al. [53] proposed an end-to-end hierarchical RNN with five branches to perform human action classification using raw skeleton sequence (RSS). They partitioned the skeleton joints into five segments and separately passed through the five branches of RNN. Finally, they combined the generated features and integrated a fully connected layer along with a softmax layer to make the decision. A two-stream RNN was introduced by Wang et al. [54] for human action recognition based on skeleton dataset. They considered both spatial and temporal information that was captured by two branches of RNN called temporal RNN and spatial RNN. Zhu et al. [55] designed a deep learning model by arranging LSTM and feed-forward fusion layer subsequently to learn co-occurrence of the human skeleton joints. They also provided a new dropout algorithm to train the model. In [56], Liu et al. explored the drawback of RNN based contextual learning to spatio-temporal learning by representing the skeleton joints into tree-structure. They again explained a novel gating technique to handle the noise and occlusion and accomplished the classification by using a spatio-temporal LSTM network. Song et al. [57] suggested an end-to-end spatio-temporal deep learning model using RNN with LSTM and got the help of spatial and temporal attention blocks to maintain the intra-frame and inter-frame relationship. A CNN-based approach for human action detection and classification was proposed by Li. et al. [58] in which they extracted discriminative features from raw skeleton joints and from the difference of skeleton joints in neighboring frames. They concatenated the extracted feature maps and applied two fully connected layers with a softmax layer to make the final prediction. Si et al. [59] separated skeleton joints of the human body into five parts and computed spatial features using a hierarchical spatial reasoning network. Then they built a temporal stack learning network consisting of velocity and position networks to capture temporal information. The dependency of RNN on the relationships between different parts of the human body inspired Zhang et al. [60] to use a simple universal geometric feature for skeleton-based action recognition. They modeled a 3-layer LSTM network to classify the geometric feature.

Some methods concentrated to map the 3D skeleton joints into a spatial format such as image format and fine-tuned the well-known deep learning models for features extraction and recognition. Du et al. [61] reorganized the skeleton sequence as a matrix by chronological order of joint coordinates in each frame along the temporal direction. Then the matrix was encoded into an image format with red, blue, and green for the coordinate values along X, Y, and Z axes. The recognition of human activity was obtained by a CNN model with spatio-temporal image representation of the skeleton joints. In [62], Liu et al. represented the skeleton sequences into image format by distinctive joint locations arrangement. They additionally mapped the relative joint velocities and used the CNN model to perform the action classification. The skeleton joint sequences were transformed into static images called skeleton optical flow guided feature (SOFGF) with determining the displacement of joints, angles, and joint distances by Ren et al. [63] and trained multi-stream CNN for discrimination. Li et al. [64] transformed the 3D skeleton sequences to color images using red, green, and blue colors which were translation and scale invariant. They developed a multi-scale CNN of image classification for human action recognition. A method for spatio-temporal information to color images of skeleton sequences called temporal pyramid skeleton motion map (TPSMM) was encoded by Chen et al. [65] in the frame to segment-wise manners. Then they conducted six CNN branches to extract features and performed classification. Li et al. [66] calculated the pair-wise distance from skeleton sequences and represented them into joint distance map (JDM) for three different views along XY, YZ, and ZX planes. Four branches of the CNN model were implemented to find out the distinctive features for action recognition using JDM. Hou et al. [67] formatted the skeleton sequences into skeleton optical spectra (SOS) using hue, saturation, and brightness colors to capture the spatial and motion information. They mapped three different views (front, side, and top) of SOS and deployed three branches of CNN to measure the classification results. Another similar method was suggested by Wang et al. [68], in which they first rotated the skeleton sequences to make view-invariant and increase the dataset. They represented the rotated skeleton sequences into joint trajectory maps (JTMs) and integrated fusion methods to get the recognition results.

3.3. Limitations of the State-of-the-Works and Our Contributions

As the hand-crafted features-based methods [43,46–52] require explicit contact by the researchers, it is always a tedious task to find out meaningful features from the skeleton sequences. It is also very hard to design a powerful deep learning model to process the raw skeleton joints to capture spatial and temporal information [53–60]. Above all, the deep learning-based human action recognition methods using raw skeleton dataset show comparatively worse performance than the deep learning-based methods using spatio-temporal representation of skeleton sequence. Sometimes, the spatio-temporal encoding cannot maintain local information as well as the global information significantly from the skeleton joints. The methods described [61–68] suffer from spatio-temporal design complexities such as view dependency, lack of motion, and deficiency of spatial and temporal information. These disadvantages lead to the low performance in skeleton-based human action recognition.

Thus, we figure out a simple yet robust, effective, and efficient way to represent the 3D skeleton joint coordinates into image format called STIF that defends both spatial and temporal variation of human behaviors specific movements. The STIF mapping facilitates to train the fine-tuned deep learning models that can automatically generate the meaningful features and performs the classification of human activity.

The major contributions of this paper are summarized as follows:

1. We idealize a novel technique to map 3D skeleton joints into spatio-temporal image. Our spatio-temporal image provides more discriminable features.
2. We adopt several fusion strategies (element-wise average, multiplication, and maximization) to expose the performance variations. Element-wise maximization shows better performance than average and multiplication.

3. We justify the robustness of our approach using both light weight and heavy weight deep learning models. We consider MobileNetV2 as the light weight and DenseNet121 and ResNet18 as the heavy weight models.
4. We compare the experimental results with prior works to show the effectiveness of our proposed method.

4. Proposed Methods

The description of the proposed methodology alongside the analysis is integrated with this section. First, the key ideas are analyzed, and then provided a detailed visual representation of the proposed system. Then we provide the spatial-temporal representation of the skeleton joint coordinates, knowledge transfer using well-known pretrained model, and finally conduct several fusion techniques to achieve better classification accuracy.

4.1. Research Motivation

With the massive advancement of modern technology, the application areas of human action recognition are spreading at a high speed in the fields of computer vision, image processing, and human–machine or human–object interaction. Various methods proposed for human action recognition using RGB [20,24,25,29–31], depth [32–35], and skeleton [42–68] dataset as described in Sections 1 and 3. However, it is still a challenging research topic to provide an effective and efficient method for human action classification. Many methods provided excellent and efficient ways of discriminating human activity using skeleton dataset. Even though the suggested methods facilitated great performance for human activity recognition using skeleton dataset, they suffered from several scarcities as illustrated in Section 3.3. The major concerns considered that most of the methods discriminated the human actions with accuracy below 90%. The weakness of the previous methods, for example, the inability to capture adequate intra-frame and inter-frame variation while representing the skeleton sequences into spatial format lessened the overall performance.

While we perform any meaningful action, for instances, drawing a circle, it has three different views along *XY*, *YZ*, and *ZX* planes that provide the spatial as well as temporal information. To capture the actual spatial views such as the circular shape of circle drawing, we connect lines between joints in adjacent frames. The temporal information is preserved by using color information that varies in every frame with the temporal changes. The line between joints in two neighboring frames with different colors defines the velocity of joints movement. By combining both spatial and temporal information, we generate spatio-temporal image which contains sufficient discriminative properties to perform the recognition.

4.2. System Architecture

Figure 5 depicts the overall architecture of the proposed system. There are four major modules: (i) spatio-temporal image formation, (ii) knowledge transfer, (iii) fusion, and (iv) classification. First, we convert the skeleton joints into a spatial format called STIF by covering both spatial information and temporal changes for the three different views *XY*, *YZ*, and *ZX* planes (front, side, and top views). Then the images are passed through the pareto frontier pretrained model referred to as the backbone network (MobileNetV2, DenseNet121, and ResNet18) trained with the ImageNet dataset. A fully connected layer is attached to each branch of the backbone network for extracting discriminative features. The generated features obtained from three different views (f_{xy}, f_{yz}, and f_{zx}) are fused in three different manners. Again, we conduct a fully connected layer to reduce the dimensionality of the features map (f_{xyz}). Finally, a fully connected (fc) and a softmax (sf) layers are added to perform the classification.

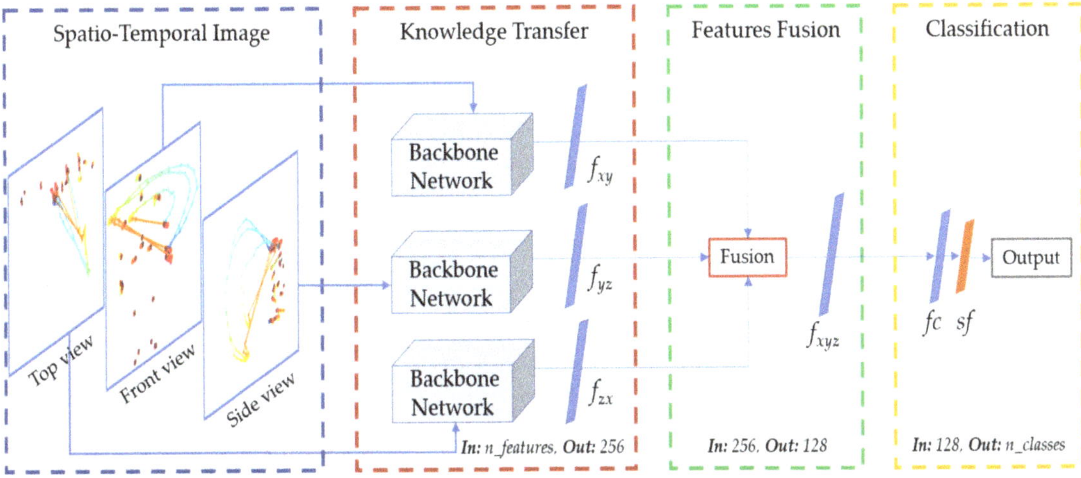

Figure 5. Architecture of the proposed system.

The input to the knowledge transfer module is the STIF representation of skeleton sequence which is indicated by n_features. Each branch including the front, side, and top of the backbone network produced 256-dimensional richer features that are fused in the fusion module. The input to the fusion module is the 256-dimensional features which are again passed through two fully connected layers and a softmax layer. The first fully connected layer reduces the 256-dimensional features to the 128-dimensional feature map. Finally, the second fully connected layer generates 27 output probabilities by following a softmax layer. The details of each module are described in the next sections.

4.2.1. Spatio-Temporal Image Formation (STIF)

Human action consists of a sequence of frames in which the spatial and temporal information changes occur over time while performing an action. The sequence of frames can be RGB, depth, skeleton, or any other format. In the proposed method, we consider only the 3D skeleton joints information for human action recognition. Skeleton joints information is usually encoded in a 3-dimensional coordinate system. Most of the devices invented for capturing skeleton joints information of the human body consider only twenty joints, specifically Microsoft Kinect captures skeleton data with twenty joint coordinates values along X, Y, and Z axes in each frame. The positions of the joints change from frame to frame as the action is rendered. We observe the changes of both the spatial and temporal information of the joint positions and represent them into spatio-temporal image. To generate the spatio-temporal image first, we map all the twenty joints in a frame with the same color in the jet color map [17] and then change the colors as the time step passed. Finally, the STIF is created by connecting lines between joints in adjacent frames subsequently.

Let us consider two joints, for example, $A(X_{i,j}, Y_{i,j})$ and $B\left(X_{(i+1),j}, Y_{(i+1),j}\right)$ along XY plane at two adjacent frames in an action where i indicates the index of frame and j represents index of joints (in our case j = 1, 2, ..., 20). The spatio-temporal image representation of an action along XY plane (front view) can be obtained by joining line between A and B using Equation (1).

$$Y - Y_{i,j} = m(X - X_{i,j}) \quad (1)$$

where m is the slope of the line passing through the points A and B given by Equation (2).

$$m = \frac{Y_{(i+1),j} - Y_{i,j}}{X_{(i+1),j} - X_{i,j}} \tag{2}$$

Similarly, we get the spatio-temporal image representation of an action along YZ plane (side view) and ZX plane (top view).

To map the spatial and temporal changes of joints information in spatial format such as image format for an action, we use the jet color map. First, we generate the jet colors information with length equal to the number of frames in an action as given in Equation (3).

$$Colors = JET(length(action)) \tag{3}$$

Figure 6 shows the jet colors in 3D space with bar chart that represents the variation of colors starting from blue to red.

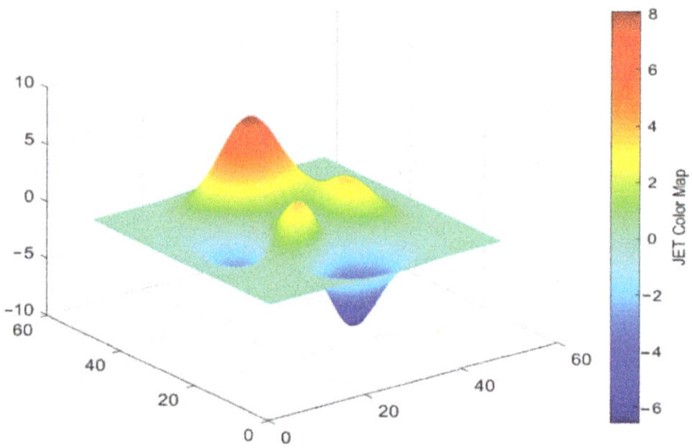

Figure 6. Jet colors generation process.

The generated colors are mapped by putting pixels for each joint as well as for the line passing through current and next frames. Equations (4) and (5) indicate the joints and line mapping with different colors to maintain spatial and temporal changes.

$$jointsMapping = putPixel(A,\ B,\ Color) \tag{4}$$

$$lineMapping = drawLine(A,\ B,\ Color) \tag{5}$$

The final spatio-temporal image is obtained by the mapping combination of joints and line as given in Equation (6).

$$spatioTemporalImage = concateMapping(jointsMapping,\ lineMapping) \tag{6}$$

The detailed of the STIF representation of human skeleton joints is summarized in Algorithm 1. The inputs to the algorithm are a sequence of frames containing twenty joint values along X, Y, and Z axes and different colors with the same length as the number of frames minus 1. The proposed method first computes the spatio-temporal image between two adjacent frames subsequently and finally combined all of them to generate the ultimate image.

Figure 7a–c shows the spatio-temporal image representation of *'Clap'* action in UTD-MHAD dataset along XY, YZ, and ZX planes (front, side, and top views) in which 1st frame indicates the mapping between frames at positions 1 and 2 in the action. The mapping of frames from positions 1 to (i + 1) be shown as ith frame and similarly (i + k)th frame is the combined mapping from frames at positions 1 to $(i + k + 1)$. The nth frame is the final spatio-temporal representation that concatenates all the previous mapping in an action. Figure 8a–c depicts the spatio-temporal visualization of the '*HighArmWave*' action in MSR-Action3D dataset along XY, YZ, and ZX planes (front, side, and top views).

Algorithm 1. Steps in spatio-temporal image representation from 3D human skeleton joints

```
spatioTemporalImage = spatioTemporalFormation(V, C)
//Input: Sequence of frames (V), Different Colors(C).
//Output: Spatio-temporal representation of an action.
f = readSequence (V)
n = f.lengthOfSequence
for i = 1:n−1 do        //Loop over all frames in a sequence.
        currentFrame = f(i)
        nextFrame = f(i+1)
        color = C(i)
        for j = 1:20 do        //Loop over 20 skeleton joints.
                jointsInCurentFrame = currentFrame(j)
                jointsInNextFrame = nextFrame (j)
                jointsMapping = putPixel(jointsInCurrentFrame, jointsInNextFrame, color)
                lineMapping = drawLine(jointsInCurrentFrame, jointsInNextFrame, color)
                spatioTemporalImage = concateMapping(jointsMapping, lineMapping)
        end for
end for
```

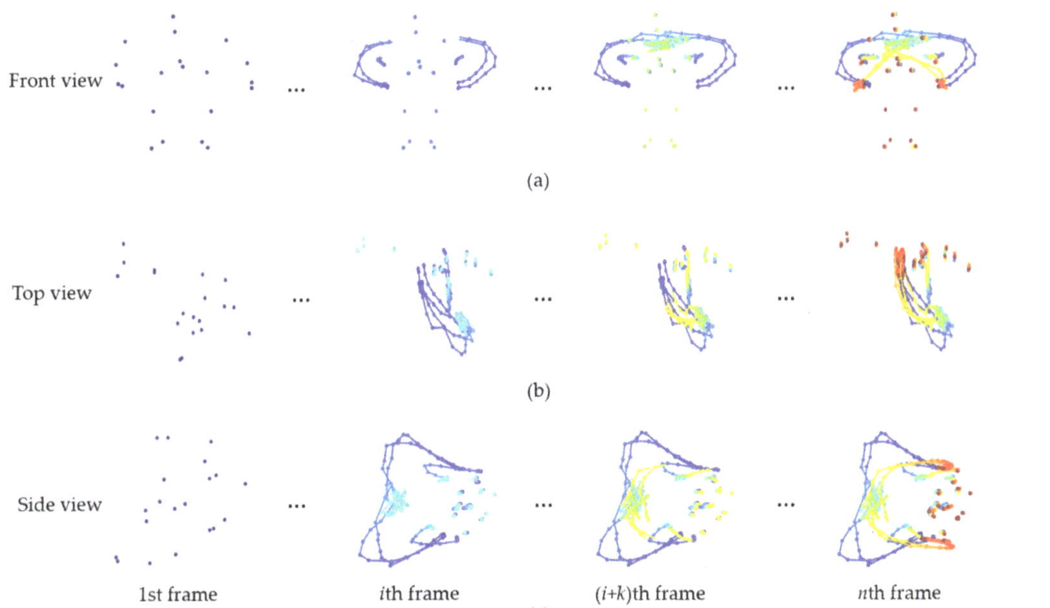

Figure 7. Spatio-temporal image formation (STIF) representation of '*Clap*' action in UTD-MHAD dataset; (**a**) front view (along XY plane), (**b**) side view (along YZ plane), and (**c**) top view (along ZX plane).

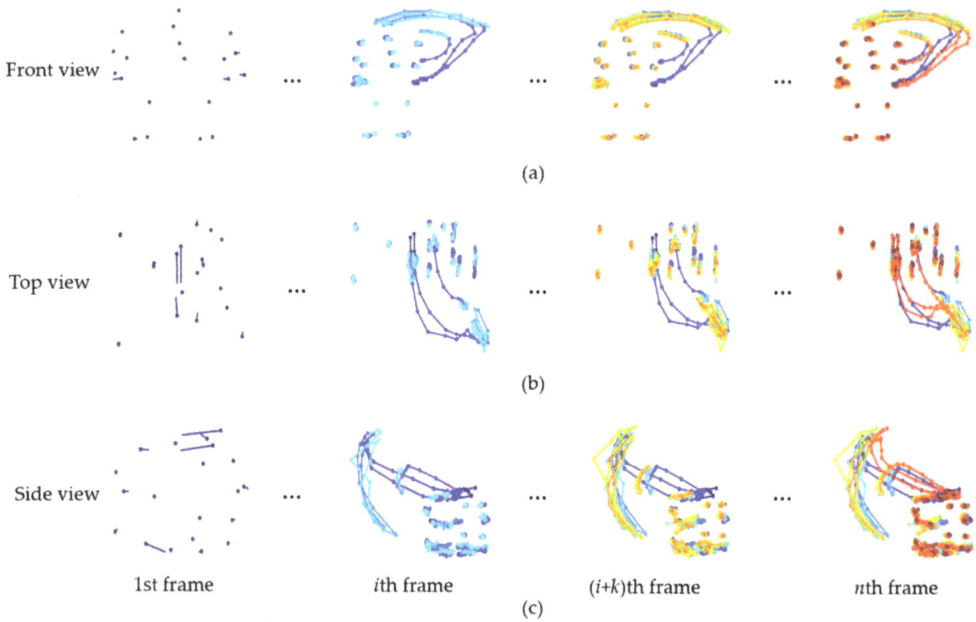

Figure 8. STIF representation of *'HighArmWave'* action in MSR-Action3D dataset; (**a**) front view (along *XY* plane), (**b**) side view (along *YZ* plane), and (**c**) top view (along *ZX* plane).

Due to the insufficiency of the dataset, we increase the dataset by using data augmentation such as rotation and scaling. Figure 9a–c visualizes the original, rotated, and scaled data along the front view of the *'SwipeLeft'* action class.

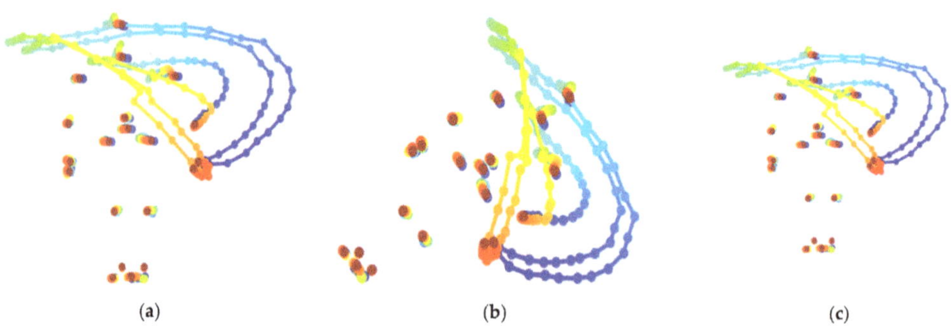

Figure 9. (**a**) Original, (**b**) rotated, and (**c**) scaled front views of *'SwipeLeft'* action data in UTD-MHAD dataset.

4.2.2. Knowledge Transfer

The benchmark human skeleton dataset is limited in volume, for instances, UTD-MHAD and MSR-Action3D datasets having fewer sequences. Even though we adopt data augmentation techniques such as scaling and rotation, it still very few in number for training deep learning models. Thus, we use the pretrained models trained with the ImageNet dataset and then the pretrained knowledge is transferred along with fully connected layers to extract the meaningful features from the UTD-MHAD and MSR-Action3D datasets for human activity recognition. We introduce three pretrained models; MobileNetV2, DenseNet121, and ResNet18 for performing the classification among the action classes. A brief description of the pretrained models used in our experiment is provided in Section 2.1.

4.2.3. Features Fusion and Classification

As described in the previous section, we consider pre-trained models to extract the discriminative features from the spatiotemporal images. We integrate a fully connected layer after the backbone model to reduce the dimensionality of the feature map. Then we apply several fusion mechanisms to compute the better classification accuracy. The fusion techniques allow us to integrate feature maps obtained from different branches of the backbone networks as shown in Figure 5.

Three different fusion techniques are introduced for comparing the performance of the proposed system. The first method is the straightforward average of the feature maps defined as f_{avg} and given in Equation (7).

$$f_{avg}(i) = \frac{1}{3}\sum[f_{xy}(i),\ f_{yz}(i),\ f_{zx}(i)] \tag{7}$$

where $i = 1, 2, \ldots, 256$ is the dimensionality of the feature map.

Each branch of the backbone network along with the fully connected layer generates 256-dimensional features namely as f_{xy}, f_{yz}, and f_{zx}. These feature maps are averaged to pass a 256-dimensional feature map through a fully connected layer to produce the feature map called f_{xyz} of size 256. This 256-dimensional feature map is then passed through a fully connected layer and a softmax layer to get the final output.

One of the most popular fusion methods is the element-wise multiplication of the feature maps defined as f_{mul} in Equation (8). Since three features with the same dimensionality (256-dimensional) are generated from the front, side, and top views images, the dimension of the resultant feature map is 256.

$$f_{mul}(i) = \prod[f_{xy}(i),\ f_{yz}(i),\ f_{zx}(i)] \tag{8}$$

where $i = 1, 2, \ldots, 256$ is the dimensionality of the feature map.

The last fusion technique is the element-wise maximization of the feature maps that yields a 256-dimensional feature map defined as f_{max} in Equation (9).

$$f_{max}(i) = \max[f_{xy}(i),\ f_{yz}(i),\ f_{zx}(i)] \tag{9}$$

where $i = 1, 2, \ldots, 256$ is the dimensionality of the feature map.

Among the above mentioned three fusion techniques, element-wise maximization is more robust compare to straight-forward average and multiplication.

5. Experimental Results

The experimental environment, training, and testing configurations, performance evaluation, and comparison are elaborately described in this section. First, we present the hardware and software used to implement the proposed system. Then, the parameters setting of training the deep learning models and performance are described in detail.

5.1. Training and Testing Configurations

We use Intel (R) Core (TM) i7 CPU, GeForce GTX 1080 GPU, Windows 10, and Linux 16.04 to accomplish the overall experiment. We conduct MATLAB 2019b and Python 3.5 as the programming language. We perform the preprocessing (spatio-temporal image representation) using MATLAB 2019b and implement deep learning using PyTorch toolbox in Python 3.5. To use the pre-trained model, we resize the STIF images generated from UTD-MHAD and MSR-Action3D datasets into 224 × 224 × 3.

We evaluate the proposed system by applying it on the UTD-MHAD and MSR-Action3D datasets as described in Section 2.2. The dataset is partitioned into training and testing data as described in Section 2.2. At the initial step, we set the learning rate as 0.001, and the learning rate decreases at the interval of 10 epoch with a factor of 0.9. We set the batch size to 16 that shuffles during the reading. The Adam optimizer is conducted for optimization purposes with a momentum value of 0.999. The training process continues until 100 epochs. Table 1 lists the hardware, software, and parameters used for training and testing the proposed system.

Table 1. Hardware, software, and parameters configurations.

Parameters	Values
Hardware	Intel (R) Core(TM) i7 CPU, GeForce GTX 1080 GPU
Software	Windows 10, Linux 16.04, MATLAB 2019b, Python 3.5
Initial learning rate	0.001
Learning rate dropping factor	0.9
Learning rate dropping period	10
Optimizer	Adam

5.2. Performance Evaluations

We calculate classification accuracies for emphasizing the effectiveness and efficiency of the proposed system using Equation (10) for each action.

$$\text{Accuracy}(\%) = \frac{\text{Total Correctly Predicted Observations}}{\text{Total Number Observations}} \times 100 \quad (10)$$

We carry out the evaluation process in six different ways to provide more clarification about the discriminability of human activity using the proposed method. We initially train the deep learning models with three different views along XY, YZ, and ZX planes separately and compute the test results. Then the features extracted from three different views are fused in three different ways and enumerate the classification results. We conduct three well-known deep learning models including MobileNetV2, DenseNet121, and ResNet18 to represent the robustness of the proposed system.

Table 2 enlists human activity recognition results with different configurations of data and models using the UTD-MHAD dataset. The proposed method achieves classification accuracies about 97.29% and 98.21% for DenseNet121 using front and top views of the UTD-MHAD dataset respectively. However, while we apply side view data, MobileNetV2 (96.06%) shows better performance than DenseNet121 (95.98%) and ResNet18 (93.17%).

Table 2. Human activity recognition with different modality and models using UTD-MHAD dataset.

Methods	MobileNetV2	DenseNet121	ResNet18
Front view (XY plane)	96.52%	97.29%	95.29%
Side view (YZ plane)	96.06%	95.98%	93.17%
Top view (ZX plane)	97.08%	98.21%	94.67%
Fusion (f_{avg})	97.98%	98.51%	95.93%
Fusion (f_{mul})	98.93%	98.89%	97.18%
Fusion (f_{max})	98.89%	99.65%	98.80%

As we mentioned in the previous discussion, we fuse the features obtained from the front, side, and top views dataset using three techniques to improve the classification accuracies in MobileNetV2, DenseNet121, and ResNet18. While fusing the features by applying the conventional average and maximization fusion techniques, the proposed method gets 98.51% and 99.65% highest classification accuracies using DeNseNet121. The element-wise multiplication of the features provides 98.93% discrimination accuracies using MobileNetV2. By considering the recognition results in Table 2, it is clear that DenseNet121 outperforms both MobileNetV2 and ResNet18 because it can generate deeper features from the STIF representation of the skeleton data.

We again investigate the performance of human activity recognition on the MSR-Action3D dataset using MobileNetV2, DenseNet121, and ResNet18 for showing the robustness of the proposed method. Table 3 shows the human action recognition results on the MSR-Action3D dataset. Like the UTD-MHAD dataset, we obtain better accuracies of about 94.83% for DenseNet121 compared with MobileNetV2 (93.83%) and ResNet18 (93.04%) on the front view dataset. A similar trend appears in the case of the top view dataset in which DenseNet121 classifies human action with an accuracy of about 93.00% while MobileNetV2 and ResNet18 secure accuracies of about 91.67% and 92.08% respectively.

Table 3. Human activity recognition with different modality and models using MSR-Action3D dataset.

Methods	MobileNetV2	DenseNet121	ResNet18
Front view (*XY* plane)	93.83%	94.83%	93.08%
Side view (*YZ* plane)	91.25%	91.25%	92.50%
Top view (*ZX* plane)	91.67%	93.00%	92.08%
Fusion (f_{avg})	95.42%	96.67%	97.50%
Fusion (f_{mul})	95.50%	96.67%	96.25%
Fusion (f_{max})	96.00%	98.75%	97.08%

ResNet18 obtains the highest accuracy about 97.50% using average fusion on the MSR-Action3D dataset. For both the multiplication and maximization fusions, DenseNet121 secures about 96.67% and 98.75% discrimination results.

To render more clarification about the performance of the proposed method for each backbone model along with each experiment, we visualize the graphical results shown in Figure 10a,b. From the Figure 10, we can argue that DenseNet121 works much better in comparison with MobileNetV2 and ResNet18 for both UTD-MHAD and MSR-Action3D datasets. At the same time, it can be stated that the three fusion techniques contribute a lot to improve the classification performance of the proposed system in case of either UTD-MHAD or MSR-Action3D dataset.

To examine the complexities such as memory consumption, operational requirements, and time complexity, we express the parameters in millions, floating-point operation per seconds (FLOPS) in giga, and time in second as listed in Table 4. The inverted-residual blocks in MobileNetV2 reduce the parameters as well as the number of operations than DenseNet121 and ResNet18. The operations increase as the features are extracted from deeper in DenseNet121. The pre-trained MobileNetV2, DenseNet121, and ResNet18 require 0.00129, 0.00235, and 0.00108 s respectively for running an action.

Table 4. Complexity analysis of different models.

Models	Parameters (M)	FLOPs (G)	Time (s)
MobileNetV2	2.56	0.33	0.00129
DenseNet121	7.22	2.90	0.00235
ResNet18	11.31	1.82	0.00108

The further inquisition of performance on different actions individually explains that the proposed method assures the recognition accuracy above 90% for the UTD-MHAD dataset. From Figure 11a, it can be said that the action classes *'SwipeLeft'*, *'SwipeRight'*, *'Clap'*, *'Throw'*, *'ArmCross'*, *'Boxing'*, *'BaseballSwing'*, *'TennisSwing'*, *'TennisServe'*, *'Push'*, *'Knock'*, *'Catch'*, *'PickUpandThrow'*, *'Jog'*, *'Walk'*, *'SitToStand'*, *'StandToSit'*, *'Lunge'*, and *'Squat'* secure highest classification accuracies with any one of the three pre-trained models.

The recognition results decrease to 58.33%, 91.67%, and 83.33% using MobileNetV2, DenseNet121, and ResNet18 in MSR-Action3D dataset for *'HandCatch'* action due to the irregularities in the dataset. On top of that, the overall discrimination performance is satisfactory for any other action classes in the MSR-Action3D dataset with three pre-trained models. Figure 11b illustrates the classification results for individual classes in the

MSR-Action3D dataset with MobileNetV2, DenseNet121, and ResNet18. The reduction of accuracies in some classes of action such as *'HandCatch'* in the MSR-Action3D dataset happens due to the irregular movement in few points of the sequences.

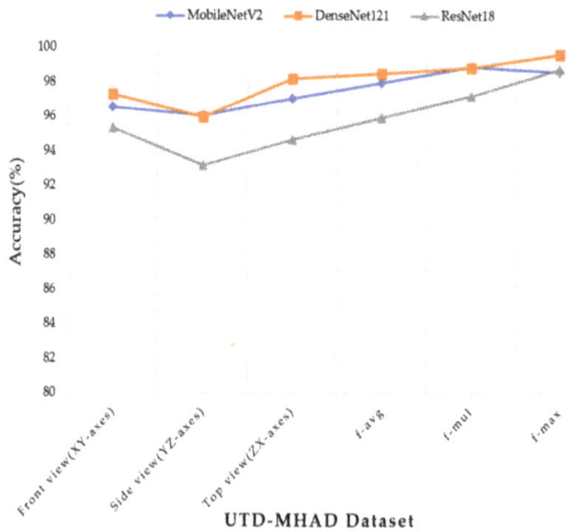

(a)

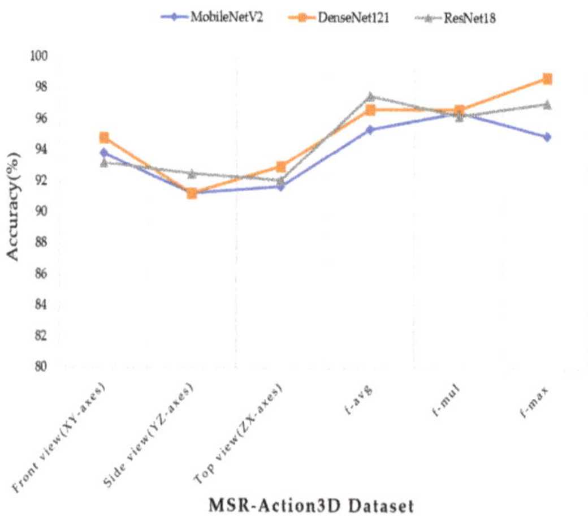

(b)

Figure 10. Visual comparisons of different views and fusions of dataset and models; (**a**) UTD-MHAD and (**b**) MSR-Action3D datasets.

5.3. State-of-the-Art Comparisons

We have already described that the proposed method fulfills the desired objective with better performance compared to the state-of-the-art methods for human action recognition using 3D skeleton dataset. For the fairness of the experimental results, we have separately mentioned the strategies with classifiers to compare the recognition accuracies between the proposed system and the prior works for UTD-MHAD and MSR-Action3D datasets as depicted in Tables 5 and 6. Among the six different experimental results (front, side, top, average, multiplication, and maximization), we only list the best one for each model.

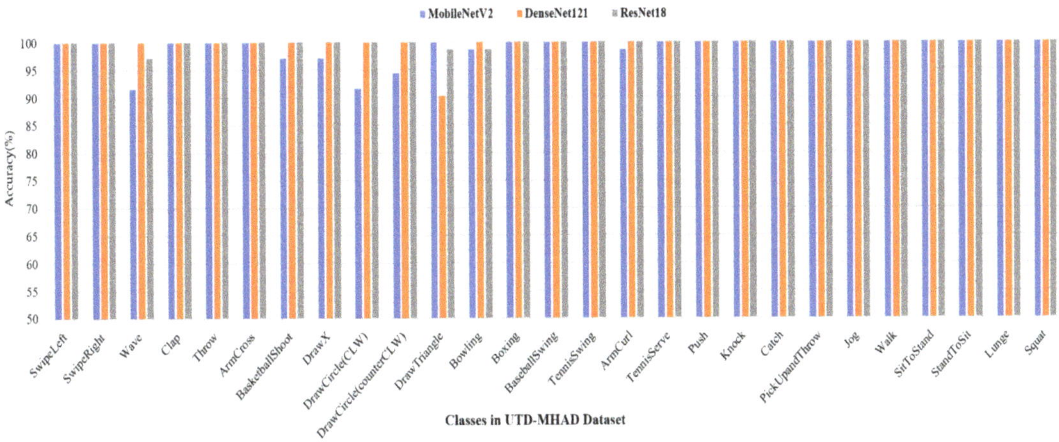

(a)

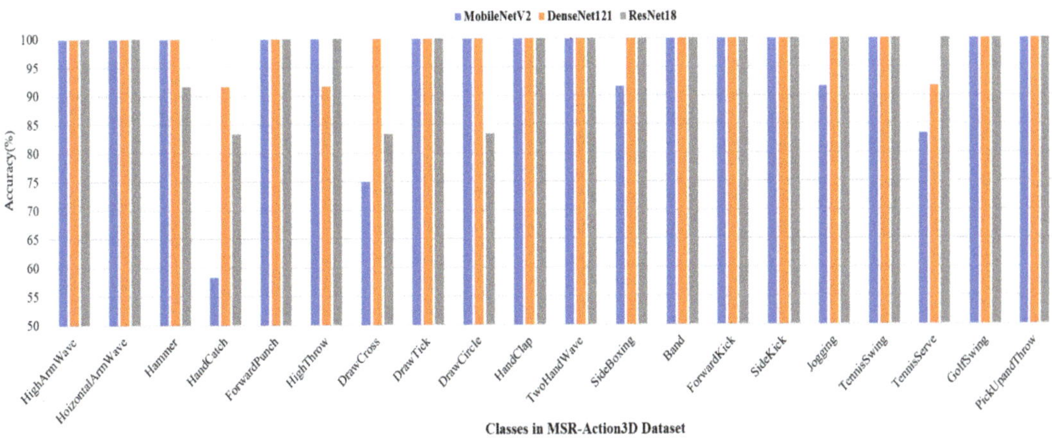

(b)

Figure 11. Classification performance of the proposed system based on individual classes in (**a**) UTD-MHAD and (**b**) MSR-Action3D datasets.

Table 5. Performance comparisons of human action recognition using UTD-MHAD dataset.

Methods	Accuracy
STIF with MobileNetV2	98.93%
STIF with DenseNet121	99.65%
STIF with ResNet18	98.80%
BOP with ELM [51]	95.30%
TPSMM with CNN [65]	88.10%
JDM with CNN [66]	93.26%
SOS with CNN [67]	86.97%
JTM with CNN [68]	85.81%

Table 6. Performance comparisons of human action recognition using MSR-Action3D dataset.

Methods	Accuracy
STIF with MobileNetV2	96.00%
STIF with DenseNet121	98.75%
STIF with ResNet18	97.08%
BOP with ELM [51]	91.90%
MIMTL with SVM [52]	93.63%
RSS with RNN [53]	94.48%
SOFGF with CNN [62]	97.25%

The previous method in [51] attains about 95.30% outcomes for human activity recognition using the UTD-MHAD dataset which is the highest among the references [51,64,66–68]. The proposed method achieves accuracy of 99.65%, the highest experimental results, with STIF representation of UTD-MHAD skeleton dataset using DenseNet121 which is around 4% greater than the methods using TPSMM [64], JDM [66], SOS [67], and JTM [68] with CNN. Even though we apply MobileNetV2 and ResNet18, our method secures about 3% better accuracy than the state-of-the-art methods.

We further provide the comparative results for MSR-Action3D skeleton dataset to boost up on the suggested system. The proposed method ensures 98.75% accuracy with DenseNet121 which is the better than the defending best accuracy of 97.25% accuracy achieved by SOFGF with CNN [62]. The other methods [51–53] classified human action with accuracies of about 4% lower than the proposed method. On the other hand, our method can obtain about 96.00% and 97.08% recognition accuracies with MobileNetV2 and ResNet18 correspondingly.

6. Analysis and Discussion

As the prior works partially lack capturing the spatial and temporal variations explicitly, we frankly provide a spatio-temporal representation of 3D skeleton joint values along the front, side, and top views. We adopt data augmentation (rotation and scaling) to increase the experimental data due to the data deficiency in UTD-MHAD and MSR-Action3D datasets. By considering the design and time complexities of the deep learning model, we fine-tune the well-known pretrained models such as MobileNetV2, DenseNet121, and ResNet18 for the human action recognition in the proposed method. We also investigate the classification results with three different fusion techniques to improve the performance.

The spatio-temporal image obtained by assembling joints mapping and line mapping between the same joints in two consecutive frames can maintain the spatial information and temporal changes with different colors in the jet color map. The variations of spatial information as well as temporal information of each action are easily distinguishable for both UTD-MHAD and MSR-Action3D datasets. The effect of the network architecture doesn't affect more on the performance of the proposed method since the lightweight MobileNetV2 works well comparable with heavy-weight DenseNet121 and ResNet18.

The fusion techniques also accelerate a bit on the performance of the proposed method. Most of the cases, element-wise maximization ensures the highest classification results compared to average and multiplication strategies.

The classification accuracies of individual classes as shown in Figure 11a,b indicate that the action performed by any parts of the human body can be classified effortlessly using deep learning with the proposed spatio-temporal image formation method.

However, the spatio-temporal representation of the 3D skeleton joints is fully confined to the regularities of the frames in an action. The irregular frames generate indiscipline spatial and temporal information that makes it more complicated to predict the correct classes of action.

7. Conclusions

In this paper, a new approach for human action recognition is suggested using deep learning with spatio-temporal image formation from 3D skeleton joints. We analyze the 3D skeleton joints and propose to encode the spatio-temporal image from 3D skeleton joints by mapping the line between the same joints in two neighboring frames. The spatial and temporal information is extracted by preserving the shape of the action and joining the line with different colors along with the temporal changes. We deploy pretrained deep learning models to evaluate the usefulness of spatio-temporal representation of the proposed method. We accomplish the experiments in two separate ways: (i) with individual views (front, side, and top views) and (ii) with fusion mechanisms (average, multiplication, and maximization). The experimental results with individual views show that the front view dataset works well. While applying the features fusion, maximization improves the recognition rate significantly.

We also compare the recognition accuracies with three deep learning models to reveal the sturdiness of our work. The features mined from the spatio-temporal image are invariant to views and speed of the action. Thus, both the pretrained lightweight and heavy weight deep learning models can individualize the actions without any difficulties.

Even though we perform the experiments with individual views along XY, YZ, and ZX planes, the discrimination accuracies of the proposed method outperform the state-of-the-art works with UTD-MHAD and MSR-Action3D benchmark skeleton datasets. The investigation of the experimental results with three different fusion methods are also conducted to bring out the most perfect approach. The overall experimental results of the proposed system using pretrained deep learning with UTD-MHAD and MSR-Action3D skeleton datasets shows better performance.

Author Contributions: Conceptualization, analysis, methodology, manuscript preparation, and experiments, N.T.; data curation, writing—review and editing, N.T., M.K.I. and J.-H.B.; supervision, J.-H.B. All authors have read and agreed to the published version of the manuscript.

Funding: This research is supported by the GRRC program of Gyeonggi province [GRRC Aviation 2017-B04, Development of Intelligent Interactive Media and Space Convergence Application System].

Institutional Review Board Statement: Not applicable.

Informed Consent Statement: Not applicable.

Data Availability Statement: Not applicable.

Acknowledgments: We would like to acknowledge Korea Aerospace University with much appreciation for its ongoing support to our research.

Conflicts of Interest: The authors declare no conflict of interest.

References

1. Eum, H.; Yoon, C.; Lee, H.; Park, M. Continuous human action recognition using depth-MHI-HOG and a spotter model. *Sensors* **2015**, *15*, 5197–5227. [CrossRef]
2. Dawar, N.; Kehtarnavaz, N. Continuous detection and recognition of actions of interest among actions of non-interest using a depth camera. In Proceedings of the IEEE International Conference on Image Processing, Beijing, China, 17–20 September 2017. [CrossRef]
3. Chu, X.; Ouyang, W.; Li, H.; Wang, X. Structured feature learning for pose estimation. In Proceedings of the IEEE Conference on Computer Vision and Pattern Recognition, Las Vegas, NV, USA, 26 June–1 July 2016. [CrossRef]
4. Ziaeefard, M.; Bergevin, R. Semantic human activity recognition: A literature review. *Pattern Recognit.* **2015**, *48*, 2329–2345. [CrossRef]
5. Chaaraoui, A.A.; Padilla-Lopez, J.R.; Ferrandez-Pastor, F.J.; Nieto-Hidalgo, M.; Florez-Revuelta, F. A vision-based system for intelligent monitoring: Human behaviour analysis and privacy by context. *Sensors* **2014**, *14*, 8895–8925. [CrossRef] [PubMed]
6. Wei, H.; Laszewski, M.; Kehtarnavaz, N. Deep Learning-Based Person Detection and Classification for Far Field Video Surveillance. In Proceedings of the 13th IEEE Dallas Circuits and Systems Conference, Dallas, TX, USA, 2–12 November 2018. [CrossRef]
7. Zhu, H.; Vial, R.; Lu, S. Tornado: A spatio-temporal convolutional regression network for video action proposal. In Proceedings of the CVPR, Venice, Italy, 22–29 October 2017. [CrossRef]

8. Wen, R.; Nguyen, B.P.; Chng, C.B.; Chui, C.K. In Situ Spatial AR Surgical Planning Using projector-Kinect System. In Proceedings of the Fourth Symposium on Information and Communication Technology, Da Nang, Vietnam, 5–6 December 2013. [CrossRef]
9. Azuma, R.T. A survey of augmented reality. *Presence: Teleoperators Virtual Environ.* **1997**, *6*, 355–385. [CrossRef]
10. Jalal, A.; Kamal, S.; Kim, D. A Depth Video Sensor-Based Life-Logging Human Activity Recognition System for Elderly Care in Smart Indoor Environments. *Sensors* **2014**, *14*, 11735–11759. [CrossRef]
11. Zheng, Y.; Ding, X.; Poon, C.C.Y.; Lo, B.P.L.; Zhang, H.; Zhou, X.; Yang, G.; Zhao, N.; Zhang, Y. Unobtrusive Sensing and Wearable Devices for Health Informatics. *IEEE Trans. Biomed. Eng.* **2014**, *61*, 1538–1554. [CrossRef] [PubMed]
12. Chen, L.; Ma, N.; Wang, P.; Li, J.; Wang, P.; Pang, G.; Shi, X. Survey of pedestrian action recognition techniques for autonomous driving. *Tsinghua Sci. Technol.* **2020**, *25*, 458–470. [CrossRef]
13. Bloom, V.; Makris, D.; Argyriou, V. G3D: A gaming action dataset and real time action recognition evaluation framework. In Proceedings of the IEEE Computer Society Conference on Computer Vision and Pattern Recognition Workshops, Providence, RI, USA, 16–21 June 2012. [CrossRef]
14. Kim, S.T.; Lee, H.J. Lightweight Stacked Hourglass Network for Human Pose Estimation. *Appl. Sci.* **2020**, *10*, 6497. [CrossRef]
15. Tasnim, N.; Islam, M.; Baek, J.H. Deep Learning-Based Action Recognition Using 3D Skeleton Joints Information. *Inventions* **2020**, *5*, 49. [CrossRef]
16. Sun, Z.; Liu, J.; Ke, Q.; Rahmani, H. Human Action Recognition from Various Data Modalities: A Review. *arXiv* **2020**, arXiv:2012.11866.
17. Pham, H.H.; Salmane, H.; Khoudour, L.; Crouzil, A.; Zegers, P.; Velastin, S.A. Spatio–temporal image representation of 3D skeletal movements for view-invariant action recognition with deep convolutional neural networks. *Sensors* **2019**, *19*, 1932. [CrossRef]
18. Arivazhagan, S.; Shebiah, R.N.; Harini, R.; Swetha, S. Human action recognition from RGB-D data using complete local binary pattern. *Cogn. Syst. Res.* **2019**, *58*, 94–104. [CrossRef]
19. Abdul-Azim, H.A.; Hemayed, E.E. Human action recognition using trajectory-based representation. *Egypt. Inform. J.* **2015**, *16*, 187–198. [CrossRef]
20. Lowe, D.G. Object recognition from local scale-invariant features Computer Vision. In Proceedings of the Seventh IEEE International Conference on Computer Vision, Corfu, Greece, 20–25 September 1999. [CrossRef]
21. Bay, H.; Ess, A.; Tuytelaars, T.; Van Gool, L. Speeded-up robust features (SURF). *Comput. Vis. Image Underst.* **2008**, *110*, 346–359. [CrossRef]
22. Yang, X.; Zhang, C.; Tian, Y. Recognizing actions using depth motion maps-based histograms of oriented gradients. In Proceedings of the 20th ACM International Conference on Multimedia, Nara, Japan, 27–31 October 2012. [CrossRef]
23. Oreifej, O.; Liu, Z. HON4D: Histogram of oriented 4D normals for activity recognition from depth sequences. In Proceedings of the Computer Vision and Pattern Recognition (CVPR), Portland, OR, USA, 23–28 June 2013. [CrossRef]
24. Mahjoub, A.B.; Atri, M. Human action recognition using RGB data. In Proceedings of the 11th International Design & Test Symposium (IDT), Tunisia, Hammamet, 18–20 December 2016. [CrossRef]
25. Al-Akam, R.; Paulus, D. Local and Global Feature Descriptors Combination from RGB-Depth Videos for Human Action Recognition. In Proceedings of the ICPRAM, Funchal, Madeira, Portugal, 16–18 January 2018. [CrossRef]
26. Li, Y.D.; Hao, Z.B.; Lei, H. Survey of convolutional neural network. *J. Comput. App.* **2016**, *36*, 2508–2515.
27. Sherstinsky, A. Fundamentals of recurrent neural network (RNN) and long short-term memory (LSTM) network. *Phys. D Nonlinear Phenom.* **2020**, *404*, 132306. [CrossRef]
28. Simonyan, K.; Zisserman, A. Two-stream convolutional networks for action recognition in videos. *arXiv* **2014**, arXiv:1406.2199. [CrossRef]
29. Zhang, B.; Wang, L.; Wang, Z.; Qiao, Y.; Wang, H. Real-time action recognition with deeply transferred motion vector cnns. *IEEE Trans. Image Proc.* **2018**, *27*, 2326–2339. [CrossRef] [PubMed]
30. Tran, D.; Bourdev, L.; Fergus, R.; Torresani, L.; Paluri, M. Learning spatiotemporal features with 3d convolutional networks. In Proceedings of the IEEE International Conference on Computer Vision, Santiago, Chile, 7–13 December 2015. [CrossRef]
31. Donahue, J.; Hendricks, L.A.; Guadarrama, S.; Rohrbach, M.; Venugopalan, S.; Saenko, K.; Darrell, T. Long-term recurrent convolutional networks for visual recognition and description. In Proceedings of the IEEE Conference on Computer Vision and Pattern Recognition, Boston, MA, USA, 7–12 June 2015.
32. Chen, C.; Jafari, R.; Kehtarnavaz, N. Action recognition from depth sequences using depth motion maps-based local binary patterns. In Proceedings of the IEEE Winter Conference on Applications of Computer Vision, Waikoloa, HI, USA, 5–9 January 2015. [CrossRef]
33. Wu, D.; Shao, L. Silhouette analysis-based action recognition via exploiting human poses. *IEEE Trans. Circuits Syst. Video Technol.* **2012**, *23*, 236–243. [CrossRef]
34. Trelinski, J.; Kwolek, B. Convolutional Neural Network-Based Action Recognition on Depth Maps. In Proceedings of the International Conference on Computer Vision and Graphics, Warsaw, Poland, 17–19 September 2018.
35. Wang, P.; Li, W.; Gao, Z.; Zhang, J.; Tang, C.; Ogunbona, P.O. Action recognition from depth maps using deep convolutional neural networks. *IEEE Trans. Hum. Mach. Syst.* **2015**, *46*, 498–509. [CrossRef]
36. Torchvision Master. Available online: https://pytorch.org/docs/stable/torchvision/models.html (accessed on 1 December 2020).

37. Sandler, M.; Howard, A.; Zhu, M.; Zhmoginov, A.; Chen, L.C. Mobilenetv2: Inverted residuals and linear bottlenecks. In Proceedings of the IEEE Conference on Computer Vision and Pattern Recognition, Salt Lake City, UT, USA, 18–22 June 2018. [CrossRef]
38. Huang, G.; Liu, Z.; Van Der Maaten, L.; Weinberger, K.Q. Densely connected convolutional networks. In Proceedings of the IEEE conference on computer vision and pattern recognition, Honolulu, HI, USA, 21–26 July 2017. [CrossRef]
39. Feichtenhofer, C.; Pinz, A.; Wildes, R.P. Spatiotemporal residual networks for video action recognition. *arXiv* **2016**, arXiv:1611.02155.
40. Chen, C.; Jafari, R.; Kehtarnavaz, N. UTD-MHAD: A Multimodal Dataset for Human Action Recognition Utilizing a Depth Camera and a Wearable Inertial Sensor. In Proceedings of the IEEE International Conference on Image Processing, Quebec City, QC, Canada, 27–30 September 2015. [CrossRef]
41. Li, W.; Zhang, Z.; Liu, Z. Action recognition based on a bag of 3D points. In Proceedings of the IEEE Computer Society Conference on Computer Vision and Pattern Recognition, San Francisco, CA, USA, 13–18 June 2010. [CrossRef]
42. Warchoł, D.; Kapuściński, T. Human Action Recognition Using Bone Pair Descriptor and Distance Descriptor. *Symmetry* **2020**, *12*, 1580. [CrossRef]
43. Vemulapalli, R.; Arrate, F.; Chellappa, R. Human action recognition by representing 3d skeletons as points in a lie group. In Proceedings of the IEEE conference on computer vision and pattern recognition, Columbus, OH, USA, 23–28 June 2014. [CrossRef]
44. Ke, Q.; Bennamoun, M.; An, S.; Sohel, F.; Boussaid, F. A new representation of skeleton sequences for 3d action recognition. In Proceedings of the IEEE conference on computer vision and pattern recognition, Honolulu, HI, USA, 21–26 July 2017. [CrossRef]
45. Presti, L.L.; Cascia, L.M. 3D skeleton-based human action classification: A survey. *Pattern Recognit.* **2016**, *53*, 130–147. [CrossRef]
46. Xia, L.; Chen, C.C.; Aggarwal, J.K. View invariant human action recognition using histograms of 3d joints. In Proceedings of the IEEE Computer Society Conference on Computer Vision and Pattern Recognition Workshops, Providence, RI, USA, 16–21 June 2012. [CrossRef]
47. Yang, X.; Tian, Y.L. Eigenjoints-based action recognition using naive-bayes-nearest-neighbor. In Proceedings of the IEEE Computer Society Conference on Computer Vision and Pattern Recognition Workshops (CVPRW), Providence, RI, USA, 16–21 June 2012. [CrossRef]
48. Zhu, Y.; Chen, W.; Guo, G. Fusing spatiotemporal features and joints for 3d action recognition. In Proceedings of the IEEE Conference on Computer Vision and Pattern Recognition Workshops, Portland, OR, USA, 23–28 June 2013. [CrossRef]
49. Evangelidis, G.; Singh, G.; Horaud, R. Skeletal quads: Human action recognition using joint quadruples. In Proceedings of the 22nd International Conference on Pattern Recognition, Stockholm, Sweden, 24–28 August 2014. [CrossRef]
50. Zanfir, M.; Leordeanu, M.; Sminchisescu, C. The moving pose: An efficient 3d kinematics descriptor for low-latency action recognition and detection. In Proceedings of the IEEE international conference on computer vision, Sydney, Australia, 1–8 December 2013. [CrossRef]
51. Agahian, S.; Negin, F.; Köse, C. Improving bag-of-poses with semi-temporal pose descriptors for skeleton-based action recognition. *Vis. Comput.* **2019**, *35*, 591–607. [CrossRef]
52. Yang, Y.; Deng, C.; Gao, S.; Liu, W.; Tao, D.; Gao, X. Discriminative multi-instance multitask learning for 3D action recognition. *IEEE Trans. Multimed.* **2017**, *19*, 519–529. [CrossRef]
53. Du, Y.; Wang, W.; Wang, L. Hierarchical recurrent neural network for skeleton based action recognition. In Proceedings of the IEEE Conference on Computer Vision and Pattern Recognition, Boston, MA, USA, 8–10 June 2015. [CrossRef]
54. Wang, H.; Wang, L. Modeling temporal dynamics and spatial configurations of actions using two-stream recurrent neural networks. In Proceedings of the IEEE Conference on Computer Vision and Pattern Recognition, Honolulu, HI, USA, 21–26 July 2017. [CrossRef]
55. Zhu, W.; Lan, C.; Xing, J.; Zeng, W.; Li, Y.; Shen, L.; Xie, X. Co-occurrence feature learning for skeleton based action recognition using regularized deep LSTM networks. In Proceedings of the AAAI Conference on Artificial Intelligence, Phoenix, AZ, USA, 12–17 February 2016.
56. Liu, J.; Shahroudy, A.; Xu, D.; Wang, G. Spatio-temporal lstm with trust gates for 3d human action recognition. In Proceedings of the European conference on computer vision, Amsterdam, The Netherlands, 8–16 October 2016. [CrossRef]
57. Song, S.; Lan, C.; Xing, J.; Zeng, W.; Liu, J. An end-to-end spatio-temporal attention model for human action recognition from skeleton data. In Proceedings of the AAAI conference on artificial intelligence, San Francisco, CA, USA, 4–9 February 2017.
58. Li, C.; Zhong, Q.; Xie, D.; Pu, S. Skeleton-based action recognition with convolutional neural networks. In Proceedings of the 2017 IEEE International Conference on Multimedia & Expo Workshops (ICMEW), Hong Kong, 10–14 July 2017.
59. Si, C.; Jing, Y.; Wang, W.; Wang, L.; Tan, T. Skeleton-based action recognition with hierarchical spatial reasoning and temporal stack learning network. *Pattern Recognit.* **2020**, *107*, 107511. [CrossRef]
60. Zhang, S.; Liu, X.; Xiao, J. On geometric features for skeleton-based action recognition using multilayer lstm networks. In Proceedings of the 2017 IEEE Winter Conference on Applications of Computer Vision (WACV), Santa Rosa, CA, USA, 24–31 March 2017. [CrossRef]
61. Du, Y.; Fu, Y.; Wang, L. Skeleton based action recognition with convolutional neural network. In Proceedings of the 2015 3rd IAPR Asian Conference on Pattern Recognition (ACPR), Kualalumpur Westin Hotel Kuala Lumpur, Malaysia, 3–6 November 2015. [CrossRef]

62. Liu, J.; Akhtar, N.; Mian, A. Skepxels: Spatio-temporal Image Representation of Human Skeleton Joints for Action Recognition. In Proceedings of the CVPR Workshops, Long Beach, CA, USA, 16–20 June 2019.
63. Ren, J.; Reyes, N.H.; Barczak, A.L.C.; Scogings, C.; Liu, M. An investigation of skeleton-based optical flow-guided features for 3D action recognition using a multi-stream CNN model. In Proceedings of the IEEE 3rd International Conference on Image, Vision and Computing (ICIVC), Chongqing, China, 27–29 June 2018. [CrossRef]
64. Li, B.; Dai, Y.; Cheng, X.; Chen, H.; Lin, Y.; He, M. Skeleton based action recognition using translation-scale invariant image mapping and multi-scale deep CNN. In Proceedings of the 2017 IEEE International Conference on Multimedia & Expo Workshops (ICMEW), Hong Kong, China, 10–14 July 2017. [CrossRef]
65. Chen, Y.; Wang, L.; Li, C.; Hou, Y.; Li, W. ConvNets-based action recognition from skeleton motion maps. *Multimed. Tools Appl.* **2020**, *79*, 1707–1725. [CrossRef]
66. Li, C.; Hou, Y.; Wang, P.; Li, W. Joint distance maps based action recognition with convolutional neural networks. *IEEE Signal Process. Lett.* **2017**, *24*, 624–628. [CrossRef]
67. Hou, Y.; Li, Z.; Wang, P.; Li, W. Skeleton optical spectra-based action recognition using convolutional neural networks. *IEEE Trans. Circuits Syst. Video Technol.* **2016**, *28*, 807–811. [CrossRef]
68. Wang, P.; Li, Z.; Hou, Y.; Li, W. Action recognition based on joint trajectory maps using convolutional neural networks. In Proceedings of the 24th ACM international conference on Multimedia, Amsterdam, The Netherlands, 15–19 October 2016. [CrossRef]

applied sciences

MDPI

Article

Action Recognition Based on the Fusion of Graph Convolutional Networks with High Order Features

Jiuqing Dong [1,†], Yongbin Gao [1,*], Hyo Jong Lee [2], Heng Zhou [1], Yifan Yao [1], Zhijun Fang [1] and Bo Huang [1]

1 International Joint Research Lab of Intelligent Perception and Control, Shanghai University of Engineering Science, No. 333 Longteng Road, Shanghai 201620, China; jiuqingdong@sues.edu.cn (J.D.); hengzhou@sues.edu.cn (H.Z.); yaoyifan1995@gmail.com (Y.Y.); zjfang@sues.edu.cn (Z.F.); huangbosues@sues.edu.cn (B.H.)

2 Division of Computer Science and Engineering, CAIIT, Jeonbuk National University, Jeonju 54896, Korea; hlee@jbnu.ac.kr

* Correspondence: gaoyongbin@sues.edu.cn; Tel.: +86-182-0190-8363

† Current address: Shanghai University of Engineering Science, No. 333 Longteng Road, Songjiang District, Shanghai 201620, China.

Received: 23 December 2019; Accepted: 18 February 2020; Published: 21 February 2020

Abstract: Skeleton-based action recognition is a widely used task in action related research because of its clear features and the invariance of human appearances and illumination. Furthermore, it can also effectively improve the robustness of the action recognition. Graph convolutional networks have been implemented on those skeletal data to recognize actions. Recent studies have shown that the graph convolutional neural network works well in the action recognition task using spatial and temporal features of skeleton data. The prevalent methods to extract the spatial and temporal features purely rely on a deep network to learn from primitive 3D position. In this paper, we propose a novel action recognition method applying high-order spatial and temporal features from skeleton data, such as velocity features, acceleration features, and relative distance between 3D joints. Meanwhile, a method of multi-stream feature fusion is adopted to fuse these high-order features we proposed. Extensive experiments on Two large and challenging datasets, NTU-RGBD and NTU-RGBD-120, indicate that our model achieves the state-of-the-art performance.

Keywords: human action recognition; graph convolution; high-order feature; spatio-temporal feature; feature fusion

1. Introduct

Action recognition is a very important task in machine vision, and it can be applied to many scenes, such as automatic driving, security, human-computer interaction, and others. Therefore, in recent years, the task of analyzing the actions of people in videos has received more and more attention. The task of action recognition has many problems which are difficult to solve by using traditional methods, such as how to deal with occlusion, illumination changes, the positioning and recognition of human actions in a single frame, and extracting the relationships of frame-wise [1]. Recent approaches in depth-based human action recognition achieved outstanding performance and proved the effectiveness of 3D representation for the classification of action classes. Meanwhile, biological observation studies have also shown that even without appearance information, the locations of a few joints can effectively represent human action [2]. For identifying human action, skeleton-based human representation has attracted more and more attention for its high level of representation and robustness in regard to position and appearance changes. Recently, graph neural networks, which generalize convolutional neural networks to graphs of arbitrary structures, have been adopted in a number of applications and have proved to be efficient for the processing of graph data [3–5]. Skeleton data also can be considered as graph structure data. Therefore, graph-based neural networks have been used for action recognition instead of the traditional CNN networks because of the successful performance. Some graph-based neural networks [6–10] are dedicated to learning both spatial and temporal features for action recognition. Meanwhile, they focus on capturing the hidden relationships among vertices in space. However, they all ignore the high-order information hidden in the skeleton data. For example, the velocity, acceleration, and relative distance information of each vertex can be extracted from the skeleton-based data. The values and directions of velocity are different for various actions. When a human is brushing his/her teeth, the hand should move up and down instead of moving back and forth. When pushing, the hand should move forward rather than backward. In a single frame, for different parts of the body, the acceleration is also varied. Additionally, there are some different actions with similar posture patterns but with different motion speeds. For example, the main difference between "grabbing another person's stuff" and "touching another person's pocket (stealing)" is the motion velocity. Therefore, taking advantage of this high-order information and extracting discriminative representations are necessary.

In this work, our main contributions are as follows:

1. We propose several high-order spatial and temporal features that are important for skeletal analysis: velocity, acceleration, and relative distance between 3D joints. Currently, the spatial features are extracted by a deep network through an adjacent matrix, while the relative distances between 3D joints are not considered in the network; we propose to use deep learning to extract the relative distances between 3D joints, which represent the postural changes of each action. Meanwhile, the widely used temporal features are extracted from the original 3D joints. The high-order motion features, such as velocity and acceleration of the joints, are nontrivial to be learned from the deep network. By explicitly calculating the high-level information as input, the deep network is able to learn higher level spatial and temporal features.
2. A multi-stream feature fusion is proposed to blend the high-order spatial and temporal features; thus, the accuracy of action recognition can be improved significantly. Our method is evaluated on the NTU-RGBD and NTU-RGBD-120 dataset, which achieves state-of-the-art performance on action detection.

2. Related Work

Recent years, NTU-RGBD [11] created a large-scale dataset for human action recognition in 2016. In 2019, NTU-RGBD has been enlarged, which is referred to NTU-RGBD-120 [12]. In addition, there are a lot of public data sets for action recognition, such as [13–19] datasets. The release of high-quality datasets have encouraged more researches on action recognition. These datasets are mainly divided into two categories, RGB-Video based and Skeleton-based. Most of the researches focus on the study of RGB video based and Skeleton-based action recognition.

2.1. RGB-Video Based Methods

In terms of video-based analysis methods, most studies consider video as a sequence of images, and then analyze the images frame by frame to learn spatial and dynamic features. Before the emergence of deep learning, the actions were identified and classified mainly by hand-designed features. [20,21] mainly introduce a method of eliminating background light flow. Their features are more focused on the description of human motion. Three hand-designed motion descriptors HOG(histogram of gradient), HOF(histogram of flow), MBH(motion boundary histograms) have been introduced, which play a very good role in the classification of motion. Since 2014, deep learning mothods have been applied to action recognition. Two-Stream Convolutional Neural Network [22] divides the convolutional neural networks into two parts, one for processing RGB images and one for processing optical flow images, which are ultimately combined and trained to extract spatial-temporal action features. The important contribution is introduced the feature of optical flow into action recognition.

After the two-stream network [22], researchers have been trying to improve its performance, such as [23–25]. Du Tran proposed that C3D [26], for the first time, applied a 3D convolution kernel to detect action and capture the motion information on the time series. After that, the 3D convolutional-based methods became popular, prestigious methods; e.g., T3D [27].

2.2. Skeleton-Based Methods

Skeleton-based analysis benefits from the development of pose estimation algorithms and the application of depth cameras. The original skeleton data are usually estimated from RGB video by a pose estimation algorithm, or directly extracted by Kinetics cameras. In the analysis of the skeleton, how to deal with the relationship among vertices in the single frame and how to deal with the interframe relationship in the skeleton sequence are very important. Some researchers believe that a certain type of action is usually only associated with and characterized by the combinations of a subset of kinematic joints. For identifying an action, not all frames in a sequence have the same importance. In order to assign different weights to different vertices of different frames, attention mechanisms and recurrent neural networks are proposed, such as STA-LSTM proposed by Sijie Song et al. [28]. A spatial attention module adaptively allocates different attentions to different joints of the input skeleton within each frame, and a temporal attention module allocates different attention levels to different frames; e.g., Inwoong Lee et al. proposed TS-LSTM [29] and Spatio-temporal LSTMs [30]. Attention-based LSTM [28] and simple LSTM networks with part-based skeleton representation have been used in [31,32]. These methods either use complex LSTM models which have to be trained very carefully or use part-based representation with a simple LSTM model. Yan et al. proposed ST-GCN [6], which was the first graph-based neural network for action recognition. They believed that the spatial configuration of the joints and their temporal dynamics were significant for action recognition. Therefore, they constructed the spatial temporal graph, which is shown in the Figure 1. This model is formulated on top of a sequence of skeleton graphs, where each node corresponds to a joint of the human body. The edges in the single-frame skeleton are composed of physical connections of the human body, and the edges of the time dimension are composed of the connections between the corresponding joins.

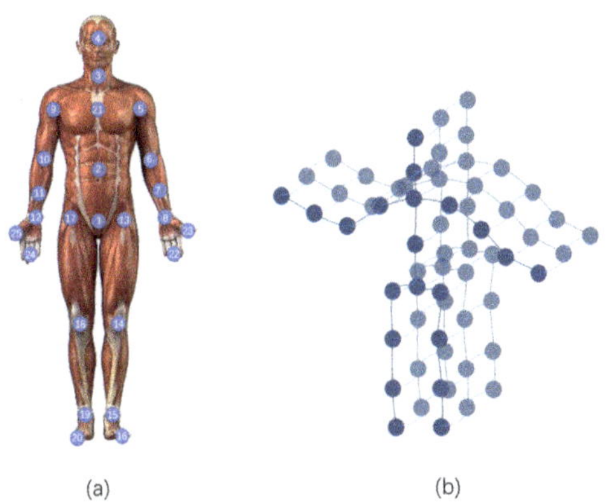

Figure 1. (**a**) The joint labeling of the NTU-RGBD and NTU-RGBD-120 datasets; the 21st node is defined as the gravity center of human. (**b**) The spatio-temporal graph used in ST-GCN [6].

Kalpit divided the skeleton graph into four subgraphs with joints shared across them and taught a recognition model using a part-based graph convolutional network [8]. AGC-LSTM [10] can not only capture features in spatial configuration and temporal dynamics but also explore the co-occurrence relationship between spatial and temporal domains.

In the previous work for action recognition task based on skeleton, only the 3D coordinate information of the joints was utilized. Nevertheless, how to effectively extract discriminative spatial and temporal features is still a challenging problem. Therefore, in this work, we put more attention on the high-order information features. The features we proposed are efficient for action recognition, and the feature fusion method we used is easy to implement.

3. Proposed Graph Convolutional Network with High-Order Features

A graph is good for representing spatial and temporal information. We can transform a frame of the skeleton data to a topological map, which contains joint and edge subsets as shown in Figure 1. A graph neural network can model joint features and structure features simultaneously, which is good method for graph data learning. As the convolution of an image is performed by a convolution kernel with a regular shape, the graph convolution layer is applied on the graph data to generate a high-level feature. Our network model is based on the 2s-AGCN [7]. The overall pipeline of our model is shown in Figure 2, where AGCN is a multi-layer graph convolution network. The networks we proposed consist of five sub-networks. Each sub-network is used to extract a variety of spatial and temporal features. Joint-coordinates, bone, and relative distance are spatial features, and velocity and acceleration of joints and bones are temporal features.

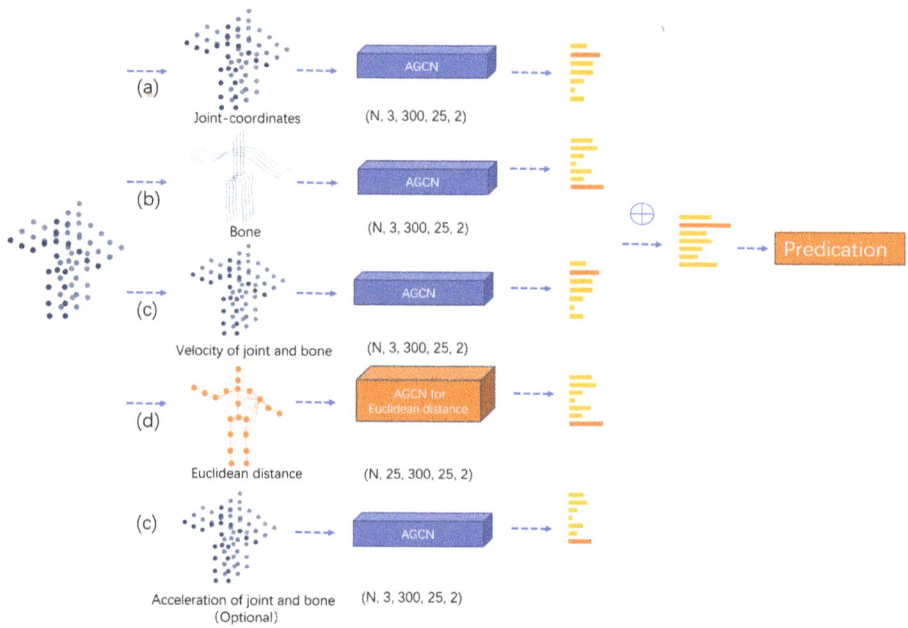

Figure 2. Illustration of the overall architecture of the MS-AGCN. The structure of the AGCN in blue is the same. The only difference between blue and orange is the number of input channels. The final score to obtain the prediction. The shape of input data is presented. (**a**) The joint feature, which is extracted from 3D coordinates of all joints. (**b**) The bone feature, which contains edge information. (**c**) The velocity feature and the acceleration feature, which are calculated from consecutive frames to obtain the temporal feature. (**d**) The relative distance feature of 3D joints; each joint contains relative distance information from others, and we only use one joint as an illustration in the figure.

3.1. Improved Graph Convolutional Network

The implementation of the graph convolution in the spatial domain is not straightforward. Concretely, the input of every layer in the network is actually a $C \times T \times N$ tensor, where C, T, and N are the number of channels, frames, and vertices, respectively. Furthermore, the edge importance matrix was proposed in ST-GCN [6], aiming to distinguish the importance of the edge of skeletons for different actions. The graph convolution operation is formulated as Equation (1) in [6]:

$$f_{out}^{n} = \sum_{s}^{S_v} W_s * (f_{out}^{n-1} * A_S) \odot M_k \tag{1}$$

where the matrix A is initial adjacency matrix proposed in [6], and S is the subset of matrix A, which is similar to the $N \times N$ adjacency matrix. W_s is the weight vector of the $C_{out}^{n} \times C_{out}^{n-1} \times 1 \times 1$ convolution operation, where $*$ denotes the matrix product. M is the edge importance matrix of $n * n$, which is dot multiplied by matrix A.

Equation (1) shows that the edge importance matrix M_k is dot multiplied to A_s. That means that if one of the elements in A_s is zero, it will always be zero, which is unreasonable. Thus, we change the computing method. We add another attention matrix M_{k1} and then multiply matrix M_k. In addition, we use the similarity matrix in 2S-AGCN [7] to estimate the similarity of two joints, and determine whether there is a connection between two vertices and how strong the connection is. Finally, Equation (1) is transformed into Equation (2):

$$f_{out}^{n} = \sum_{s}^{S_v} W_s * (f_{out}^{n-1} * (A_S \oplus M_{k1} \oplus S_k)) \odot M_k \tag{2}$$

where $\oplus$ denotes matrix addition. S_k is the similarity matrix proposed in 2s-AGCN [7]. M_{k1} is a new attention matrix we added.

For the temporal domain, since the number of neighbors for each vertex is fixed as two (corresponding joints in the two consecutive frames), it is straightforward to perform the graph convolution similar to the classical convolution operation. Concretely, we perform $K_t * 1$ convolution on the output feature map calculated above, where K_t is the kernel size of temporal convolution. Spatial convolution is combined with temporal domain convolution into a graph convolution module. The details are shown in Figure 3:

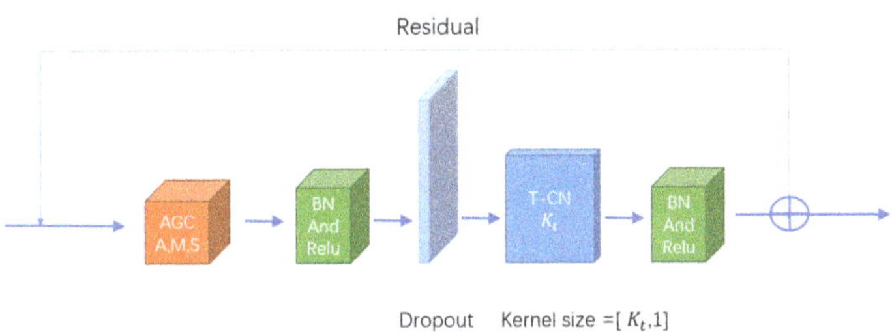

Figure 3. An AGCN block consists of spatial GCN(AGC), temporal GCN(T-CN), and other operations: batch normalization (BN), Relu, dropout, and the residual block. A, M, and S in AGC represent the adjacency matrix, edge importance matrix, and similarity matrix, respectively.

3.2. High-Order Spatial Features

For spatial features in a single frame, we propose combining the bone feature with the relative distance feature of 3D joint. From the Figure 2b,d we can directly get the information contained by these two features.

Bone feature: Shi et al. [7] argued that the coordinate information of the joints could not represent the action of the human body well. Therefore, they proposed the second-order information, which is referred to bone feature, as a feature to enhance the performance on action recognition. The bone feature is extracted from bone data, which includes the length and the direction. Each bone is a human physical connection between joints; Shi defined the person's center of gravity as the target joint; and all directions of the bone are centripetal. Each bone is connected to two joints. The distance from joint $j_1(x_1, y_1, z_1)$ to center of gravity is farther than $j_2(x_2, y_2, z_2)$. The vector representation of bone between j_1 and j_2 is $e_{j_1,j_2} = (x_1 - x_2, y_1 - y_2, z_1 - z_2)$. The direction is from j_1 to j_2.

The number of bones is always one less than the number of joints because each bone is connected to two joints. In order to keep the quantity consistent, we set the empty bone at the center of gravity. The input dimension of the bone network thereby can be the same as the joint network.

Relative distance feature of 3D Joints: We find that the feature extracted from relative distance between 3D joints is useful for skeleton data. For example, nodding requires only a head movement. The acceleration/velocity values of all vertices are zero, except for those of head-related joints. However, the relative distance from the head to the other joints must be changing at all frames and it can not be zero. In addition, we set the distance between the vertex and itself as zero, so the relative distance information of one vertex is 25-dimensional. For a single frame skeleton, we can use a 25×25 matrix to represent it. This matrix is a diagonal matrix, and the principal diagonal elements are zeros. The shape of relative distance information is $(N, 25, T, 25, 2)$, while the shape of other information is $(N, 3, T, 25, 2)$, where N denotes the batch-size we set and T denotes the length of one action sequence.

3.3. High-Order Temporal Features

For temporal features in a single frame, we propose the velocity feature and the acceleration feature. From the Figure 2c, we can directly get the information contained by these two features.

Velocity feature: Velocity features of an action are very crucial for action recognition. Learning velocity features can be relatively complemented with learning features of the joint and bone. For skeleton data, we calculate the motion velocity information of each vertex. The velocity of vertex v_1 is equal to the coordinate of v_1 in the next frame minus the current frame. We can obtain the velocity in three directions (x, y, z), which is helpful for analyzing the action. Velocities of different orientations correspond to different changes. Therefore, velocity analysis in each orientation of the vertex is effective for the final prediction. $j_1^t(x_1^t, y_1^t, z_1^t)$ denotes the coordinates of joint j_1 at t frame. $j_1^{t+1}(x_1^{t+1}, y_1^{t+1}, z_1^{t+1})$ denotes the coordinates of joint j_1 at $T+1$ frame. The velocity of $v_1^t(v_{x1}^t, v_{y1}^t, v_{z1}^t)$ at t frame can be written as:

$$v_1^t(v_{x1}^t, v_{y1}^t, v_{z1}^t) = j_1^{t+1} - j_1^t = (x_1^{t+1} - x_1^t, y_1^{t+1} - y_1^t, z_1^{t+1} - z_1^t) \tag{3}$$

For all joints, Equation (3) is transformed into Equation (4):

$$v^t(v_x^t, v_y^t, v_z^t) = j^{t+1} - j^t = (x^{t+1} - x^t, y^{t+1} - y^t, z^{t+1} - z^t) \tag{4}$$

where v denotes the velocity of all joints in a single frame. Moreover, we calculate the velocity of the edge between the two joints, which is the velocity of the bone. The calculation method of velocity of the bone is the same as that of the joints. We use the 3D velocity of the bone as a feature and feed it into the network. More details of the training results and comparison experiments are provided in Section 4.

Acceleration feature: Acceleration is a physical quantity used to describe the change in velocity. Acceleration is helpful for analyzing action. In one skeleton sequence, the velocities of joints may have different changes. Some joints move at a constant velocity, while other joints accelerate. The acceleration of the joint is equal to the velocity of the current frame minus the corresponding joint of the previous frame. Its feature dimensions are also three-dimensional. Basically, that means that the calculation method of acceleration information is the same as that of the velocity information. Therefore, the features extracted from velocity and acceleration information are similar, while the acceleration uses more frames to calculate the high-order motion. We can calculate acceleration information based on Equation (5) as follows:

$$a_1^t = v_1^{t+1} - v_1^t = (v_{x1}^{t+1} - v_{x1}^t, v_{y1}^{t+1} - v_{y1}^t, v_{z1}^{t+1} - v_{z1}^t) \tag{5}$$

For all joints, Equation (5) is transformed into Equation (6):

$$a^t = v^{t+1} - v^t = (v_x^{t+1} - v_x^t, v_y^{t+1} - v_y^t, v_z^{t+1} - v_z^t) \tag{6}$$

where a_1^t denotes the acceleration of joint j_1 at t frame. v_1^{t+1} and v_1^t denote the velocity of joint j_1 at $t+1$ and t frames, respectively, and a^t denotes the acceleration of all joints in t frame.

3.4. High-Order Features Fusion

Joint Feature: For both of NTU-RGBD and NTU-RGBD-120 datasets, the joint features are extracted from the 3D coordinates of the skeleton sequence. Joint features are fundamental and important features for the skeleton data. Joints coordinates contain abundant spatial and temporal information. Our baseline is a single stream of 3D joint. We also put the joint data into our neural networks to extract joint feature as shown in Figure 2a.

Features extracted only by 3D joints are not enough for action recognition. We propose several pieces of high-order information as input which is effective for action recognition. In front of the input layer, a batch normalization layer is added to normalize the input data. A global average pooling layer is added at the end of the network to pool feature maps of different samples to the same size. Both the input and output of the network are graph-structures data in the graph convolution. The last graph convolution layer generates a discriminative feature and puts it into the standard soft-max classifier. The final score is the weighted summation of the scores of five streams, which is used to predict the action label. We believe that the information contained in the joints, bones, and relative distance is the most fundamental and important. Therefore, these features should be set large weights. The velocity and acceleration information are auxiliary features that strengthen the temporal relationship. These features should be set small weights. The weighted summation method can be formulated as Equation (7):

$$S_f = S_a W_a + S_b W_b + S_c W_c + S_d W_d \tag{7}$$

where S_a, S_b, S_c, and S_d denote the score of joint, bone, joint and bone velocity, and relative distance, respectively. S_f denotes the final score. W_* denotes the weights of scores.

4. Experiments

4.1. Datasets

NTU-RGBD [11] contains 56,880 video clips of 60 actions. The samples were taken from 40 different people by using a Kinect v2 camera. The ages of subjects are between 10 and 35. They used three cameras simultaneously to capture three different horizontal views from the same action. For the camera position setting: the three cameras were at the same height but three different horizontal angles: −45°, 0°, +45° [11]. The dataset provides two methods to evaluate the performance of action classification: cross-subject and cross-view. The training set of cross-subject includes 40,320 samples, which consists of actions performed by 20 subjects. The testing set contains 16,560 samples, which consists of samples taken by another 20 subjects [11]. The cross-subject training set includes 37,920 samples taken by Cameras 2 and 3, and testing set contains 18,960 samples taken by Camera 1.

NTU-RGBD-120 [12] is an extension of NTU-RGBD, which is much larger and provides much more variation of environmental conditions, subjects, camera views, etc. It contains 114,480 video clips of 120 actions. The ages of subjects are between 10 and 57, and heights are between 1.3 m and 1.9 m. The dataset provides two criteria to evaluate the performance of action classification: cross-subject and cross-setup. The training set of cross-subject includes 63,026 samples, which consists of actions performed by 53 subjects. The testing set contains 50,919 samples taken by another 53 subjects [12]. The cross-setup training set includes 54,468 samples consisting of the samples with even collection setup IDs. Testing set contains 59,477 samples, which consists of samples with odd setup IDs. Different setup IDs correspond to changeable vertical heights of the cameras and their distances to the subjects.

4.2. Data Augmentation

During the experiment, we performed the data analysis and gathered statistics on the samples of incorrect recognition. Experiments show that the graph convolution is efficient for the large displacement. However, we also found that the fine-grained actions were more likely to predict incorrectly. Thus, we made a data augmentation for these action categories, which consists of 16 categories. They are drinking water, eating a meal/snack, brushing teeth, clapping, reading, writing, wearing a shoe, taking off a shoe, making a phone call, playing with the phone/tablet, typing on the keyboard, pointing to something with a finger, taking a selfie, sneezing, coughing, touching the head (headache), and touching the neck (neckache). Considering that the datasets were collected in-three-dimensions, and in order to maintain the relative position of the joints unchanged, we performed the rotation of the skeleton data with angles of $\pm 2^{\circ}$.

4.3. Training Detail

All experiments were conducted on the Pytorch deep learning framework. Stochastic gradient descent (SGD) with Nesterov momentum (0.9) was applied as the optimization strategy. The batch size was 64. Cross-entropy was selected as the loss function to backpropagate gradients. The weight decay was set to 0.0001. For both the NTU-RGBD [11] and NTU-RGBD-120 [12] datasets, there are at most two people in each sample of the dataset. If the number of bodies in the sample was less than two, we padded the second body with 0. The maximum number of frames in each sample is 300. For samples with less than 300 frames, we repeated the samples until it reached 300 frames. The learning rate was set as 0.1 and was divided by 10 at the 30th epoch and 40th epoch. The training process was ended at the 50th epoch.

4.4. Ablation Experiment

In Section 3, we add the joints feature, bones feature, joint-velocity feature, bone-velocity feature, and relative distance feature for action recognition. Since the acceleration feature is similar to the velocity feature, the accuracy after fusion is not significantly improved. The ablation studies of different features are shown in Tables 1 and 2, where J, B, JV, BV, and RD denote that features of joint, bone, joint-velocity, bone-velocity and relative-distance, respectively. Obviously, the multi-feature fusion method outperforms the single-feature-based methods on two benchmark evaluations.

Table 1. Comparisons of the validation accuracy with different input modalities on a cross-subject benchmark of the NTU-RGBD dataset.

Methods	Accuracy (%) (No Augmentation)	Accuracy (%) (Augmentation)
Joint	86.7	87.7
Bone	87.0	88.1
Joint-Velocity	86.1	86.8
Bone-Velocity	85.4	86.9
Relative-Distance	87.1	87.5
J+B+JV+BV+RD	**90.5**	**91.7**

Table 2. Comparisons of the validation accuracy with different input modalities on a cross-view benchmark of the NTU-RGBD dataset.

Methods	Accuracy (%) (No Augmentation)	Accuracy (%) (Augmentation)
Joint	93.0	93.8
Bone	93.4	94.3
Joint-Velocity	93.0	93.5
Bone-Velocity	92.7	93.4
Relative-Distance	93.2	94.0
J+B+JV+BV+RD	**95.8**	**96.8**

Tables 3 and 4 are the results on NTU-RGBD-120 dataset. The results also illustrate that the multi-feature fusion method is more effective. The recognition accuracy of our model in NTU-RGBD-120 is slightly lower than the accuracy of NTU-RGBD. The major reasons leading to this result were: (1) NTU-RGBD-120 adds some fine-grained object-related individual actions. For these actions, the body movements are not significant, and the sizes of the objects involved are relatively small; e.g., when "counting money" and "playing magic cube". (2) Some fine-grained hand/finger motions are added in NTU-RGBD-120. Most of the actions in the NTU-RGBD dataset have significant body and hand motions, while the NTU-RGBD-120 dataset contains some actions that have fine-grained hand and finger motions, such as "making an ok sign" and "snapping fingers". (3) The third limitation is the large number of action categories. When only a small set of classes is available, each can be very distinguishable by finding a simple motion pattern or even by the appearance of an interacted object. However, when the number of classes increases, similar motion patterns and interacted objects will be shared among different classes, which makes the action recognition much more challenging.

Table 3. Comparisons of the validation accuracy with different input modalities on cross-subject benchmark of NTU-RGBD-120 dataset.

Methods	Accuracy (%) (No Augmentation)
Joint	80.7
Bone	81.2
Joint-Velocity	78.5
Bone-Velocity	79.2
Relative-Distance	81.5
J+B+JV+BV+RD	**86.4**

Table 4. Comparisons of the validation accuracy with different input modalities on cross-setup benchmark of NTU-RGBD-120 dataset.

Methods	Accuracy (%) (No Augmentation)
Joint	84.3
Bone	84.5
Joint-Velocity	81.4
Bone-Velocity	82.3
Relative-Distance	84.5
J+B+JV+BV+RD	**89.2**

4.5. Comparison with the State-of-the-Art

We compare the final model with the state-of-the-art skeleton-based action recognition methods on NTU-RGBD dataset and NTU-RGBD-120 dataset. The results of the comparison are shown in Tables 5 and 6. The methods used for comparison include the handcraft-feature-based methods [33], RNN-based methods [28,29,34,35], CNN-based methods [36,37], and GCN-based methods [6–10]. From Table 5, we can see that our proposed method achieves the best performances of 96.8% and 91.7% in terms of two criteria on the NTU-RGBD dataset.

Since the NTU-RGBD-120 dataset was released in 2019, there are no related works on this dataset yet. Therefore, we only cite the result of relevant methods mentioned in the original paper of this dataset. As shown in the Table 6, our method is significantly better than the others.

Table 5. Comparisons of the validation accuracy with state-of-the-art methods on the NTU-RGBD dataset.

Methods	Cross-Subject (%)	Cross-View (%)
Lie Group(2014) [33]	50.1	82.8
Trust Gate ST-LSTM(2016) [29]	69.2	77.7
Two-stream RNN(2017) [34]	71.3	79.5
STA-LSTM(2017) [28]	73.4	81.2
VA-LSTM(2017) [35]	79.4	87.6
SR-TSL(2018) [37]	84.8	92.4
HCN(2018) [36]	86.5	91.1
ST-GCN (2018) [6]	81.5	88.3
AS-GCN(2018) [9]	86.8	94.2
PB-GCN (2018) [8]	87.5	93.2
2s-AGCN(2019) [7]	88.5	95.1
AGC-LSTM(2019) [10]	89.2	95.0
ours	**91.7**	**96.8**

Table 6. The results of different methods, which are designed for 3D human activity analysis, using the cross-subject and cross-setup evaluation criteria on the NTU RGB+D 120 dataset.

Methods	Cross-Subject (%)	Cross-Setup (%)
ST-LSTM(2016) [29]	55.7	57.9
Internal Feature Fusion(2017) [38]	58.2	60.9
GCA-LSTM(2017) [30]	58.3	59.2
Multi-Task Learning Network(2017) [39]	58.4	57.9
FSNet(2018) [40]	59.9	62.4
Skeleton Visualization (Single Stream)(2017) [41]	60.3	63.2
Two-Stream Attention LSTM(2018) [38]	61.2	63.3
Multi-Task CNN with RotClips(2018) [42]	62.2	61.8
Body Pose Evolution Map(2018) [43]	64.6	66.9
ours	**86.4**	**89.4**

5. Conclusions

In this work, we propose several spatial and temporal features which are more effective for skeleton-based action recognition. By blending these high-order features, the deep network highlights the spatial changes and temporal changes of the 3D joints, which are crucial for action recognition. It is worth mentioning that the multi-feature fusion method outperforms the single-feature-based method. For each high-order feature added, the accuracy of the final result is improved by about 1%. On the cross-subject and cross-view evaluation criteria of the NTU-RGBD dataset, blending high-order features can improve the accuracy by 3.8% and 2.8%, respectively. What is more, for the cross-subject and cross-setup evaluation criteria of the NTU-RGBD-120 dataset, blending high-order features can improve the accuracy by 5.7% and 4.9%, respectively. The results prove the efficiency of the high-order features and indicate that the performance of our model is the state-of-the-art. In future work, we will add visual information to solve the problems caused by object-related individual actions, and prepare to add some part-based features to solve the problem of fine-grained actions.

6. Patents

Using the method we proposed in this article, we published an invention patent. There is some information about our invention patent. More details can be searched for publication number (CN110427834A) from the official website of the State Intellectual Property Office of China.

China Patent: Jiuqing Dong, Yongbin Gao, Yifan Yao, Jia Gu, and Fangzheng Tian. Behavior recognition system and method based on skeleton data [P]. CN110427834A,2019-11-08.

Author Contributions: Conceptualization, J.D., Y.G. and H.J.L.; methodology, J.D. and B.H.; software, J.D. and H.Z.; validation, J.D., Y.Y. and H.Z.; formal analysis, B.H.; investigation, Y.G.; resources, Z.F.; data curation, J.D.; writing—original draft preparation, J.D.; writing—review and editing, H.J.L. and Y.G.; visualization, H.Z.; supervision, Y.G.; project administration, Z.F.; funding acquisition, Y.G. and Z.F. All authors have read and agreed to the published version of the manuscript.

Funding: This work is supported in part by the Youth Program of National Natural Science Foundation of China (Grand No.:61802253), the National Natural Science Foundation of China (Grand No.:61831018, 61772328). In part by the Chenguang Talented Program of Shanghai under Grand 17Cg59. In part by Basic Science Research Program through the National Research Foundation of Korea(NRF) funded by the Ministry of Education (GR2019R1D1A3A03103736).

Acknowledgments: We thank LEE from Jeonbuk National University for his great help. We also thank anonymous reviewers for their careful reading and insightful comments.

Conflicts of Interest: The authors declare no conflict of interest.

References

1. Dai, R.; Gao, Y.; Fang, Z.; Jiang, X.; Wang, A.; Zhang, J.; Zhong, C. Unsupervised learning of depth estimation based on attention model and global pose optimization. *Signal Process. Image Commun.* **2019**, *78*, 284–292. [CrossRef]
2. Johansson, G. Visual perception of biological motion and a model for its analysis. *Percept. Psychophys.* **1973**, *14*, 201–211. [CrossRef]
3. Defferrard, M.; Bresson, X.; Vandergheynst, P. Convolutional neural networks on graphs with fast localized spectral filtering. In *Advances in Neural Information Processing Systems 29*; Lee, D.D., Sugiyama, M., Luxburg, U.V., Guyon, I., Garnett, R., Eds.; Curran Associates, Inc.: Red Hook, NY, USA, 2016; pp. 3844–3852. Available online: http://papers.nips.cc/paper/6081-convolutional-neural-networks-on-graphs-with-fast-localized-spectral-filtering (accessed on 20 February 2020).
4. Bruna, J.; Zaremba, W.; Szlam, A.; LeCun, Y. Spectral networks and locally connected networks on graphs. *arXiv* **2013**, arXiv:1312.6203.
5. Kipf, T.N.; Welling, M. Semi-supervised classification with graph convolutional networks. *arXiv* **2016**, arXiv:1609.02907.
6. Yan, S.; Xiong, Y.; Lin, D. Spatial temporal graph convolutional networks for skeleton-based action recognition. In Proceedings of the Thirty-Second AAAI Conference on Artificial Intelligence, New Orleans, LA, USA, 2–7 February 2018.
7. Shi, L.; Zhang, Y.; Cheng, J.; Lu, H. Two-Stream Adaptive Graph Convolutional Networks for Skeleton-Based Action Recognition. In Proceedings of the IEEE Conference on Computer Vision and Pattern Recognition, Long Beach, CA, USA, 16–20 June 2019; pp. 12026–12035.
8. Thakkar, K.; Narayanan, P. Part-based graph convolutional network for action recognition. *arXiv* **2018**, arXiv:1809.04983.
9. Li, M.; Chen, S.; Chen, X.; Zhang, Y.; Wang, Y.; Tian, Q. Actional-Structural Graph Convolutional Networks for Skeleton-based Action Recognition. In Proceedings of the IEEE Conference on Computer Vision and Pattern Recognition, Long Beach, CA, USA, 16–20 June 2019; pp. 3595–3603.
10. Si, C.; Chen, W.; Wang, W.; Wang, L.; Tan, T. An Attention Enhanced Graph Convolutional LSTM Network for Skeleton-Based Action Recognition. In Proceedings of the IEEE Conference on Computer Vision and Pattern Recognition, Long Beach, CA, USA, 16–20 June 2019; pp. 1227–1236.
11. Shahroudy, A.; Liu, J.; Ng, T.T.; Wang, G. NTU RGB+ D: A large scale dataset for 3D human activity analysis. In Proceedings of the IEEE Conference on Computer Vision and Pattern Recognition, Las Vegas, NV, USA, 27–30 June 2016; pp. 1010–1019.
12. Liu, J.; Shahroudy, A.; Perez, M.L.; Wang, G.; Duan, L.Y.; Chichung, A.K. NTU RGB+ D 120: A Large-Scale Benchmark for 3D Human Activity Understanding. *IEEE Trans. Pattern Anal. Mach. Intell.* **2019**. [CrossRef] [PubMed]
13. Li, W.; Zhang, Z.; Liu, Z. Action recognition based on a bag of 3D points. In Proceedings of the 2010 IEEE Computer Society Conference on Computer Vision and Pattern Recognition-Workshops, San Francisco, CA, USA, 13–18 June 2010; pp. 9–14.
14. Sung, J.; Ponce, C.; Selman, B.; Saxena, A. Human activity detection from RGBD images. In Proceedings of the Workshops at the Twenty-Fifth AAAI Conference on Artificial Intelligence, San Francisco, CA, USA, 7–11 August 2011.
15. Koppula, H.S.; Gupta, R.; Saxena, A. Learning human activities and object affordances from rgb-d videos. *Int. J. Robot. Res.* **2013**, *32*, 951–970. [CrossRef]
16. Bloom, V.; Makris, D.; Argyriou, V. G3D: A gaming action dataset and real time action recognition evaluation framework. In Proceedings of the 2012 IEEE Computer Society Conference on Computer Vision and Pattern Recognition Workshops, Providence, RI, USA, 16–21 June 2012; pp. 7–12.
17. Liu, C.; Hu, Y.; Li, Y.; Song, S.; Liu, J. PKU-MMD: A large scale benchmark for continuous multi-modal human action understanding. *arXiv* **2017**, arXiv:1703.07475.
18. Jhuang, H.; Garrote, H.; Poggio, E.; Serre, T.; Hmdb, T. A large video database for human motion recognition. In Proceedings of the IEEE International Conference on Computer Vision, Barcelona, Spain, 6–13 November 2011; Volume 4, p. 6.

19. Soomro, K.; Zamir, A.R.; Shah, M. UCF101: A dataset of 101 human actions classes from videos in the wild. *arXiv* **2012**, arXiv:1212.0402.
20. Wang, H.; Kläser, A.; Schmid, C.; Liu, C.L. Dense trajectories and motion boundary descriptors for action recognition. *Int. J. Comput. Vis.* **2013**, *103*, 60–79. [CrossRef]
21. Wang, H.; Schmid, C. Action recognition with improved trajectories. In Proceedings of the IEEE International Conference on Computer Vision, Sydney, Australia, 1–8 December 2013; pp. 3551–3558.
22. Simonyan, K.; Zisserman, A. Two-stream convolutional networks for action recognition in videos. Advances in Neural Information Processing Systems (NIPS). 2014. Available online: http://papers.nips.cc/paper/5353-two-stream-convolutional (accessed on 20 February 2020).
23. Wang, L.; Xiong, Y.; Wang, Z.; Qiao, Y.; Lin, D.; Tang, X.; Van Gool, L. Temporal segment networks: Towards good practices for deep action recognition. In *European Conference on Computer Vision*; Springer: Berlin, Germany, 2016; pp. 20–36.
24. Lan, Z.; Zhu, Y.; Hauptmann, A.G.; Newsam, S. Deep local video feature for action recognition. In Proceedings of the IEEE Conference on Computer Vision and Pattern Recognition Workshops, Honolulu, HI, USA, 21–26 July 2017; pp. 1–7.
25. Zhou, B.; Andonian, A.; Oliva, A.; Torralba, A. Temporal relational reasoning in videos. In Proceedings of the European Conference on Computer Vision (ECCV), Munich, Germany, 8–14 September 2018; pp. 803–818.
26. Tran, D.; Bourdev, L.; Fergus, R.; Torresani, L.; Paluri, M. Learning spatiotemporal features with 3D convolutional networks. In Proceedings of the IEEE International Conference on Computer Vision, Santiago, Chile, 7–13 December 2015; pp. 4489–4497.
27. Diba, A.; Fayyaz, M.; Sharma, V.; Karami, A.H.; Arzani, M.M.; Yousefzadeh, R.; Van Gool, L. Temporal 3D convnets: New architecture and transfer learning for video classification. *arXiv* **2017**, arXiv:1711.08200.
28. Song, S.; Lan, C.; Xing, J.; Zeng, W.; Liu, J. An end-to-end spatio-temporal attention model for human action recognition from skeleton data. In Proceedings of the Thirty-First AAAI Conference on Artificial Intelligence, San Francisco, CA, USA, 4–9 February 2017.
29. Liu, J.; Shahroudy, A.; Xu, D.; Wang, G. Spatio-temporal lstm with trust gates for 3D human action recognition. In *European Conference on Computer Vision*; Springer: Berlin, Germany, 2016; pp. 816–833.
30. Liu, J.; Wang, G.; Hu, P.; Duan, L.Y.; Kot, A.C. Global context-aware attention LSTM networks for 3D action recognition. In Proceedings of the IEEE Conference on Computer Vision and Pattern Recognition, Honolulu, HI, USA, 21–26 July 2017; pp. 1647–1656.
31. Du, Y.; Wang, W.; Wang, L. Hierarchical recurrent neural network for skeleton based action recognition. In Proceedings of the IEEE Conference on Computer Vision and Pattern Recognition, Boston, MA, USA, 7–12 June 2015; pp. 1110–1118.
32. Tao, L.; Vidal, R. Moving poselets: A discriminative and interpretable skeletal motion representation for action recognition. In Proceedings of the IEEE International Conference on Computer Vision Workshops, Santiago, Chile, 7–13 December 2015; pp. 61–69.
33. Vemulapalli, R.; Arrate, F.; Chellappa, R. Human action recognition by representing 3D skeletons as points in a lie group. In Proceedings of the IEEE Conference on Computer Vision and Pattern Recognition, Columbus, OH, USA, 23–28 June 2014; pp. 588–595.
34. Wang, H.; Wang, L. Modeling temporal dynamics and spatial configurations of actions using two-stream recurrent neural networks. In Proceedings of the IEEE Conference on Computer Vision and Pattern Recognition, Honolulu, HI, USA, 21–26 July 2017; pp. 499–508.
35. Zhang, P.; Lan, C.; Xing, J.; Zeng, W.; Xue, J.; Zheng, N. View adaptive recurrent neural networks for high performance human action recognition from skeleton data. In Proceedings of the IEEE International Conference on Computer Vision, Honolulu, HI, USA, 21–26 July 2017; pp. 2117–2126.
36. Li, C.; Zhong, Q.; Xie, D.; Pu, S. Co-occurrence feature learning from skeleton data for action recognition and detection with hierarchical aggregation. *arXiv* **2018**, arXiv:1804.06055.
37. Si, C.; Jing, Y.; Wang, W.; Wang, L.; Tan, T. Skeleton-based action recognition with spatial reasoning and temporal stack learning. In Proceedings of the European Conference on Computer Vision (ECCV), Munich, Germany, 8–14 September 2018; pp. 103–118.
38. Liu, J.; Shahroudy, A.; Xu, D.; Kot, A.C.; Wang, G. Skeleton-based action recognition using spatio-temporal LSTM network with trust gates. *IEEE Trans. Pattern Anal. Mach. Intell.* **2017**, *40*, 3007–3021. [CrossRef]

39. Ke, Q.; Bennamoun, M.; An, S.; Sohel, F.; Boussaid, F. A new representation of skeleton sequences for 3D action recognition. In Proceedings of the IEEE Conference on Computer Vision and Pattern Recognition, Honolulu, HI, USA, 21–26 July 2017; pp. 3288–3297.
40. Liu, J.; Shahroudy, A.; Wang, G.; Duan, L.Y.; Chichung, A.K. Skeleton-based online action prediction using scale selection network. *IEEE Trans. Pattern Anal. Mach. Intell.* **2019**. [CrossRef]
41. Liu, M.; Liu, H.; Chen, C. Enhanced skeleton visualization for view invariant human action recognition. *Pattern Recognit.* **2017**, *68*, 346–362. [CrossRef]
42. Ke, Q.; Bennamoun, M.; An, S.; Sohel, F.; Boussaid, F. Learning clip representations for skeleton-based 3D action recognition. *IEEE Trans. Image Process.* **2018**, *27*, 2842–2855. [CrossRef] [PubMed]
43. Liu, M.; Yuan, J. Recognizing human actions as the evolution of pose estimation maps. In Proceedings of the IEEE Conference on Computer Vision and Pattern Recognition, Salt Lake City, UT, USA, 18–22 June 2018; pp. 1159–1168.

applied sciences

MDPI

Article

Action Recognition Algorithm of Spatio–Temporal Differential LSTM Based on Feature Enhancement

Kai Hu [1,2,*], Fei Zheng [1,3], Liguo Weng [1,2], Yiwu Ding [1] and Junlan Jin [1]

1 School of Automation, Nanjing University of Information Science & Technology, Nanjing 210044, China; zhengfei@nuist.edu.cn (F.Z.); 002311@nuist.edu.cn (L.W.); dyw_0909@nuist.edu.cn (Y.D.); jjl0610@nuist.edu.cn (J.J.)
2 Jiangsu Provincial Collaborative Innovation Center for Atmospheric Environment and Equipment Technology, Nanjing University of Information Science & Technology, Nanjing 210044, China
3 China Telecom Ningbo Branch, Ningbo 315000, China
* Correspondence: 001600@nuist.edu.cn; Tel.: +86-137-7056-9871

Abstract: The Long Short-Term Memory (LSTM) network is a classic action recognition method because of its ability to extract time information. Researchers proposed many hybrid algorithms based on LSTM for human action recognition. In this paper, an improved Spatio–Temporal Differential Long Short-Term Memory (ST-D LSTM) network is proposed, an enhanced input differential feature module and a spatial memory state differential module are added to the network. Furthermore, a transmission mode of ST-D LSTM is proposed; this mode enables ST-D LSTM units to transmit the spatial memory state horizontally. Finally, these improvements are added into classical Long-term Recurrent Convolutional Networks (LRCN) to test the new network's performance. Experimental results show that ST-D LSTM can effectively improve the accuracy of LRCN.

Keywords: action recognition; Long Short-Term Memory; spatio–temporal differential

Citation: Hu, K.; Zheng, F.; Weng, L.; Ding, Y.; Jin, J. Action Recognition Algorithm of Spatio–Temporal Differential LSTM Based on Feature Enhancement. *Appl. Sci.* **2021**, *11*, 7876. https://doi.org/10.3390/app11177876

Academic Editor: Hyo Jong Lee

Received: 6 August 2021
Accepted: 24 August 2021
Published: 26 August 2021

1. Introduction

Human action recognition involves many fields, such as computer vision, image processing, deep learning, etc. It is widely used in human–computer interaction [1], video surveillance [2], intelligent transportation, sports analysis, smart home, etc. It has both academic significance and practical value. Human action recognition aims to identify action categories of moving objects and predict further actions. Its research methods are divided into two categories: one is based on manual feature extraction [3–7], and the other is based on deep learning.

The manual feature extraction method uses a traditional machine learning model to extract features from the video, then it encodes the features, standardizes the encoding vectors, trains the model, and finally carries out prediction and classification. Its advantage lies in its need-based feature extraction, strong pertinence, and simple implementation. There are noises [8] in the datasets, such as illumination, similar actions (like jogging and running), dynamic backgrounds, etc. These noises make manually extracted features ineffective in classification, so its related research is limited. Improved Dense Trajectories [9] (iDT) algorithm is one of the best algorithms based on traditional methods, and its stability is high. Many researchers combined iDT with deep learning methods to achieve higher recognition accuracy. However, the calculation speed of the iDT algorithm is very slow and it can not meet real-time requirements.

Most existing deep learning methods for action recognition are developed from convolutional neural networks. Compared with a single image, the video, which is the target of action recognition, has time-series information. Therefore, the action recognition algorithm based on deep learning pays more attention to time-series features.

In deep networks [10,11], LSTM is often applied in action recognition. It is a kind of time recurrent neural network, which is specially designed to solve the long-term dependence problem of a general Recurrent Neural Network (RNN). Ng et al. [12] proposed a two-stream convolutional network model combined with LSTM, which can reduce computational cost and learn global video features. The two-stream convolutional network uses the CNN network (AlexNet or GoogLeNet) on ImageNet to extract image features and optical flow features of the video frames. Although the accuracy achieved by this network is only fair, it provides a new idea for the research of action recognition. Even if there is a lot of noise in optical flow images, the network combined with LSTM is helpful in classification. Du et al. [13] proposed an end-to-end recurrent pose-attention network (RPAN). The RPAN combines the attention mechanism with the LSTM network to represent more detailed actions. Long et al. [14] proposed an RNN framework with multimodal keyless attention fusion. The network divides visual features (including RGB image features and optical flow features) and acoustic features into equal-length segments, and inputs them to LSTM. The network's advantage is that it reduces computation cost and improves computation speed. The LSTM is applied to extract different features in this network. Wang et al. [15] put forward the I3D-LSTM model by combining Inflated 3D ConvNets (I3D) and LSTM network; it can learn low-level and high-level features well. He et al. [16] proposed the DB-LSTM (Densely-connected Bi-directional LSTM) model; it uses dense hopping connections of Bi-LSTM (Bi-directional Long Short-Term Memory) to strengthen the feature propagation and reduce the number of parameters. This network is also an extended form of the two-stream network. Song et al. [17] used skeleton information to train the LSTM, and divided the network into two sub-networks: a temporal attention sub-network and a spatial attention sub-network.

In general, the deep learning networks of action recognition are mainly based on three types: the two-stream convolutional network, 3D convolutional network, and the LSTM network. Because the data in many practical application scenarios are generated in non-Euclidean space, the deep learning algorithm [18] meets great challenges in graph data, because the data in many practical scenarios are generated in non-Euclidean space. Therefore, action recognition algorithms based on the graph convolutional network are born. With the birth of skeletal datasets such as NTU RGB+D, action recognition algorithms based on the graph convolutional network are further developed. Most of the existing research on deep learning action recognition is based on the basic LSTM model, and many hybrid models are derived.

An action provides information in both the time domain and the space domain, and hence there are time change characteristics and space change characteristics. Although LSTM can deal with time-series information very well, it cannot deal with spatial features and features of temporal and spatial change. To make up for this shortcoming, researchers mostly increase the extraction and processing of spatial features by integrating other deep learning modules. Wang et al. [19] proposed a Spatio–Temporal LSTM (ST-LSTM) for spatio–temporal sequence prediction, which can extract spatio–temporal information. This paper further studies the ST-LSTM structure and considers its internal structure from the point of view of control theory: the ST-LSTM unit has proportional (P) and integral (I) links in the convolutional calculation and forgets temporal and spatial memory states. Compared with the typical PID control architecture, the ST-LSTM lacks the differential (D) unit. From the point of view of practical programming, the weights of gated units are always positive, and the differential calculation cannot be generated inside units. Therefore, this paper introduces the corresponding differential calculation and improves its stacked mode, to improve the feature processing on both time and space at the same time. From the point of view of robot control, the first-order differential in time represents the action speed information, and the first-order differential in space represents the position change information. The contributions of this paper are as follows:

(1) Feature enhancement is carried out. A spatio–temporal differential LSTM unit is proposed, which combines the concept of differential control in PID into the deep learning network. This modification not only considers the influence of time series and spatial position relationship on action recognition, but also increases the influence

of action speed and position change. For ST-LSTM units, a differential part is added for the temporal memory state and spatial memory state. A new LSTM unit named ST-D LSTM is designed.

(2) Feature enhancement is carried out. Due to differential calculation in ST-D LSTM units, the transfer of the two spatial states across time steps is required. Therefore, this paper designs a stacking method, that is, the horizontal transmission of spatial memory states is added. In this paper, the accuracy and stability of the stacked ST-D LSTM units are tested on different datasets; the influence of the number of stacked layers on the accuracy is studied by comparisons with other behavior recognition algorithms.

This paper is divided into five sections. Section 1 introduces the development of action recognition research. Section 2 introduces the methodology of ST-D LSTM. Section 3 introduces the ST-D LSTM unit model. Section 4 tests the performance of the ST-D LSTM model. Section 5 summarizes the work of this paper.

2. Methodology

PID control is the abbreviation of proportional integral and differential control; it has good robustness and high reliability. In the control system, the PID controller calculates the control error according to the given value and the actual output value, and then carries on proportional, integral, and differential operations on the error; finally, it combines the three operation results to obtain the control signal. Generally speaking, PID control is a linear control algorithm based on the estimation of error "past", "present", and "future" information.

Conventional PID control has three correction links: proportional, integral and differential. Their specific functions are as follows: the proportional link reflects control error proportionally, and controls the "present" error of the system. The integral controller produces the control effect at the fastest speed. It reflects the rapidity of PID control. The integral link can memorize error. In view of the "past" error of the system, the integral controller is mainly to eliminate the steady-state error. The strength of the integral function mainly depends on the integral time constant Ti. The larger Ti, the weaker the integral action. The integral function decides the accuracy of the PID control. The differential link can reflect the trend of the error (change rate). Aiming at the "future" error of the system, the differential controller improves the dynamic characteristics of the closed-loop system by acting in advance, which reflects the stability of the PID control.

After the analysis of the classic LSTM model, it is found that the recurrent memory network retains the results of the previous video frame h_{t-1} and inputs the information of current video frame x_t. The network uses different weights w_f and w_i to express the relationship between them. Moreover, it is found that when w_f and w_i are positive, it is a kind of integral (I) relation; when w_f and w_i are negative, it is a kind of differential (D) relation. Due to the weight added to video frames, this is also a proportional (P) relationship. When referring to the code of the ST-LSTM on the Github, it is found that w_f and w_i are positive, so for the ST-LSTM, its internal temporal memory state and spatial memory state have a proportional (P) and integral (I) relationship. From the point of view of PID control, the differential link in the ST-LSTM is missing, so we try to add a differential (D) to the ST-LSTM. From the perspective of deep learning, adding differential is also an idea of feature enhancement.

From the point of view of robot kinematics, action characteristics include posture, position, speed, etc. Taking the manipulator of a robot as an example, the action of the arm includes the translation of the center of mass and the rotation around the centroid. When the manipulator is analyzed by the Newton–Euler equation, the dynamic equation is as follows:

$$\tau = M(\theta)\ddot{\theta} + V(\theta,\dot{\theta}) + G(\theta) \quad (1)$$

In the above formula, $M(\theta)$ is the $n \times n$ mass matrix of the operating arm, $V(\theta,\dot{\theta})$ is the centrifugal force and the Gordian force vector of $n \times 1$. $G(\theta)$ is the gravity vector of $n \times 1$, which depends on the position and velocity. $M(\theta)$ and $G(\theta)$ are complex functions about positions of all joints of the operating arm θ. $\dot{\theta}$ represents the angle velocity. $\ddot{\theta}$ represents the acceleration. Therefore, in the control theory, the control of the robot needs a differential state.

The action recognition network based on deep learning pays attention to the extraction of action posture information. Enhancing the information extraction of limb speed and position changes can improve the final performance of the network. Velocity and position changes are the first-order differential of action temporal state and spatial state, respectively. Therefore, the differential of PID control is introduced into the ST-LSTM to extract more information such as gesture and velocity position changes.

Moreover, although the ST-LSTM increases the influence of the spatial series on the gesture, the time series taken into account by a unit is only the current time series and the last time series. Due to the proportional relationship in the forgetting gate, only part of the previous time series is retained. However, for a complete action, the action is continuous, a complete action cannot be completed in only two short time series. A simple action (such as bowing) needs at least 3–4 time series to complete, and there are actions which are more complex and need more time series to complete. Therefore, it is necessary to retain more time-series information.

Based on the above ideas, the Spatio–Temporal Differential LSTM unit is proposed, it combins the ST-LSTM with a differential module. Moreover, a basic and a multi-layer LSTM are built, to show the performance of the improved differential LSTM network. It is shown that the ST-D LSTM can improve the recognition performance and can capture more action information. The ST-D LSTM can be flexibly embedded into different networks to achieve different applications.

This paper uses the idea of differential control in PID control. The input differential can capture the speed information, and the temporal state differential can capture the change information of action position. The improved ST-D LSTM unit can improve the accuracy of action recognition, and increase the stability of the network.

3. ST-D LSTM

Although researchers made some progress in accuracy, the framework of most algorithms is too complex. The improvement of accuracy depends on the network depth and the number of parameters. This paper proposes the ST-D LSTM structure based on spatio–temporal differential and the suitable stacking method. In order to better demonstrate its performance and usage, we used ST-D LSTM to replace LSTM in the classic LRCN. The network structure can simultaneously take into account temporal and spatial information and complete the transmission of spatial information changes across time steps. In the process of information transmission, the horizontal structure pays attention to the feature extraction on the time flow, and the vertical structure pays attention to the feature extraction on the spatial flow. Moreover, the input differential increases the feature extraction of the limb movement speed. The spatial differential information across video frames can increase the feature extraction of the position changes in different frames. The combination of horizontal and vertical information transmission mode enables the network to combine temporal and spatial features and corresponding features, to make the final judgment. This method can extract more action features without adding other deep learning modules, achieve better recognition accuracy and avoid increasing the network complexity.

3.1. The Internal Structure of the ST-D LSTM

Wang et al. [19] proposed the ST-LSTM structure for spatio–temporal sequence prediction; it can realize information transmission between different layers of LSTM units.

ST-LSTM is improved based on the ConvLSTM [20] structure. Vertically, spatial information memory states between the LSTM units at different layers are similar to the horizontal memory states of the ConvLSTM unit, and the spatio–temporal memory module is added based on the original horizontal memory state. The ST-LSTM transmits the information of hidden layers, and increases the transmission of spatial information in the vertical direction, to realize the transmission of memory information between different layers in this time step. ST-LSTM is the core part of the PredRNN algorithm.

For action recognition, limb position change is a vital feature; that is, the time change and position change should be considered at the same time. The zigzag transfer method enables the stacked ST-LSTM unit to transfer the spatial state longitudinally at each time step. Although the PredRNN algorithm considers both temporal and spatial features through the zigzag cross-layer connection, it ignores changes of temporal and spatial features. For this reason, the SpatioTemporal Differential LSTM (ST-D LSTM) unit is proposed, with the idea of spatio–temporal variation based on the spatial memory state of the ST-LSTM unit.

The ST-D LSTM is similar to the LSTM. It also contains the forgetting gate, the input gate, and the output gate. Furthermore, the ST-D LSTM unit also contains two cell states: the temporal memory module C_{t-1}^{l} and the spatial memory module S_{t}^{l-1}. The temporal memory module stores the temporal characteristic information of the previous $t-1$ moments in the same layer units, while the spatial memory module stores the spatial characteristic information of different layer units. x_t represents input in the ST-D LSTM unit; h_{t-1}^{l} is the hidden layer state. k_t, i_t and f_t are the conversion mechanism, the input gate and the output door of temporal memory, respectively. k_t', i_t' and f_t' are the conversion mechanism, the input, and the output door of the spatial memory, respectively. The output gate o_t combines temporal memory and spatial memory.

Similarly to the differential part in the PID control, the differential module of spatial memory state is added to the original LSTM unit according to the connection mode of the input gate. The "future" error, that is the characteristic change information, is introduced into the present state by integral calculation, so that the network can improve the accuracy and stability. In addition, the input differential module is added at the same time to increase the propagation of spatial features in the same layer of the LSTM unit along the horizontal time step, so that the network can take into account the temporal information, the limb moving speed and trajectory. The ST-D LSTM internal structure diagram is shown in Figure 1.

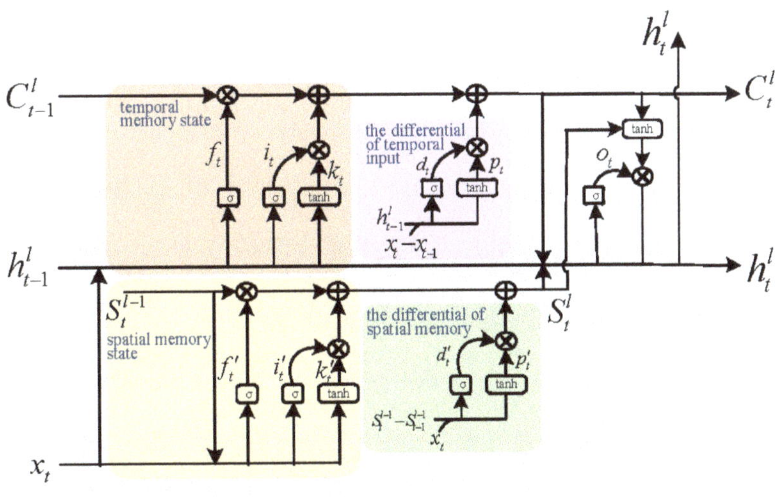

Figure 1. The internal structure diagram of the ST-D LSTM.

In the mathematical model, t is a small value, so the input differential $\frac{dx(t)}{dt}$ is approximated to $x_t - x_{t-1}$, that is $\frac{dx(t)}{dt} \approx x_t - x_{t-1}$. Similarly, the spatial memory differential can be expressed as $S_{t}^{l-1} - S_{t-1}^{l-2}$. Approximation can make the calculation easier while realizing the differentiation of the input and spatial state. The differential processing is similar to the optical flow method in image processing. The input differentiation provides information on the image speed change and the spatial memory differentiation provides the position change information of the image.

In this paper, the LRCN network framework is used for subsequent experiments, and the input to the ST-D LSTM unit is features extracted by the CNN, so convolutions are not used in the ST-D LSTM unit, and each gate can be considered a fully connected connection. The temporal memory state equations of the forgetting gate, input gate, and input differentiation in the ST-D LSTM unit are shown in Equations (2) and (3):

$$\begin{pmatrix} f_t \\ i_t \\ k_t \end{pmatrix} = \begin{pmatrix} \sigma \\ \sigma \\ \tanh \end{pmatrix} \left(W \cdot \left[x_t, h^l_{t-1} \right] \right) \tag{2}$$

$$\begin{pmatrix} d_t \\ p_t \end{pmatrix} = \begin{pmatrix} \sigma \\ \tanh \end{pmatrix} \left(W \cdot \left[x_t - x_{t-1}, h^l_{t-1} \right] \right) \tag{3}$$

The spatial memory equations of the forgetting gate, input gate and differentiation in the ST-D LSTM unit are shown in Equations (4) and (5):

$$\begin{pmatrix} f'_t \\ i'_t \\ k'_t \end{pmatrix} = \begin{pmatrix} \sigma \\ \sigma \\ \tanh \end{pmatrix} \left(W \cdot \left[x_t, S^{l-1}_t \right] \right) \tag{4}$$

$$\begin{pmatrix} d'_t \\ p'_t \end{pmatrix} = \begin{pmatrix} \sigma \\ \tanh \end{pmatrix} \left(W \cdot \left[x_t, S^{l-1}_t - S^{l-1}_{t-1} \right] \right) \tag{5}$$

When $l = 1$, $S^{l-1}_t = S^L_t, S^{l-1}_{t-1} = S^L_{t-2}$.

The updated temporal cell state and spatial cell state are:

$$C^l_t = f_t \circ C^l_{t-1} + i_t \circ k_t + d_t \circ p_t \tag{6}$$

$$S^l_t = f'_t \circ S^{l-1}_t + i'_t \circ k'_t + d'_t \circ p'_t \tag{7}$$

The equation of the output gate in the ST-D LSTM unit is:

$$O_t = \sigma(w_O \cdot [h^l_{t-1}, C^l_t, S^l_t, x_t] + b_O) \tag{8}$$

$$\mathrm{h}^l_t = O_t \circ \tan(C^l_t, S^l_t) \tag{9}$$

3.2. The Stacked Mode of the ST-D LSTM Unit

The differential calculation of spatial states in ST-D LSTM units requires the transmission of spatial memory in the same layer across two steps. To cooperate with the spatial state differentiation, an improved transfer method of state memories is proposed. The spatial memory at each step is divided into horizontal and vertical transmission after output, and the differential calculation is carried out outside the unit. This method will not increase the amount of data in transmission, so the speed of the network will not be too slow. The connection is shown in Figure 2.

As shown in Figure 2, based on the traditional LSTM cell stacked mode, and with reference to the vertical propagation of the PredRNN spatial memory state, the split propagation is carried out to increase the horizontal transmission of the spatial memory. Moreover, the differential calculation is carried out outside the unit; that is, the differentiation between the spatial memory of the upper layer at this time step S^{l-1}_t and S^{l-1}_{t-1} the spatial memory at the previous step is added. In this connection mode, the temporal memory state is only transmitted horizontally, and the temporal information features extracted by each layer are partially retained and input to the next layer. The horizontal transmission of the spatial memory state makes the location feature changes with the same precision rate to be transmitted. For the unit in the first layer at time t, the differentiation between the spatial memory state of the previous time step S_{t-1}^{l} and that of the time step S_{t-2}^{l} is added; that is, $S_{t-1}^{l}-S_{t-2}^{l}$. The spatial memory state output of the unit is divided into two directions, one direction continues the longitudinal spatial memory transmission, and the other direction performs the differential calculation. This connection mode can increase the information of position change without affecting the calculation speed, and subsequent experiments will verify its effectiveness.

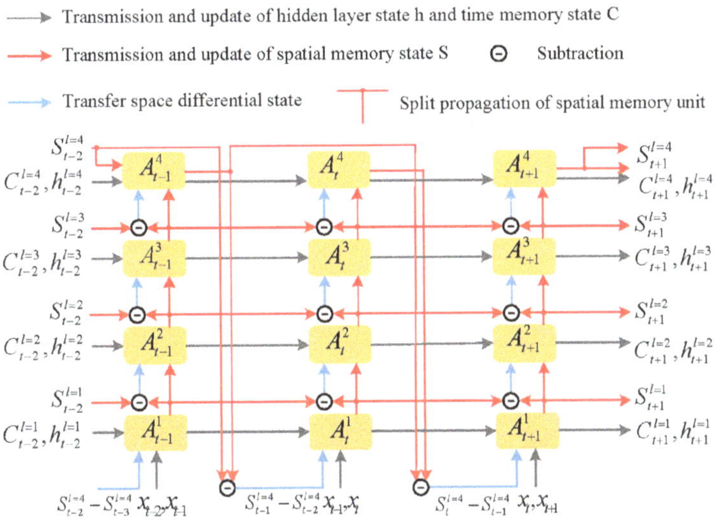

Figure 2. The connection mode between ST-D LSTM units.

4. Experiments

In order to show the performance of the ST-D LSTM unit, this section carries out experiments on the three datasets, UCF-101, HMDB-51, and Hollywood2. The results directly prove its advantages in accuracy, and the influence of the stack number of ST-D LSTM units on recognition accuracy is further studied. Finally, this section compares the recognition accuracy of the ST-D LSTM unit with other algorithms on UCF-101 and HMDB-51.

4.1. Datasets

Research teams, both overseas and domestic, usually use human action datasets in algorithm training to detect the algorithm's accuracy and robustness. The dataset has at least the following two essential functions:

(1) The researchers do not have to consider the process of collection and pretreatment.
(2) It is able to compare different algorithms under the same standard.

The KTH dataset [21] was released in 2004. The KTH dataset includes six kinds of actions (including strolling, jogging, running, boxing, waving, and clapping) performed by 25 people in 4 different scenes. The dataset has 2391 video samples and includes scale transformation, clothing transformation, and lighting transformation. However, the shooting camera is fixed, and the background is similar.

The Weizmann dataset [22] was released in 2005 and includes nine people completing ten kinds of actions (bending, stretching, high jump, jumping, running, standing, hopping, walking, waving1, and waving). In addition to category tags, the dataset contains silhouettes of people in the foreground and background sequences to facilitate background extraction. However, the dataset has a fixed perspective and simple backgrounds.

The above two datasets are released early. The citation rate of these datasets is high. However, with the rapid development of action recognition, there are shortcomings: the background is simple, the angle is fixed, and each video has only one person. The above two datasets already cannot satisfy actual action recognition requirements, so they are rarely used now.

The Hollywood2 dataset [23] was released in 2009. The video data in the dataset are collected from Hollywood movies. There are 3669 video clips in total, including 12 action categories (such as: answering the phone, eating, driving, etc.) extracted from 69 movies and 10 scenes (outdoor, shopping mall, kitchen, etc.). The dataset is close to real situations.

The University of Central Florida released the UCF-101 dataset [24] in 2012. The dataset samples include various action samples collected from TV stations and video samples saved from YouTube. There are 13,320 videos, including five types of actions (human–object interaction, human–human interaction, limb movements, body movement, and playing musical instruments), and 101 class-specific small actions.

Brown University released the HMDB-51 dataset [25] in 2011. The samples come from video clips of YouTube. There are 51 types of sample actions and 6849 videos in total. Each type of sample action in the dataset contains at least 101 videos.

The UCF-101 dataset and the HMDB-51 dataset have many action types and a wide range of actions. The scenes in the Hollywood2 dataset are more complex and closer to real life. To comprehensively verify the ST-D LSTM unit's performance, three datasets, UCF-10, HMDB-51, and Hollywood2, were chosen for training and testing. Furthermore, the ST-D LSTM unit's performance was tested in the above three databases, respectively. The UCF-101 and HMDB-51 datasets are commonly used in deep learning algorithms, so these two datasets were used when the ST-D LSTM unit was compared with other deep learning-based algorithms.

4.2. Method

To test the accuracy of the ST-D LSTM, a simple Long-term Recurrent Convolutional Network [26] (LRCN) is adopted in experiments.

The LRCN connects the stacked LSTM model directly with the CNN; it extracts the spatial features of the pre-trained CNN and inputs spatial features to the LSTM model to learn the temporal and spatial features at the same time. The framework of LRCN is shown in Figure 3. The model first converts the video to frame images, then uses the pre-trained CNN to extract the spatial features of the frame images; next, it inputs the extracted features into the ST-D LSTM network to extract the temporal and spatial information. As a result, the network learns the temporal relationship from spatial features of frame images. Finally, the result is classified by Softmax.

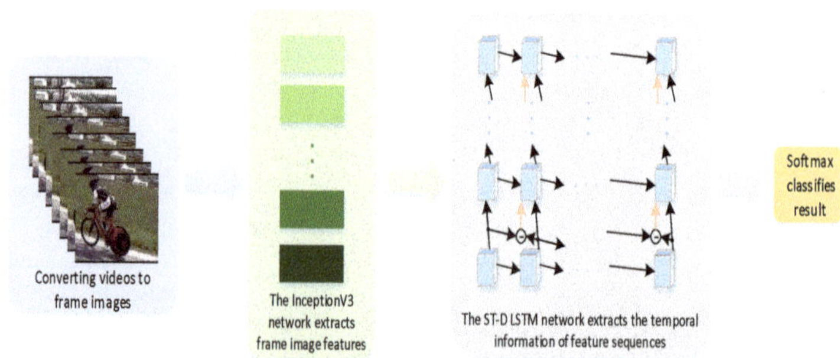

Figure 3. The LRCN network framework based on the ST-D LSTM.

In the experiment, the convolutional network is used to extract spatial features and the LSTM network is used to extract temporal features. However, it is slightly different from the original LRCN. In CNN feature extraction, the InceptionV3 with less computation but high performance is used to extract image features. In the LSTM network, the number of hidden layers is defined according to the requirements of computer performance, and the LSTM unit uses the ST-D LSTM unit.

The ST-D LSTM unit is applied to the network model in Figure 3, and is evaluated in terms of accuracy, loss and standard deviation. To better show the improved LSTM units' performance, experiments were carried out on three datasets of HMDB-51, UCF-101 and Hollywood2, respectively. The experiments use only a single variable of the LSTM unit. The input data model, training parameters, and other parameters are consistent. The batch_size is 32, the number of the hidden layers is 5, the hidden layers' parameter is 1024, the full connection layers' parameter is 512, and the loss function is the classic cross-entropy function. In the follow-up experiments, one layer, two layers, three layers, four layers, and five hidden layers are used to study the influence of the number of hidden layers on recognition accuracy.

The assessment method is the direct hold-out method. To avoid the data division influencing the result and increase the final evaluation result's fidelity, the training set and the testing set are divided in the same way at each type of action in every dataset in the experiment. The training set accounts for 70% of the total dataset, and the testing set accounts for 30% of the total dataset. Simultaneously, to make the results more stable and reliable, this paper uses multiple hold-outs to take the average of the results. Each LSTM unit uses the hold-out method to divide the dataset. After an experiment is concluded, the dataset is re-divided, and the experiment is performed again, and this is then repeated. The experiments were performed using three datasets of five different LSTM units, each repeated three times. At last, the average accuracy of three experimental results is the result of the LSTM unit.

The experiment's hardware configuration is an Intel I7-9700K CPU, two Nvidia GeForce GTX2080Ti graphics cards, 4 × 16 G total 64 GB memory. The software environment was configured as Ubuntu 16.04, CUDA 8.0, Cudnn 6.0 for CUDA 8.0, TensorFlow 1.4, and Python 3.5.

4.3. Experimental Results and Analysis

4.3.1. The Influence of Internal Structure on Accuracy

In this experiment, the LRCN network was selected as the basic network framework. The basic LSTM unit, ST-LSTM unit and ST-D LSTM unit were used in the stacking part of the LSTM, and the common connection mode; the zigzag connection mode and the differential connection mode corresponding to each unit were selected. The number of hidden layers was 5 and the parameter was set to 1024. Figures 4 and 5 show the comparison of accuracy and loss optimization of the basic LSTM unit, ST-LSTM unit, and ST-D LSTM unit in the three datasets, respectively.

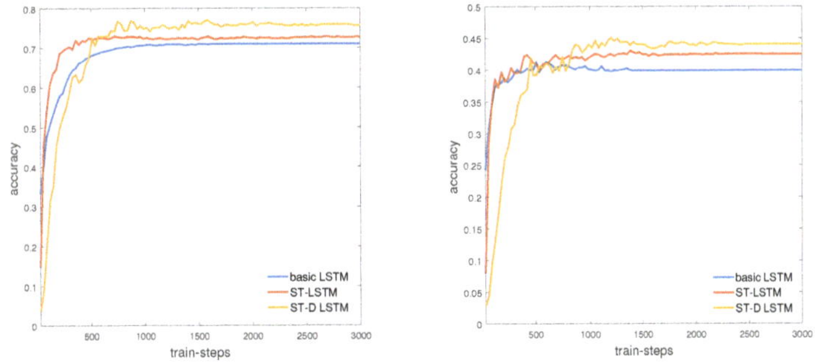

(a) UCF-101 (b) HMDB-51

Figure 4. *Cont.*

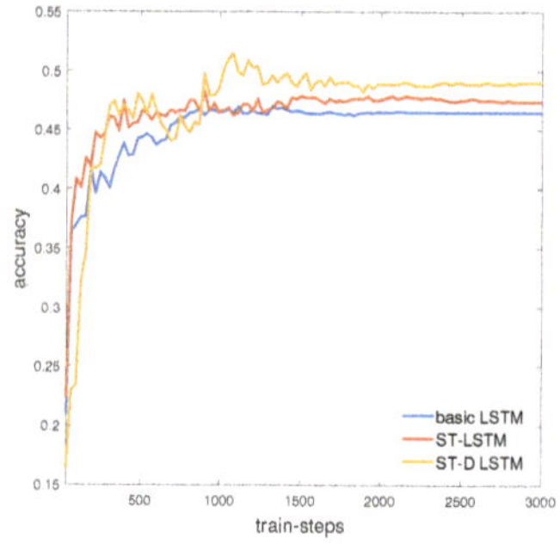

(c) Hollywood2

Figure 4. The comparison of different LSTM units on three datasets in accuracy.

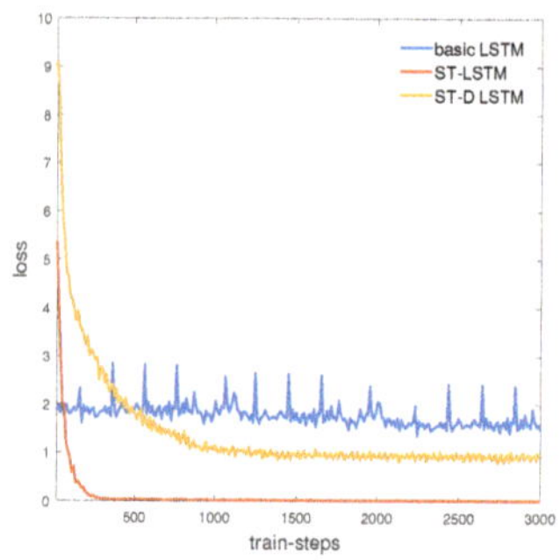

(a) UCF-101

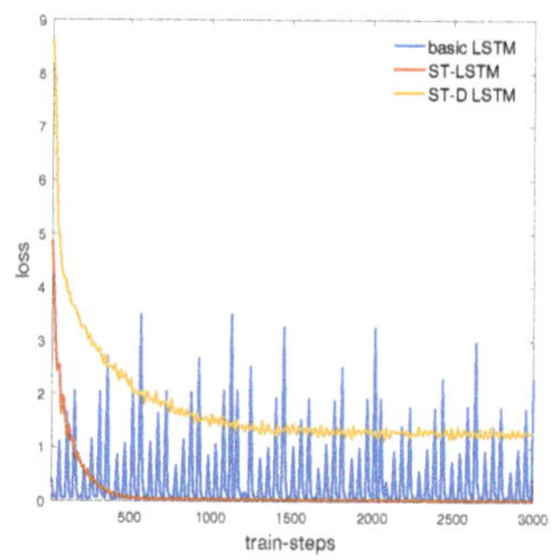

(b) HMDB-51

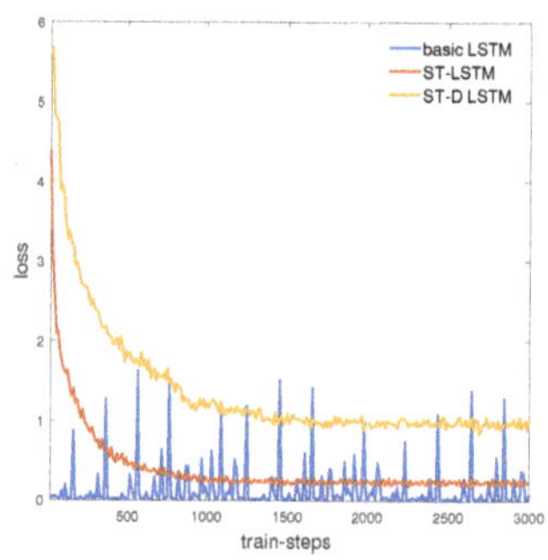

(c) Hollywood2

Figure 5. The comparison of different LSTM units on three datasets in loss.

Figure 4 shows the accuracy of the basic LSTM unit, ST-LSTM unit, and ST-D LSTM unit. Table 1 shows the final accuracy when the accuracy reaches a stable stage. As shown in Figure 4 and Table 1, due to the differential transmission, the accuracy of the ST-D LSTM unit is the slowest to reach the stable stage, but its final recognition accuracy is the highest. Thus, the temporal state differential and input differential modules can increase the extraction and improve the accuracy.

As shown in Figure 5, the loss of the ST-D LSTM can finally converge to a stable stage, but the convergence rate and the final convergence value are slightly lower than those of the ST-LSTM, which may be caused by the differential module. To objectively compare the loss optimization processes, the same loss function and optimizer are used in different

LSTM units. It can be found that the loss value of ST-D LSTM unit still has room to be optimized, and the loss function can be further designed and optimized.

Table 1. The accuracy of different LSTM units on three datasets.

	UCF-101	**HMDB-51**	**Hollywood2**
basic LSTM	71.15%	39.99%	46.49%
ST-LSTM	72.73%	42.53%	47.41%
ST-D LSTM	75.70%	44.11%	49.02%

4.3.2. The Influence of the Number of Stacking Layers

In the performance verification and comparison experiment, the recognition accuracy obtained by stacking five-layer ST-D LSTM units was used. However, in the actual process of parameter adjustment, it can be found that the performance of stacking different layers of ST-D LSTM units is different in accuracy and training speed. Therefore, the ST-D LSTM units are stacked one layer, two layers, three layers, four layers, and five layers, respectively, and the LRCN network is applied for experiments. In this experiment, only the number of layers varies, the other parameters such as batch size, parameters of the hidden layer, training steps and so on are consistent. The process of accuracy climbing is shown in Figure 6, and the stable accuracy is shown in Table 2.

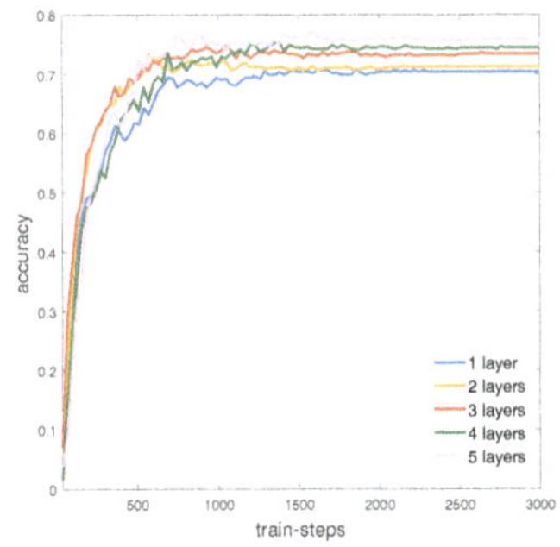

(**a**) UCF-101

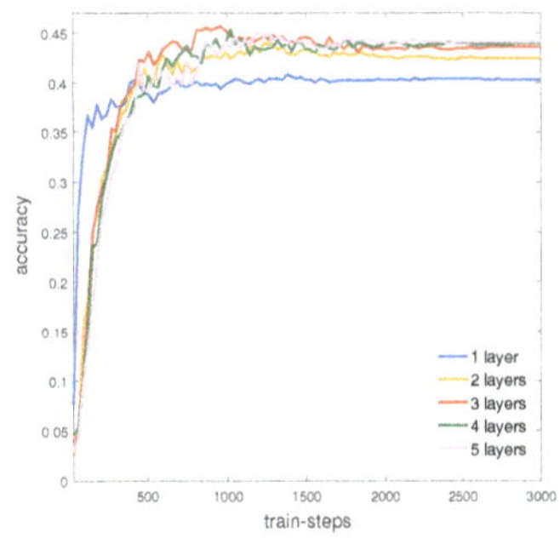

(**b**) HMDB-51

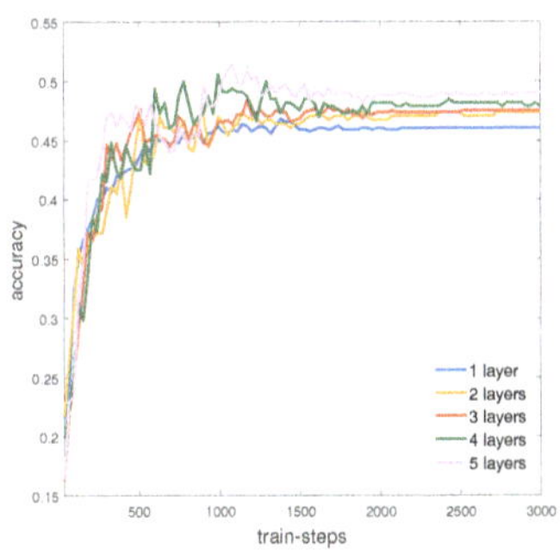

(**c**) Hollywood2

Figure 6. The comparison of the accuracy increasing process of stacked ST-D LSTM units with different layers.

When ST-D LSTM units with different layers are stacked, there is a significant difference in training speed. The impacts are studied from two aspects of accuracy and training speed. The network training speed is shown in Table 3. In the speed experiment, the fps index is used, that is, the number of video frames processed in one second.

Table 2. The accuracy comparison of stacked ST-D LSTM units with different layers.

	UCF-101	HMDB-51	Hollywood2
1 layer	70.47%	40.39%	46.12%
2 layers	71.32%	42.51%	47.41%
3 layers	73.44%	43.61%	47.54%
4 layers	74.48%	44.01%	48.21%
5 layers	75.70%	44.11%	49.02%

Table 3. The training speed comparison of stacked ST-D LSTM units with different layers (in frames per second).

	UCF-101	HMDB-51	Hollywood2
1 layer	38	35	56
2 layers	31	24	42
3 layers	27	17	34
4 layers	16	10	23
5 layers	11	10	14

Through experiments, it can be found that increasing the number of layers can improve the accuracy. When five layers are stacked, ST-D LSTM units perform the best on the HMDB-51, UCF-101, and Hollywood2 datasets. However, increasing layers will also increase the time needed for reading data and training. Stacking too many layers will slow down the training. When studying the translation task based on LSTM, Wu et al. [27] found that the network can work well by simply stacking four layers of LSTM units, and six layers is the limit. Stacking more than eight layers makes the network fail. Table 2 shows that when ST-D LSTM units are stacked to layers 4 and 5 on the HMDB-51 dataset, the recognition accuracy only increases slightly. Therefore, although stacked LSTM layers can increase network performance, in general, the LSTM units can better balance the training speed and accuracy with 4–5 stacked layers.

4.3.3. Comparison of ST-LSTM and ST-D LSTM in Terms of Stability and Accuracy

For stability experiments, the ST-LSTM and ST-D LSTM units, which are both stacked five-layers, were applied to the LRCN network for three repeated experiments. The average accuracy was calculated as the final result. The standard deviation was calculated to compare the stability of the ST-LSTM unit and ST-D LSTM unit. The average accuracy and standard deviation of the three repeated experiments are plotted. As shown in Figure 7, in three different datasets, the accuracy of the ST-D LSTM unit is higher than that of the ST-LSTM unit, but the standard deviation is not higher than that of the ST-LSTM unit. Therefore, the ST-D LSTM unit has good stability.

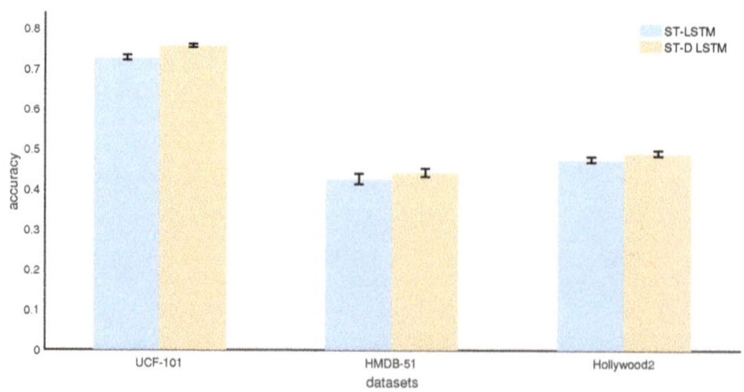

Figure 7. The comparison of accuracy and standard deviation between the ST-LSTM and ST-D LSTM.

In order to further verify the performance of the ST-D LSTM unit, the ST-D LSTM unit is compared with other deep learning algorithms. The experiment is performed on the UCF-101 and HMDB-51 datasets and results are shown in Table 4.

Table 4. The accuracy comparison of various deep learning algorithms on UCF-101 and HMDB-51 datasets.

		UCF-101	HMDB-51
Two-stream Convolutional Network [28]		73.00%	40.50%
LRCN	basic LSTM	71.15%	39.99%
	ST-LSTM	72.73%	42.53%
	BiLSTM [22]	70.00%	39.81%
	LSTM+attention	72.40%	41.50%
	ST-D LSTM	75.70%	44.11%

The ST-D LSTM is compared with the two-stream convolutional network, the LRCN network with an attention mechanism, and the LRCN network with BiLSTM. Due to differential calculation, the ST-D LSTM unit is more sensitive to action changes and can achieve high accuracy on the UCF-101 and the HMDB-51 datasets.

5. Conclusions and Prospect

Human action recognition has many applications in today's society. Although existing networks can achieve good accuracy, many have limitations in application scenarios. In this paper, the internal structure of the LSTM unit is improved. A ST-D LSTM unit with high accuracy and high reliability is proposed and applied to action recognition. The ST-D LSTM unit updates and transmits action spatial feature change information: the differential operation of the spatial memory state is carried out in the process of transmission, and hence the ST-D LSTM has proportional, integral and differential operations. The ST-D LSTM can satisfy the requirements of rapidity, accuracy, and stability. In the verification experiments, the accuracy of the ST-D LSTM unit is better than that of the ST-LSTM unit in the UCF-101, HMDB-51, and Hollywood2 datasets, and its stability is no less than that of the ST-LSTM unit. However, due to the methods of data reading and transferring in deep learning, the differential calculation leads to a double increase in the amount of data. Therefore, the speed of the ST-D LSTM network cannot be guaranteed, and the amount of parameters needs to be further optimized. Compared with other action recognition algorithms based on deep learning, the ST-D LSTM unit shows good accuracy in the UCF-101 and HMDB-51 datasets. The ST-D LSTM unit is applied to the LRCN network in the experiments. Because the LRCN algorithm extracts features before processing them, the LRCN network applying in the ST-D LSTM unit does not achieve the end-to-end training. In the follow-up research, the ST-D LSTM unit can use convolutional calculations in the internal structure. The ST-D LSTM unit can be applied to other network frameworks to achieve the end-to-end training. Moreover, the ST-D LSTM unit can also be applied to other scenarios, such as attitude estimation, sequence prediction, and so on.

Author Contributions: Conceptualization, K.H. and F.Z.; methodology, K.H.; software, F.Z.; validation, F.Z., Y.D. and J.J.; formal analysis, F.Z., J.J.; investigation, L.W.; resources, K.H.; data curation, K.H.; writing—original draft preparation, F.Z.; writing—review and editing, F.Z., L.W. and K.H; visualization, F.Z.; supervision, K.H.; project administration, K.H.; funding acquisition, K.H. All authors have read and agreed to the published version of the manuscript

Funding: The research in this paper is supported by the National Natural Science Foundation of China (42075130), Industry prospect and key core technology key projects of Jiangsu Province (BE2020006-2), the key special project of the National Key R&D Program (2018YFC1405703), NUIST Students' Platform for Innovation and Entrepreneurship Training Program (202010300050Z). I would like to express my heartfelt thanks to the reviewers who submitted valuable revisions to this article.

Institutional Review Board Statement: Ethical review and approval were waived for this study, due to the data being provided publicly.

Informed Consent Statement: Not applicable.

Data Availability Statement: The code used to support the findings of this study are available from the corresponding author upon request. The data are from the open dataset of HMDB-51 (https://serre-lab.clps.brown.edu/resource/hmdb-a-large-human-motion-database/, accessed on 23 August 2021), UCF-101 (www.crcv.ucf.edu/data/UCF101.php, accessed on 23 August 2021), Hollywood2 (www.di.ens.fr/~laptev/actions/hollywood2/, accessed on 23 August 2021).

Conflicts of Interest: The authors declare no conflict of interest.

References

1. Roitberg, A.; Perzylo, A.; Somani, N.; Giuliani, M.; Rickert, M.; Knoll, A. Human activity recognition in the context of industrial human-robot interaction. In Proceedings of the Signal and Information Processing Association Annual Summit and Conference (APSIPA), Chiang Mai, Thailand, 9–12 December 2014; pp. 1–10.
2. Wang, H.; Kläser, A.; Schmid, C.; Liu, C.L. Dense trajectories and motion boundary descriptors for action recognition. *Int. J. Comput. Vis.* **2013**, *103*, 60–79. [CrossRef]
3. Yang, X.; Tian, Y.L. Action Recognition Using Super Sparse Coding Vector with Spatio-temporal Awareness. In Proceedings of the European Conference on Computer Vision, Zurich, Switzerland, 6–12 September 2014; pp. 727–741.
4. Peng, X.; Zou, C.; Qiao, Y.; Peng, Q. Action Recognition with Stacked Fisher Vectors. In Proceedings of the European Conference on Computer Vision, Zurich, Switzerland, 6–12 September 2014; pp. 581–595.
5. Peng, X.; Wang, L.; Wang, X.; Yu, Q. Bag of visual words and fusion methods for action recognition: Comprehensive study and good practice. *Comput. Vis. Image Underst.* **2016**, *150*, 109–125. [CrossRef]
6. Arandjelovic, R.; Zisserman, A. All about VLAD. In Proceedings of the IEEE Conference on Computer Vision and Pattern Recognition, Portland, OR, USA, 23–28 June 2013; pp. 1578–1585.
7. Duta, I.C.; Ionescu, B.; Aizawa, K.; Sebe, N. Spatio-temporal vlad encoding for human action recognition in videos. In Proceedings of the International Conference on Multimedia Modeling, Reykjavik, Iceland, 4–6 January 2017; pp. 365–378.
8. Zhu, H.; Zhu, C.; Xu, Z. Research advanves on human activity recognition datasets. *Acta Autom. Sin.* **2018**, *44*, 978–1004.
9. Wang, H.; Schmid, C. Action recognition with improved trajectories. In Proceedings of the IEEE International Conference on Computer Vision, Sydney, Australia, 1–8 December 2013; pp. 3551–3558.
10. Chen, B.; Xia, M.; Huang, J. Mfanet: A multi-level feature aggregation network for semantic segmentation of land cover. *Remote Sens.* **2021**, *13*, 731. [CrossRef]
11. Xia, M.; Zhang, X.; Weng, L.; Xu, Y. Multi-stage feature constraints learning for age estimation. *IEEE Trans. Inf. Forensics Secur.* **2020**, *15*, 2417–2428. [CrossRef]
12. Ng, Y.H.; Hausknecht, M.; Vijayanarasimhan, S.; Vinyals, O.; Toderici, G. Beyond short snippets: Deep networks for video classification. In Proceedings of the IEEE Conference on Computer Vision and Pattern Recognition, Boston, MA, USA, 7–12 June 2015; pp. 4694–4702.
13. Du, W.W.; Wang, Y.; Qiao, Y. Rpan: An end-to-end recurrent pose-attention network for action recognition in videos. In Proceedings of the IEEE International Conference on Computer Vision, Venice, Italy, 22–29 October 2017; pp. 3725–3734.
14. Long, X.; Gan, C.; De, M.G.; Liu, X.; Li, Y.; Li, F.; Wen, S. Multimodal keyless attention fusion for video classification. In Proceedings of the Thirty-Second AAAI Conference on Artificial Intelligence, New Orleans, LA, USA, 2–7 February 2018.
15. Wang, X.; Miao, Z.; Zhang, R.; Hao, S. I3d-lstm: A new model for human action recognition. In Proceedings of the IOP Conference Series: Materials Science and Engineering, Jakarta, Indonesia, 29–31 March 2014; p. 32035.
16. He, J.Y.; Wu, X.; Cheng, Z.Q.; Yuan, Z.; Jiang, Y.G. DB-LSTM: Densely-connected Bi-directional LSTM for human action recognition. *Neurocomputing* **2021**, *444*, 319–331. [CrossRef]
17. Song, S.; Lan, C.; Xing, J.; Zeng, W.; Liu, J. An end-to-end spatio-temporal attention model for human action recognition from skeleton data. In Proceedings of the AAAI Conference on Artificial Intelligence, San Francisco, CA, USA, 4–9 February 2014.
18. Xia, M.; Wang, T.; Zhang, Y.; Liu, J.; Xu, Y. Cloud/shadow segmentation based on global attention feature fusion residual network for remote sensing imagery. *Int. J. Remote Sens.* **2021**, *42*, 2022–2045. [CrossRef]
19. Wang, Y.; Long, M.; Wang, J.; Gao, Z.; Yu, P.S. Predrnn: Recurrent neural networks for predictive learning using spatiotemporal lstms. In Proceedings of the 31st International Conference on Neural Information Processing Systems, Long Beach, CA, USA, 4–9 December 2017; pp. 879–888.
20. Shi, X.; Chen, Z.; Wang, H.; Yeung, D.Y.; Wong, W.K.; Woo, W.C. Convolutional LSTM network: A machine learning approach for precipitation nowcasting. In Proceedings of the Advances in Neural Information Processing Systems, Palais des Congrès de Montréal, Montréal, QC, Canada, 7–10 December 2015; pp. 802–810.
21. Schulst, C.; Laptev, I.; Caputo, B. Recognizing human actions: A local SVM approach. In Proceedings of the 17th International Conference on Pattern Recognition, Cambridge, UK, 23–25 August 2004; pp. 32–36.

22. Moshe, B.; Lena, G.; Eli, S.; Michal, I.; Ronen, B. Actions as Space-Time Shapes. In Proceedings of the Tenth IEEE International Conference on Computer Vision, Beiging, China, 17–20 October 2005; pp. 1395–1402.
23. Marszalek, M.; Laptev, I.; Schmid, C. Actions in context. In Proceedings of the IEEE Conference on Computer Vision and Pattern Recognition, Miami, FL, USA, 20–25 June 2009; pp. 2929–2936.
24. Soomro, K.; Zamir, A.R.; Shah, M. UCF101: A dataset of 101 human actions classes from videos in the wild. *arXiv* **2012**, arXiv:1212.0402.
25. Kuehne, H.; Jhuang, H.; Garrot, E.; Poggio, T.; Serre, T. HMDB: A large video database for human motion recognition. In Proceedings of the International Conference on Computer Vision, Barcelona, Spain, 6–13 November 2011; pp. 2556–2563.
26. Donahue, J.; Anne, H.L.; Guadarrama, S.; Rohrbach, M.; Venugopalan, S.; Saenko, K.; Darrell, T. Long-term recurrent convolutional networks for visual recognition and description. In Proceedings of the IEEE Conference on Computer Vision and Pattern Recognition, Boston, MA, USA, 7–12 June 2015; pp. 2625–2634.
27. Wu, Y.; Schuster, M.; Chen, Z.; Le, Q.V.; Norouzi, M.; Macherey, W.; Macherey, W. Google's neural machine translation system: Bridging the gap between human and machine translation. *arXiv* **2016**, arXiv:1609.08144.
28. Simonyan, K.; Zisserman, A. Two-stream convolutional networks for action recognition in videos. *arXiv* **2014**, arXiv:1406.2199.

applied sciences

MDPI

Article

Learning Class-Specific Features with Class Regularization for Videos

Alexandros Stergiou *, Ronald Poppe and Remco C. Veltkamp

Department of Information and Computing Sciences, Utrecht University, Princetonplein 5, 3584 CC Utrecht, The Netherlands; r.w.poppe@uu.nl (R.P.); r.c.veltkamp@uu.nl (R.C.V.)

* Correspondence: a.g.stergiou@uu.nl

Received: 28 July 2020; Accepted: 4 September 2020; Published: 8 September 2020

Abstract: One of the main principles of Deep Convolutional Neural Networks (CNNs) is the extraction of useful features through a hierarchy of kernels operations. The kernels are not explicitly tailored to address specific target classes but are rather optimized as general feature extractors. Distinction between classes is typically left until the very last fully-connected layers. Consequently, variances between classes that are relatively similar are treated the same way as variations between classes that exhibit great dissimilarities. In order to directly address this problem, we introduce *Class Regularization*, a novel method that can regularize feature map activations based on the classes of the examples used. Essentially, we amplify or suppress activations based on an educated guess of the given class. We can apply this step to each minibatch of activation maps, at different depths in the network. We demonstrate that this improves feature search during training, leading to systematic improvement gains on the *Kinetics*, *UCF-101*, and *HMDB-51* datasets. Moreover, *Class Regularization* establishes an explicit correlation between features and class, which makes it a perfect tool to visualize class-specific features at various network depths.

Keywords: class regularization; 3D-CNN; spatiotemporal activations; class-specific features

1. Introduction

Video-based action recognition has seen tremendous progress since the introduction of Convolutional Neural Networks (CNNs) [1,2]. The hierarchical application of 3D convolutional operations has been shown to effectively capture descriptive spatiotemporal features.

CNNs include multiple layers that are stacked in a single, hierarchical architecture. Features are calculated by successive convolutions. Kernels in early layers focus on simple textures and patterns, while deeper layers focus on more complex parts of objects or scenes. As these features become more dependent on the different weighting of neural connections in previous layers, only a portion of them becomes descriptive for a specific class [3,4]. Yet, all kernels are learned in a class-agnostic way. Consequently, much of the discriminative nature of CNNs is achieved only in the very last fully-connected layers. This hinders easy interpretation of the part of the network that is informative for a specific class.

We aim at forcing the network to propagate class-specific activations throughout the network. We propose a method named *Class Regularization* that relates class information to extracted features of network blocks. This information is added back to the network by amplifying and suppressing activation values with respect to predicted classes. *Class Regularization* has a beneficial effect on the nonlinearities of the network by modulating the effects of the activations. Owing to this, the architecture can effectively distinguish between the most class-informative kernels in each part of the network given a selected class. This procedure reduces the dependency on many uncorrelated features in the last fully-connected layers that are responsible for the final class predictions, essentially penalizing overfitting.

Our contributions are the following:

- We propose *Class Regularization*, a regularization method applied in spatiotemporal CNNs. The method does not change the overall structure of the architecture, but can be used as an additional step after each operation or block.
- We demonstrate that the relationships between classes and features can be visualized by propagating class-based feature information through normalized neural weights for each of the model's building blocks.
- We report performance gains for benchmark action recognition datasets *Kinetics, UCF-101*, and *HMDB-51* by including *Class Regularization* in convolution blocks.

We discuss advances in vision-based action recognition in Section 2. A detailed overview of *Class Regularization* appears in Section 3. Experiments are presented in Section 4. We conclude in Section 5.

2. Related Work

Significant advancements have been made in the recognition of actions in videos with the introduction of deep neural approaches that are based on the hierarchical discovery of informative features [5,6]. These architectures provide the basis to further accommodate temporal information.

Due to the indirect relationship between temporal and spatial information, one of the first attempts on video recognition with neural models was the use of Two-stream networks [7]. These networks contain two separate models that use still video frames and optical flow as inputs. Class predictions are made after combining the extracted features of the separate networks. Two-stream networks were also used as a base method for works such as Temporal Segment Networks (TSN, [8]) which use scattered snippets from the video and fuse their predictions. This approach sparked research on the selection of informative frames [9,10]. Other extensions of Two-stream networks include the use of residual connections [11,12] that share spatiotemporal information across multiple layers.

An alternative approach to capture temporal information in CNNs is through 3D convolutions [13]. 3D convolutions include an additional time dimension to the two spatial dimensions. 3D convolutions have been shown to outperform standard image-based networks [14] in video classification. A fusion of Two-stream networks and 3D convolutions has been explored with the I3D architecture [15]. The two spatiotemporal models are trained in parallel on frame and optical flow data, with the benefit of also processing temporal-only information. Further structures that have been explored with 3D convolutions include Residual Networks [16].

Inspired by depthwise and pointwise convolutions performed over image channels [17], researchers have introduced ways of splitting the 3D convolutions in subsequent 2D convolution operations. This can be achieved through either using spatial filters followed by temporal-only filters [18,19] or convolutions in groups [20,21]. Alternative implementations include the use of long-sequence and short-sequence kernels [22] and dimension-based iterations of time-width and time-height [23,24]. Others have worked on the minimization of the computational requirements by shifting activations along time [25]. Previous works that have touched upon regularization were aimed more at the overall temporal consistency of features (e.g., [26]) without the consideration of feature combinations that are most informative for some classes.

Although these techniques have shown great promise in terms of accuracy and computational performance, there is still a lack of better spatiotemporal representations for intermediate network layers. Currently, there is no standardized way of processing the temporal information. Our proposed method, named *Class Regularization*, fills this void as it can be added to virtually all network architectures with minimum additional computational cost in order to enhance the correlation between specific spatiotemporal features and the action class.

3. Regularization for Convolutional Blocks

In CNNs, explicitly adding class information through regularization is challenging, given the increasing level of ambiguity of the model's inner workings with respect to the network depth.

We make the assumption that, when testing a video in a trained CNN, a speculative guess can be made on which class is represented by observing the activations produced at a certain layer. The underlying idea is that different kernels focus on different spatiotemporal patterns that appear in different classes. These guesses depend on the layer depth. Deeper network layers can distinguish class-specific features better because of the larger feature complexity. Therefore, guesses at different parts of the model should have consequently greater or lesser effect. To include the layer feature complexity, we define an *affection rate* ($\mathfrak{A}$) that specifies how the class predictions affect the network's activation maps. The values are chosen given the layer depth and the level of uncertainty of their class estimates. The discovery of feature correspondence between the layer's feature space and the class's feature space is implemented through pointwise convolutions and by incorporating their kernel updates as part of the training procedure. A complete workflow of the regularization method appears in Figure 1.

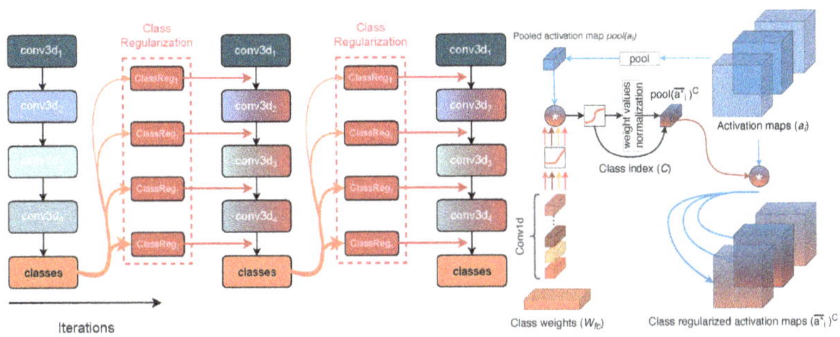

(**a**) Network view (**b**) Block view

Figure 1. Class Regularization. An additional pathway between class weights and intermediate features is added (**a**) connecting different parts of the network across iterations. In *Class Regularization* blocks, activation maps are pooled to vector volumes (pool(a_i)) and multiplied by the class weights in order to select the resulting highest class activation (pool$(\overline{a_i^*})^C$). A computational overview of the in-block operations (**b**) appears in Algorithm 1.

3.1. Layer Fusion with Class Predictions

We first discuss the main process for finding class estimates based on extracted features from the convolutional block at depth i in the network. We start by creating a vector representation of the features as distributed between the activation's channels for a spatiotemporal activation map input of size ($FxHxW$). Considering the produced activation map of the ith block (denoted as a_i), a global feature representation of the activations is created through a spatiotemporal sampling operation: $pool(a_i)$ (Equation (1)). The produced volume can be interpreted as a single vector descriptor containing a combination of the feature intensity values in the form of their average activations in a significantly lower dimensional space.

$$pool(a_i) = \frac{1}{F \times H \times W} \sum_{f=1}^{F} \sum_{h=1}^{H} \sum_{w=1}^{W} a_{(f,h,w,i)}. \tag{1}$$

To include class-based information in intermediate feature layers, we use the weights of the network's final prediction layer W_{fc}. This allows to establish a relationship between previous and current iterations in a recurrent fashion, by taking class-specific features into account. To obtain feature alignment given the channel sizes for the current activation and the prediction weights, a one-dimensional convolutional operation ($conv = W_{fc} * \mathcal{W}$) is applied to the class weights with a kernel ($\mathcal{W}$) size of 1 followed by a *relu* activation function.

Next, we use a standard matrix-to-matrix multiplication to create an association between the channel descriptor and each of the classes. The volume Z_i is of size $\{Z_i^{[1]}, ..., Z_i^{[CL]}\}$, for CL classes, with each unit $Z_i^{[j]}$ having the same channel dimensionality as a_i. This operation allows an early estimate for the indexes of the most relevant features for each class.

One could use the outputs of the multiplication as a separate loss function but we have two main reasons for refraining from doing so. First, due to the limited feature complexity, it is significantly more difficult to make educated class predictions in early layers of the network. This also corresponds to the notion of hierarchical feature extraction in CNNs. Second, through these multiple output points, multiple error derivatives are to be calculated which will slow down the training process significantly.

3.2. Fusion of Class Weight Vector and Spatiotemporal Activation Maps

Our aim is to let the produced class-based activations Z_i become indicators for the most informative class features. With this, we aim at obtaining the maximum class probability through a standard softmax activation function applied on the activations. This converts the weighed sum logit score to a probabilistic distribution over all classes: $S(Z_i)$. Based on (5) in Algorithm 1, the index of the maximum class activation is selected (C) to define the class weight vector with the highest correspondence based on the class features of the layer.

Algorithm 1 Class regularization overview

1: **Inputs:**
Values of activation map a_i for ith layer.
2: **Outputs:**
Class-regularized activations $(\overline{a_i^*})^C$
3: $W_i \leftarrow relu(W_{fc} * W)$
▷ Weight dimensionality conversion
4: $Z_i \leftarrow W_i * pool(a_i)$
▷ Weighted sum per class neuron
5: $C \leftarrow \underset{j}{\text{argmax}}\{S(Z_i^{[1]}), ..., S(Z_i^{[CL]})\} \; \forall \; S(Z_i^{[j]}) = \frac{\epsilon^{Z_i^{[j]}}}{\sum_{c \in 1,...,CL} \epsilon^{Z_i^{[c]}}}$
▷ Largest softmax activation class search
6: $\widehat{W}_i \leftarrow \mathfrak{A} * \frac{(W_i - min\{W_i\})*(1-\mathfrak{A})}{max\{W_i\} - min\{W_i\}}$
▷ Weight scaling
7: $(\overline{a_i^*})^C \leftarrow \widehat{W}_i^{[c]} * a_i$
▷ Final weight regularization.

As we want to amplify activation features, we need to normalize weight vector W_i before fusing it with the layer's activation maps. Layer features that are less informative for a specific class should be scaled down, while informative ones should be scaled up. To this end, we set a value ($\mathfrak{A}$) which will be referred to as the *affection rate*. It determines the bounds that the weight vector will be normalized to ($\widehat{W}_i$)—see step (6) in Algorithm 1. We are not using a standardization method as in *batch normalization* [27] that guarantees a zero-mean output. This is because we use a multiplication operation for including the class weight information to the activation maps. Therefore, zero-mean normalization will remove part of the information because values below one will decrease in the feature intensity. It also hinders performance as it effectively contributes to the occurrence of *vanishing gradients* because the produced activation map values would be reduced at each iteration.

In our final step, we inflate the normalized weight vector $\widehat{W}^{[c]}$ to match the dimensions of the spatiotemporal activation maps. We then perform a matrix-to-matrix multiplication between the activation maps of the layer a_i and the normalized weights with the axis of symmetry being the depth or channel's dimension.

3.3. Performing Updates to Regularized Volumes

Class Regularization is performed on activation maps in the network to manipulate the activation values of the upcoming operations. We underline that the value of the *affection rate* $\mathfrak{A}$ used in the normalization can be trained through a separate objective function. In addition, our method is independent of the training iteration or layer number that it is applied to, and can process examples independently for online learning. Therefore, the discovery of class connections can also be performed in minibatches, to then return regularized volumes over single or multiple class weights.

The feature representation that is captured $(\overline{a_i^*})^C$ depends on the specific example a_i, the layer or block i of the architecture that the regularization was applied to, the chosen class C that was selected, and the affection rate $\mathfrak{A}$. The distribution inside $\widehat{W}_i^{[C]}$ has an expected value of 1 and a distribution of $2(1-\alpha)$. Due to the low computational overhead to back-propagate through the proposed method, the computation times are not significantly affected by including *Class Regularization* in a network.

3.4. Class Regularization for Visualizations

As *Class Regularization* is based on the injection of class-based information inside the feature-extraction process, a direct correlation between classes and features is made at each block in which the method is applied. Being able to represent the class features given a different feature space improves the overall explainability capabilities of the model. Through feature correlation, the method alleviates the curse of dimensionality problem of previous visualization methods that rely on back-propagating from the predictions to a particular layer [28]. Since the classes are represented in the same feature space as the activation maps of the block, we can discover regions in space and time that are informative over multiple network layers. To the best of our knowledge, this is the first method to visualize spatiotemporal class-specific features at each layer of the network.

By extending the proposed algorithm to include an adaptation of *Saliency Tubes* [29] in each block, we can create visual representations of the features with the highest activations per class. We create two visual examples of the class *bowling* in Kinetics-400 to demonstrate class activations in different layers of the network (Figure 2). As observed in the clip segment, in both cases, early layer features are significantly less deterministic of the class and mostly target the distinction between foreground and background. In later layers, the focus shifts from the actor to the background in the predictions layer of the first clip. Ball and bowling pins are present at a wallpaper, which demonstrates that a strong still-frame visual signal is favored in case the action or objects within the action are occluded. In the second clip, the main field of focus of the network in the final layers is towards the area between the actor's hand and the ball. We note that, by design, all network architectures used fuse all temporal activations to a single frame at their final convolution block, which only allows the visualization of spatial extension of the activations.

The amplification of layer features can also be visualized as in Figure 3. The top-3 kernels to be amplified for a baseball hit example correspond to spatiotemporal features such as the appearance of the bat, the field, and the movement of the bat during a swing. In addition, these amplifications are also propagated to deeper layers in the network through the connections of the most informative kernels.

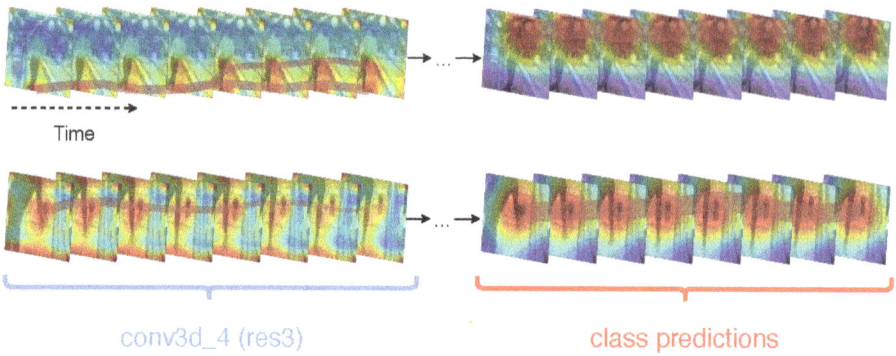

Figure 2. Layerwise class feature correspondence. Each of the *Saliency Tubes* [29] represents the class activations of a *Class Regularized* Wide-ResNet50 model in layers *res3* and *predictions (fc)*. Both clips correspond to visualizations for class *bowling* from Kinetics-400 [30] for two different examples with variations in their spatiotemporal regions.

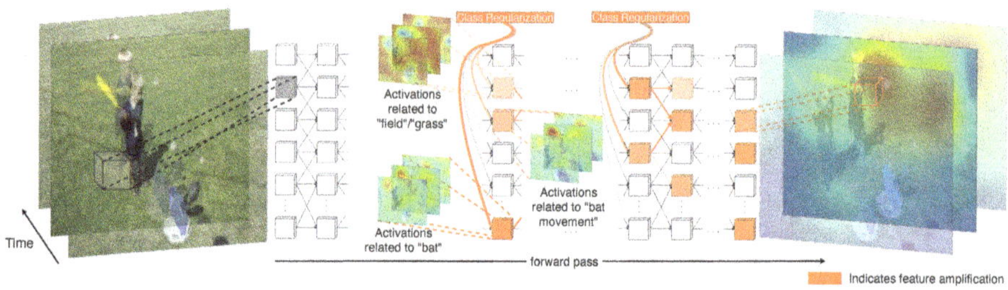

Figure 3. Visualization of feature amplification. As class-specific activations are reused by the network, informative spatiotemporal features for specific classes during an iteration are amplified. The effect of this amplification is propagated to deeper layers in the network through the connections of the layers in which *Class Regularization* is applied.

4. Experiments

We demonstrate the merits of *Class Regularization* on the action recognition classification performance on three benchmark datasets, and using a number of widely used CNN architectures. Results are summarized in Table 1. We further statistically compare the classification performance between predictions from different architectures and from different blocks within the same architecture.

For our experiments, we consider the widely used Kinetics-400 [30] dataset as a baseline. Then, each of the selected models is further fine-tuned on both UCF-101 [31] and HMDB-51 [32] by training the 1D convolutions inside *Class Regularization* to ensure a dimensionality correspondence between the new class weight vectors and the activation maps of each layer that the method is applied to. We further allow updates on the final two convolution blocks of the selected networks.

Training. The models trained on Kinetics are initialized with a standard Kaiming initialization [33] without inflating the 3D weights. This allows for a direct comparison between architectures with and without *Class Regularization* and to compare the respective accuracy rates. For all the experiments, we use an *SGD* optimizer with 0.9 momentum. *Class Regularization* is added at the end of each bottleneck block in the ResNets and at the end of each mixed block in I3D.

Table 1. Architectures of 3D convolution models with and without class regularization. We note that in the *Class Regularized* networks, only an eighth of the additional parameters are trainable, with the rest corresponding to nontrainable class weights tensor duplicates.

Layer Name	3D ResNet101 (w/o CN)	3D Wide ResNet 50 (w/o CN)	I3D (w/o CN)
$con3d_1$		$7 \times 7 \times 7, 64$ conv	
$con3d_2$	$\begin{bmatrix} 1\times1\times1 \\ 3\times3\times3 \\ 1\times1\times1 \end{bmatrix} (\times 64) \times 3$	$\begin{bmatrix} 1\times1\times1 \\ 3\times3\times3 \\ 1\times1\times1 \end{bmatrix} (\times 128) \times 3$	$\begin{bmatrix} 1\times1\times1 & 1\times1\times1 & 1\times1\times1 & 3\times3\times3, (pool) \\ & 3\times3\times3 & 3\times3\times3 & 1\times1\times1 \end{bmatrix} (\times 480) \times 2$
$ClassReg_1$	- / $\alpha = 0.9$	- / $\alpha = 0.9$	- / $\alpha = 0.8$
$con3d_2$	$\begin{bmatrix} 1\times1\times1 \\ 3\times3\times3 \\ 1\times1\times1 \end{bmatrix} (\times 128) \times 4$	$\begin{bmatrix} 1\times1\times1 \\ 3\times3\times3 \\ 1\times1\times1 \end{bmatrix} (\times 256) \times 4$	$\begin{bmatrix} 1\times1\times1 & 1\times1\times1 & 1\times1\times1 & 3\times3\times3, (pool) \\ & 3\times3\times3 & 3\times3\times3 & 1\times1\times1 \end{bmatrix} (\times 832) \times 5$
$ClassReg_2$	- / $\alpha = 0.8$	- / $\alpha = 0.8$	- / $\alpha = 0.7$
$con3d_2$	$\begin{bmatrix} 1\times1\times1 \\ 3\times3\times3 \\ 1\times1\times1 \end{bmatrix} (\times 256) \times 23$	$\begin{bmatrix} 1\times1\times1 \\ 3\times3\times3 \\ 1\times1\times1 \end{bmatrix} (\times 512) \times 6$	$\begin{bmatrix} 1\times1\times1 & 1\times1\times1 & 1\times1\times1 & 3\times3\times3, (pool) \\ & 3\times3\times3 & 3\times3\times3 & 1\times1\times1 \end{bmatrix} (\times 1024) \times 2$
$ClassReg_3$	- / $\alpha = 0.7$	- / $\alpha = 0.7$	- / $\alpha = 0.6$
$con3d_2$	$\begin{bmatrix} 1\times1\times1 \\ 3\times3\times3 \\ 1\times1\times1 \end{bmatrix} (\times 512) \times 3$	$\begin{bmatrix} 1\times1\times1 \\ 3\times3\times3 \\ 1\times1\times1 \end{bmatrix} (\times 1024) \times 3$	-
$ClassReg_4$	(-) / $\alpha = 0.6$	- / $\alpha = 0.6$	-
predictions		global average pool, softmax unit group	

We use transformations for both spatial and temporal dimensions. In the temporal dimension, we use clips of 16 frames that are randomly extracted from different subsequences of the video. For the validation sets, we only use the center 16 frames. Spatially, we use a cropping size of 112×112 and 256×256 for fine-tuning. All models are trained for 170 epochs, as no further improvements were observed afterwards in our initial experiments with all three networks on Kinetics. We also used a step-based learning rate reduction of 10% every 50 epochs.

Datasets. Kinetics-400 consists of roughly 240 K training videos and 20 k validation videos of 400 different human actions. We report the top-1 accuracy alongside the computational cost (FLOPs) for each of the networks using spatiotemporally cropped clips. UCF-101 and HMDB-51 have 13 k and 9 k videos, respectively. They are used to demonstrate the transfer abilities of the proposed *Class Regularization* as well as the usability of our method in smaller datasets.

4.1. Main Results

A comparison between results of models trained from scratch on Kinectics-400 appears in Table 2. Existing networks consider a complete change in the overall architecture or convolution operations in models, which is computationally challenging given the large memory requirements of spatiotemporal models. New models need to be trained for a significant number of iterations in order to achieve mild improvements, while additionally using large datasets [34,35] for pretraining. In contrast, the proposed *Class Regularization* method is used on top of existing architectures and only requires fine-tuning the dimensionality correspondence between the number of features in a specific layer and the features that are used for class predictions. Overall, the largest improvements were observed on networks with larger number of layers, such as ResNet101-3D, in comparison to networks with lower number of layers, with the best performing architectures being I3D and ResNet101-3D with *Class Regularization* with 67.8% top-1 accuracy and 67.7% top-1 accuracy, respectively.

Table 2. Accuracy rates of different spatiotemporal convolutional architectures on the Kinetics-400 dataset.

Model	Backbone	Depth	# Params (M)	GFLOPS	Top-1 (%)
ResNet50-3D [16]	ResNet	50	36.72	80.32	0.636
ResNet101-3D [16]	ResNet	101	69.06	110.98	0.652
ResNeXt101-3D [16]	ResNet	101	69.06	148.91	0.667
Wide ResNet50-3D [16]	ResNet	50	140.94	72.32	0.640
I3D [15]	Inception	48	12.07	55.79	0.664
R(2+1)D-ResNet50 [19]	Resnet	50	34.86	89.14	0.645
R(2+1)D-ResNet101 [19]	ResNet	101	67.22	159.21	0.668
MF-Net [36]	ResNet	50	8.03	22.7	0.653
ResNet101-3D w/*Class Regularization*	ResNet	101 + 4	37.10	126.13	0.677
Wide ResNet50-3D w/*Class Regularization*	ResNet	50 + 4	141.32	82.67	0.653
I3D w/*Class Regularization*	Inception	48 + 3	69.44	62.96	**0.678**

4.2. Direct Comparisons with and without Class Regularization

In Table 3, we pairwise compare three architectures after training directly on Kinetics, and after adding *Class Regularization* blocks and fine-tuning. For each architecture and dataset, networks with *Class Regularization* outperform those without. The largest gain is observed in the 101-layer 3D Resnet (+2.45%), while we obtained improvements of +1.37% and +1.43% on the Wide-ResNet50 and I3D, respectively. Our approach appears to be especially useful for deeper networks. Since the effective description of classes is achieved through large feature spaces, *Class Regularization* significantly benefits models that include complex and large class weight spaces. The set of influential class features can be better distinguished with minimal computational costs, as shown in Figure 4.

Table 3. Direct comparison of architectures with (in orange) and without (in black) *Class Regularization* blocks. Latency in msec.

Model	Added Latency	Kinetics	UCF101	HMDB51
ResNet101	-	65.29%	88.23%	62.47%
ResNet101	+ 98.786	67.74%	88.84%	63.31%
Wide ResNet50	-	63.96%	87.52%	61.62%
Wide ResNet50	+102.995	65.33%	89.11%	63.24%
I3D	-	66.42%	91.80%	64.27%
I3D	+68.34	67.85%	93.17%	65.83%

When transferring weights, the retraining process does not change for the *Class Regularization* networks as the additional training phase is only performed in order for the model to learn the feature correspondence.

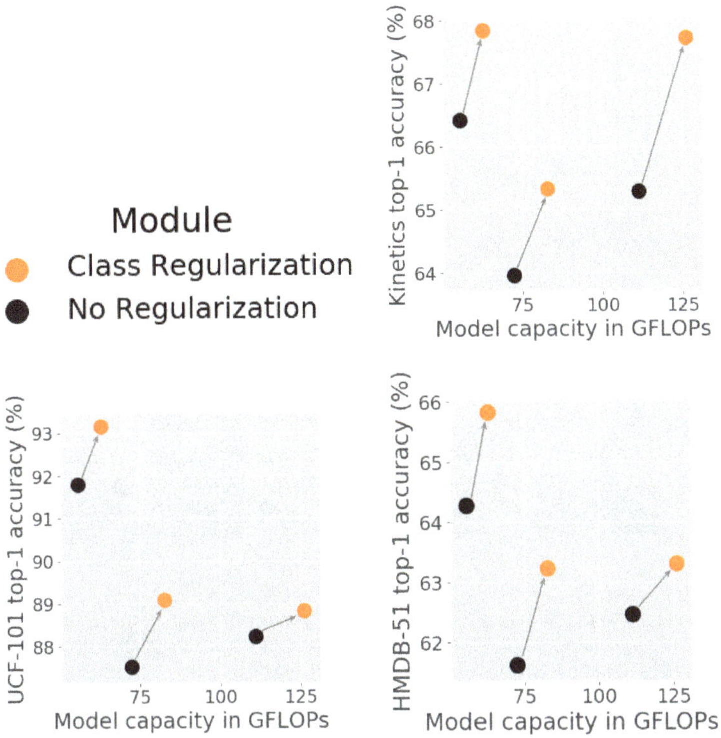

Figure 4. Class Regularization accuracy/computation trade-off.

Advancements can also be achieved in transfer learning, as shown by the rates for UCF-101 (split 1) and HMDB-51 (split 1) in Table 3. Accuracy for non-*Class-Regularized* networks are recalculated in order to ensure the same training setting. The largest gain from the original implementation in UCF-101 is on the Wide ResNet50 with +1.59% followed by I3D with +1.26% and ResNet101 +0.61%. For the HMDB-51 dataset, the model pair that exhibits the largest gap in performance is Wide ResNet50 with a +1.62% improvement, I3D with +1.56%, and ResNet101 with +0.84%. Overall, the minor deterioration of the accuracy gains in transfer learning could be contributed to the fact that kernels have been already trained in conjunction with class information from a different dataset.

As shown in Figure 5, *Class Regularization* demonstrates improvements for most classes in Kinetics. The greatest improvements are observed for classes that can be better defined based on their execution instead of their appearance. Examples are "bench pressing" (9.1% improvement), "high jump" (8.9% improvement), and "jogging" (15.6% improvement). Amplifying features that are characteristic of those classes improves the model's recognition capabilities. In contrast, classes that are more likely to contain significant variation in feature values are more prone to be classified wrongly. This is particularly true for classes that exhibit large intraclass variation, such as "parkour", where there are no standard actions performed. Other examples are "garbage collection" in Figure 5 which could be performed either mechanically (top), by a single person (mid), or by multiple people (bottom). Other examples include either oscillations as in "pumping gas", or require contextual information as in "sniffing".

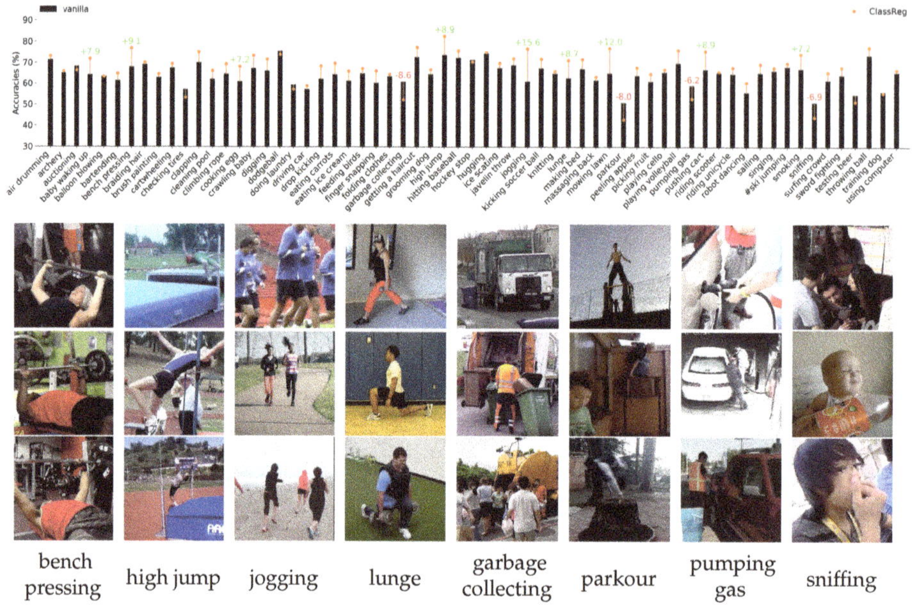

Figure 5. Per-class performance for ResNet-101 w/o Class Regularization with illustrative examples of classes with large performance gains or losses in Kinetics-400 [30].

4.3. Resulting Statistical Significance

To analyze whether the classification performance with the addition of *Class Regularization* to a network is statistically relevant, we perform a McNemar's test on UCF-101 and HMDB-51. We study if the difference in accuracy between a network with and without *Class Regularization* is to be attributed to *marginal homogeneity* (i.e., the result of sampling distribution) or if the variance is significant enough to conclude that the two models indeed perform differently. (Table 4a–c) summarize the McNemar tests, defined in Equation (2), where b are the cases in which the nonregularized model ($\neg Reg$) is correct (+) and the regularized (Reg) is wrong (-), and c are the cases in which the nonregularized model is correct while the regularized is wrong.

$$x^2 = \frac{(b-c)^2}{b+c} \qquad \begin{cases} H_0 : P_b = P_c \\ H_1 : P_b \neq P_c \end{cases} \tag{2}$$

We notice that in each of the tested architectures, the c values were significantly larger than those of b. In particular, a standard 3D ResNet-101 model with class regularization is different than that of a model without, with x^2 values of 2.48 and 2.84 for UCF-101 and HMDB-51, respectively. For both datasets, the average probability that the difference between the model with and without is attributed to statistical error is approximately 10%. The difference is also evident in architectures that include a larger number of extracted features per layer. This can be seen from the results of the Wide-ResNet-50 model, in which the x^2 values are 3.40 and 3.08 for UCF-101 and HMDB-51, respectively. The difference is statistically significant at the $p = 0.05$ level. The final comparison considers I3D as the base network to also account for cases of convolution operations being split to multiple and differently sized cross-channeled convolutions. The probability that the measured difference is not due to a systematic difference in performance is approximately 10%. The sum of (b, c)

values in every case is reasonably large in order to sufficiently approximate x^2 and to conclude that marginal probabilities for models with and without *Class Regularization* are not homogeneous.

Table 4. McNemar's statistical significance test on the first split of UCF-101 and HMDB-51. Results with and without regularization for (**a**) ResNet101, (**b**) Wide-ResNet, and (**c**) I3D models. (**d–f**) present results for class-regularized ResNet101 across different blocks in the architecture.

a ResNet101-3D

	Reg (+)	*Reg* (-)
¬*Reg* (+)	773 / 924	23 / 35
¬*Reg* (-)	36 / 47	83 / 527

b Wide-ResNet50-3D

	Reg (+)	*Reg* (-)
¬*Reg* (+)	764 / 895	34 / 49
¬*Reg* (-)	51 / 68	65 / 518

c I3D

	Reg (+)	*Reg* (-)
¬*Reg* (+)	824 / 887	15 / 96
¬*Reg* (-)	27 / 118	51 / 429

d ResNet101-3D (over different depths)

	Reg^2 (+)	Reg^2 (-)
Reg^1 (+)	18 / 30	0 / 2
Reg^1 (-)	35 / 89	861 / 1411

e ResNet101-3D (over different depths)

	Reg^3 (+)	Reg^3 (-)
Reg^2 (+)	52 / 104	1 / 5
Reg^2 (-)	730 / 829	131 / 592

f ResNet101-3D (over different depths)

	Reg^4 (+)	Reg^4 (-)
Reg^3 (+)	772 / 930	10 / 3
Reg^3 (-)	37 / 38	95 / 559

Finally, we compare the class predictions found by the *Class Regularization* method in different blocks within the architecture in (Table 4d–f). These panels provide an overview of how different depths of the network perform for the given task by gradually ablating blocks of convolutions. As observed for both datasets, the largest change in performance is in the third convolution block (Table 4e), with the transition of information from the third to the fourth block of convolutions (Table 4f) only accounting for a small change in performance based on c. We use this to further demonstrate the merits of *Class Regularization* as a quantitative way of understanding the informative parts of the overall network, complementing the class-specific feature visualizations.

5. Conclusions

We have introduced *Class Regularization*, a method that focuses on class-specific features rather than treating each convolution kernel as class-agnostic. *Class Regularization* allows the network to strengthen or weaken layer activations based on the batch data. The method can be added to any layer or block of convolutions in pretrained models. It is lightweight, as the class weights from the prediction layer are shared throughout *Class Regularization* blocks. To avoid the vanishing gradient problem and the possibility of negatively influencing activations, the weights are normalized between a range given an *affection rate* value.

We evaluated the proposed method on three benchmark datasets: Kinetics-400, UCF-101, and HMDB-51; and three models: ResNet101, Wide ResNet50, and I3D. We consistently show improvement when using *Class Regularization*, with a performance gain of up to 2.45%. The achieved improvements were done with minimal additional computational cost over the original architectures. We also perform a statistical significance test to demonstrate that the outcomes of the different models are indeed based on the additional regularization, rather than being the result of incidental sampling.

Class Regularization can also aid in improving the explainability of 3D-CNNs. Qualitative visualizations reveal which spatiotemporal features are strongly correlated with specific classes. Such analyses can be made for specific layers and, as such, provide insight into the discriminative patterns that specific features represent.

Future works based on *Class Regularization* should aim towards addressing the feature variations of actions within the same classes, with a greater focus towards the temporal domain. The inclusion of global information and the calibration of local spatiotemporal patterns based on such information, as with *Squeeze and Recursion* blocks [26], shows a promising direction towards creating spatiotemporal features that can better treat the variations of human actions and interactions.

Author Contributions: A.S. was responsible for the methodology formalization. He has performed the software implementation and the experimentation. He is also the main contributor for the article. R.P. took part in the conceptualization of the work and edited the article. He also provided the funding for the work. R.C.V. supervised the work. All authors have read and agreed to the published version of the manuscript.

Funding: This research is supported by the Netherlands Organization for Scientific Research (NWO) with a TOP-C2 grant for "Automatic recognition of bodily interactions" (ARBITER).

Conflicts of Interest: The authors declare no conflict of interest.

References

1. Herath, S.; Harandi, M.; Porikli, F. Going deeper into action recognition: A survey. *Image Vis. Comput.* **2017**, *60*, 4–21. [CrossRef]
2. Stergiou, A.; Poppe, R. Analyzing human-human interactions: A survey. *Comput. Vis. Image Underst.* **2019**, *188*, 102799. [CrossRef]
3. Bau, D.; Zhu, J.Y.; Strobelt, H.; Zhou, B.; Tenenbaum, J.B.; Freeman, W.T.; Torralba, A. Visualizing and understanding generative adversarial networks. *arXiv* **2019**, arXiv:1901.09887.
4. Gilpin, L.H.; Bau, D.; Yuan, B.Z.; Bajwa, A.; Specter, M.; Kagal, L. Explaining explanations: An overview of interpretability of machine learning. In Proceedings of the International Conference on Data Science and Advanced Analytics (DSAA), Turin, Italy, 1–3 October 2018; pp. 80–89.
5. He, K.; Zhang, X.; Ren, S.; Sun, J. Deep residual learning for image recognition. In Proceedings of the Computer Vision and Pattern Recognition (CVPR), Las Vegas, NV, USA, 27–30 June 2016; pp. 770–778.
6. Simonyan, K.; Zisserman, A. Very deep convolutional networks for large-scale image recognition. *arXiv* **2014**, arXiv:1409.1556.
7. Simonyan, K.; Zisserman, A. Two-stream convolutional networks for action recognition in videos. In Proceedings of the Advances in Neural Information Processing Systems (NIPS), Montreal, QC, Canada, 8–13 December 2014; pp. 568–576.
8. Wang, L.; Xiong, Y.; Wang, Z.; Qiao, Y.; Lin, D.; Tang, X.; Van Gool, L. Temporal segment networks: Towards good practices for deep action recognition. In Proceedings of the European Conference on Computer Vision (ECCV), Amsterdam, The Netherlands, 8–16 October 2016; pp. 20–36.
9. Diba, A.; Sharma, V.; Van Gool, L. Deep temporal linear encoding networks. In Proceedings of the Computer Vision and Pattern Recognition (CVPR), Honolulu, HI, USA, 21–26 July 2017; pp. 2329–2338.
10. Wang, Y.; Song, J.; Wang, L.; Van Gool, L.; Hilliges, O. Two-Stream SR-CNNs for Action Recognition in Videos. In Proceedings of the British Machine Vision Conference (BMVC), York, UK, 19–22 September 2016.
11. Feichtenhofer, C.; Pinz, A.; Wildes, R. Spatiotemporal residual networks for video action recognition. In Proceedings of the Advances in Neural Information Processing Systems (NIPS), Barcelona, Spain, 5–10 December 2016; pp. 3468–3476.
12. Wang, X.; Girshick, R.; Gupta, A.; He, K. Non-Local Neural Networks. In Proceedings of the Conference on Computer Vision and Pattern Recognition (CVPR), Salt Lake City, UT, USA, 18–22 June 2018.
13. Ji, S.; Xu, W.; Yang, M.; Yu, K. 3D convolutional neural networks for human action recognition. *Trans. Pattern Anal. Mach. Intell.* **2013**, *35*, 221–231. [CrossRef]
14. Tran, K.N.; Gala, A.; Kakadiaris, I.A.; Shah, S.K. Activity analysis in crowded environments using social cues for group discovery and human interaction modeling. *Pattern Recognit. Lett.* **2014**, *44*, 49–57. [CrossRef]
15. Carreira, J.; Zisserman, A. Quo vadis, action recognition? A new model and the Kinetics dataset. In Proceedings of the Computer Vision and Pattern Recognition (CVPR), Honolulu, HI, USA, 21–26 July 2017; pp. 4724–4733.
16. Hara, K.; Kataoka, H.; Satoh, Y. Can spatiotemporal 3D CNNs retrace the history of 2D CNNs and ImageNet? In Proceedings of the Computer Vision and Pattern Recognition (CVPR), Salt Lake City, UT, USA, 18–22 June 2018; pp. 18–22.
17. Chollet, F. Xception: Deep Learning with Depthwise Separable Convolutions. In Proceedings of the Conference on Computer Vision and Pattern Recognition (CVPR), Honolulu, HI, USA, 21–26 July 2017; pp. 1800–1807.

18. Qiu, Z.; Yao, T.; Mei, T. Learning spatio-temporal representation with pseudo-3d residual networks. In Proceedings of the International Conference on Computer Vision (ICCV), Venice, Italy, 22–29 October 2017; pp. 5534–5542.
19. Tran, D.; Wang, H.; Torresani, L.; Ray, J.; LeCun, Y.; Paluri, M. A Closer Look at Spatiotemporal Convolutions for Action Recognition. In Proceedings of the Conference on Computer Vision and Pattern Recognition (CVPR), Salt Lake City, UT, USA, 18–22 June 2018; pp. 6450–6459.
20. Lee, M.; Lee, S.; Son, S.; Park, G.; Kwak, N. Motion Feature Network: Fixed Motion Filter for Action Recognition. In Proceedings of the European Conference on Computer Vision (ECCV), Munich, Germany, 8–14 September 2018.
21. Tran, D.; Wang, H.; Torresani, L.; Feiszli, M. Video Classification With Channel-Separated Convolutional Networks. In Proceedings of the International Conference on Computer Vision (ICCV), Seoul, Korea, 27 October–2 November 2019.
22. Feichtenhofer, C.; Fan, H.; Malik, J.; He, K. SlowFast networks for video recognition. In Proceedings of the International Conference on Computer Vision (ICCV), Seoul, Korea, 27 October–2 November 2019.
23. Li, C.; Zhong, Q.; Xie, D.; Pu, S. Collaborative Spatiotemporal Feature Learning for Video Action Recognition. In Proceedings of the Conference on Computer Vision and Pattern Recognition (CVPR), Long Beach, CA, USA, 16–20 June 2019; pp. 7872–7881.
24. Stergiou, A.; Poppe, R. Spatio-Temporal FAST 3D Convolutions for Human Action Recognition. In Proceedings of the International Conference On Machine Learning and Applications (ICMLA), Boca Raton, FL, USA, 16–19 December 2019; pp. 183–190.
25. Lin, J.; Gan, C.; Han, S. TSM: Temporal Shift Module for efficient video understanding. In Proceedings of the International Conference on Computer Vision (ICCV), Seoul, Korea, 27 October 2019–2 November 2019; pp. 7083–7093.
26. Stergiou, A.; Poppe, R. Learn to cycle: Time-consistent feature discovery for action recognition. *arXiv* **2020**, arXiv:2006.08247.
27. Ioffe, S.; Szegedy, C. Batch normalization: Accelerating deep network training by reducing internal covariate shift. In Proceedings of the International Conference on Machine Learning (ICML), Lille, France, 6–11 July 2015; pp. 448–456.
28. Stergiou, A.; Kapidis, G.; Kalliatakis, G.; Chrysoulas, C.; Poppe, R.; Veltkamp, R. Class Feature Pyramids for Video Explanation. In Proceedings of the IEEE International Conference on Computer Vision Workshops (ICCVW), Seoul, Korea, 27–28 October 2019.
29. Stergiou, A.; Kapidis, G.; Kalliatakis, G.; Chrysoulas, C.; Veltkamp, R.; Poppe, R. Saliency Tubes: Visual Explanations for Spatio-Temporal Convolutions. In Proceedings of the International Conference on Image Processing (ICIP), Taipei, Taiwan, 22–25 September 2019.
30. Kay, W.; Carreira, J.; Simonyan, K.; Zhang, B.; Hillier, C.; Vijayanarasimhan, S.; Viola, F.; Green, T.; Back, T.; Natsev, P. The Kinetics human action video dataset. *arXiv* **2017**, arXiv:1705.06950.
31. Soomro, K.; Zamir, A.R.; Shah, M. UCF101: A dataset of 101 human actions classes from videos in the wild. *arXiv* **2012**, arXiv:1212.0402.
32. Kuehne, H.; Jhuang, H.; Garrote, E.; Poggio, T.; Serre, T. HMDB: A large video database for human motion recognition. In Proceedings of the International Conference on Computer Vision (ICCV), Barcelona, Spain, 6–13 November 2011; pp. 2556–2563.
33. He, K.; Zhang, X.; Ren, S.; Sun, J. Delving deep into rectifiers: Surpassing human-level performance on imagenet classification. In Proceedings of the International Conference on Computer Vision (ICCV), Santiago, Chile, 7–13 December 2015; pp. 1026–1034.
34. Deng, J.; Dong, W.; Socher, R.; Li, L.J.; Li, K.; Li, F.-F. Imagenet: A large-scale hierarchical image database. In Proceedings of the Computer Vision and Pattern Recognition (CVPR), Miami, FL, USA, 20–25 June 2009; pp. 248–255.

35. Karpathy, A.; Toderici, G.; Shetty, S.; Leung, T.; Sukthankar, R.; Fei-Fei, L. Large-scale video classification with convolutional neural networks. In Proceedings of the Computer Vision and Pattern Recognition (CVPR), Columbus, OH, USA, 23–28 June 2014; pp. 1725–1732.
36. Chen, Y.; Kalantidis, Y.; Li, J.; Yan, S.; Feng, J. Multi-Fiber networks for Video Recognition. In Proceedings of the European Conference on Computer Vision (ECCV), Munich, Germany, 8–14 September 2018.

applied sciences

MDPI

Article

Low-Cost Embedded System Using Convolutional Neural Networks-Based Spatiotemporal Feature Map for Real-Time Human Action Recognition

Jinsoo Kim and Jeongho Cho *

Department of Electrical Engineering, Soonchunhyang University, Asan 31538, Korea; js.kim@sch.ac.kr
* Correspondence: jcho@sch.ac.kr; Tel.: +82-41-530-4960

Abstract: The field of research related to video data has difficulty in extracting not only spatial but also temporal features and human action recognition (HAR) is a representative field of research that applies convolutional neural network (CNN) to video data. The performance for action recognition has improved, but owing to the complexity of the model, some still limitations to operation in real-time persist. Therefore, a lightweight CNN-based single-stream HAR model that can operate in real-time is proposed. The proposed model extracts spatial feature maps by applying CNN to the images that develop the video and uses the frame change rate of sequential images as time information. Spatial feature maps are weighted-averaged by frame change, transformed into spatiotemporal features, and input into multilayer perceptrons, which have a relatively lower complexity than other HAR models; thus, our method has high utility in a single embedded system connected to CCTV. The results of evaluating action recognition accuracy and data processing speed through challenging action recognition benchmark UCF-101 showed higher action recognition accuracy than the HAR model using long short-term memory with a small amount of video frames and confirmed the real-time operational possibility through fast data processing speed. In addition, the performance of the proposed weighted mean-based HAR model was verified by testing it in Jetson NANO to confirm the possibility of using it in low-cost GPU-based embedded systems.

Keywords: CNN; human action recognition; spatiotemporal feature; embedded system; real-time

Citation: Kim, J.; Cho, J. Low-Cost Embedded System Using Convolutional Neural Networks-Based Spatiotemporal Feature Map for Real-Time Human Action Recognition. *Appl. Sci.* **2021**, *11*, 4940. https://doi.org/10.3390/app11114940

Academic Editor: Hyo Jong Lee

Received: 23 April 2021
Accepted: 25 May 2021
Published: 27 May 2021

1. Introduction

Human action recognition (HAR) in video is one of the most challenging tasks in the field of computer vision as it requires simultaneous consideration of spatial and temporal representations of motion [1]. Unlike image classification [2] and object detection [3], which utilize spatial expression extracted by convolutional neural network (CNN) based on an image, HAR recognizes action through spatiotemporal feature extracted from time-varying motions, as well as the appearance of the person extracted from the video, which is a series of images [4]. During the early stage, research [5,6] that recognized actions through a two-dimensional CNN, which learns existing spatial expressions, was performed; however, action recognition was difficult as there were limitations in terms of learning the conditions and temporal features, such as the scale and pose of the human appearance, the similarity of movement, and the change of camera's point of view [7]. Therefore, CNN-based models that learn motion representations through spatiotemporal features are being proposed [8–10], because action in video is recognized based on extracted spatiotemporal features by identifying the connectivity of movements that change over time. The models that recognize action by learning time-varying motion patterns based on CNN have achieved significant performance improvements in the field of HAR [11,12]. The latest CNN-based models for HAR learn time-varying motion representations via an extended CNN structure that combines CNNs applied to static images with networks that extract temporal features [13]. Typically, there are 3D CNNs that simultaneously extract spatiotemporal features, which represent human appearance and motion representations through filters, two-stream CNN, which combines spatial CNN network and temporal CNN network, and convolution recurrent neural network (CRNN), which combines CNN and recurrent neural networks (RNNs) [14].

3D CNN is a structure that applies a convolution layer consisting of 3D kernels to a 3D image stack generated by stacking sequential images over time, simultaneously learning the spatial and temporal features of the input data through a 3D kernel [15]. Therefore, it has the advantage of recognizing action by directly generating hierarchical representations of spatiotemporal information through motion information encoded from the sequential image, and there is a disadvantage [16] that computationally-heavy 3D CNN-based models have high computation and memory cost.

Two-stream CNN recognizes action through a model that fuses spatial CNN network, which extracts appearance information—a spatial feature of motion—from images, and temporal CNN network, which extracts temporal features from motion vectors changing over time, such as optical flow, in parallel structures [17]. Such parallel structures overcome the limitations that conventional CNNs are difficult to learn temporal representations and efficiently fuse spatial features and temporal features extracted from optical flows through late fusion. However, due to the fusion of two CNN networks into a parallel structure, it has a limitation [18] that it is difficult to apply to real-time HAR systems with a long data processing time compared to single-stream CNNs.

CRNNs are single-stream CNN structures and recognize action by identifying context for time-varying motions through RNNs after extracting spatial feature maps from sequential image stacks to CNNs [19]. Long short-term memory (LSTM) is used as a network to analyze time information, and a model that fuses CNN and LSTM is called ConvLSTM. ConvLSTM is an LSTM that learns time-varying motion patterns from images, so it has the advantage of identifying the connectivity of spatial features that change over time and of extracting temporal features well. However, ConvLSTM has the disadvantage that the more images that are extracted from the video, the more the complexity of the model increases.

The CNN-based HAR models that were mentioned can be used in various fields that require actual HAR through the action recognition performance that improved compared with existing techniques, but owing to the complexity of the model and long data processing time, there are still limitations to be operated in real-time in low-cost embedded systems [20]. Recently, HAR has been applied to integrated monitoring systems, which detect emergency situations with CCTVs. Most of integrated monitoring systems receive multiple CCTV images through network-based digital video communication and extract monitoring information [21] as shown in Figure 1. The video data transmission process is used to recognize behavior in a server PC equipped with a GPU that has strong computational efficiency owing to the complexity of the HAR model. Therefore, HAR models have a limitation in recognizing behavior in real-time, being installed in a single embedded system directly connected to the CCTV.

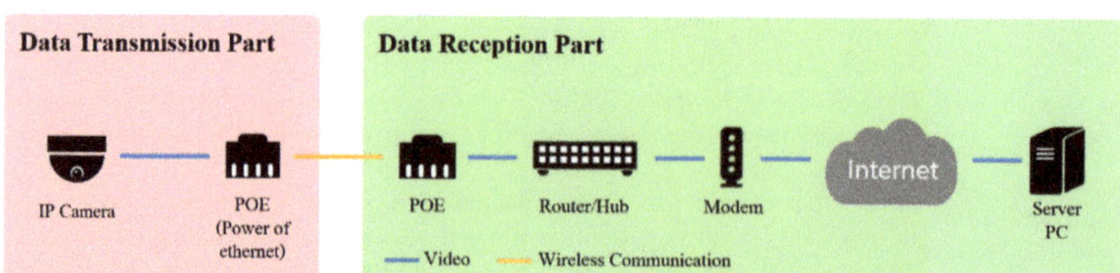

Figure 1. An example of a wireless communication network for recognizing behavior through CCTV using a complex HAR model.

In particular, it is essential for the system to recognize human action quickly in urgent situations related to the safety, such as violence and theft on the streets, and in emergencies, such as detecting urgent situations or abnormal behaviors of the elderly and single-person households [22]. When the action is recognized through the HAR model in the embedded system where the camera is installed, time and cost in the process of transmitting high-capacity video data can be reduced [23].

Therefore, in this paper, we propose a weighted mean-based single-stream CNN model that recognizes action faster than conventional models. As the proposed method has a simpler structure than the existing CNN-based models with high complexity, recognizing behavior in a single embedded system connected to CCTV without transmitting image data through network communication is possible. The main idea is to build a lightweight CNN-based HAR model that can be applied to low-cost embedded systems with low-end GPUs to apply HAR to CCTV-based surveillance fields and to recognize instantaneous motions in emergency situations at high speed. The proposed system extracts spatial and temporal features by weight and averaged the change rate of frames according to time on a spatial feature map extracted by CNN. The weighted mean is calculated sequentially by the change rate of each frame extracted according to any time interval from the video and the change rate of frame at the corresponding time is weighted in the process of averaging feature maps, so the spatial and temporal features are created in the contexts of the integrated time-varying motion representation. Spatial feature maps at a specific time where the amount of motion change is high are weighted more than the point of time when the amount of change is low and the context of the motion representation is created; spatial representations are extracted efficiently at the point of time affecting the recognition performance of systems. The weighted mean also recognizes action by processing data at a rapid rate through a lighter structure than conventional CNN-based HAR models. 3D CNNs extract temporal features using spatiotemporal filters, two-stream CNNs utilize optical flows, and ConvLSTMs utilizes LSTMs to extract temporal features. Such models take a long time in the calculation process with an extended CNN structure combining networks that extracts temporal features in the existing two-dimensional CNN structure. However, the proposed model recognizes action by inputting a one-dimensional spatiotemporal feature vector into FC layers, where the vector is generated by spatial feature maps and frame change rate using weighted mean. Therefore, video data are processed at high speed while maintaining the existing two-dimensional CNN structure.

2. Related Works

In ImageNet Challenge 2012 [24], a deep learning-based CNN with superior performance than the algorithm that was to be used in the existing computer vision has been proposed, and in recent studies, the deep learning-based CNN shows high level of usability for HARs. HAR models applied by CNN automatically extract and learn motion features from video and utilize the motion features extracted from different modalities of data to enhance the performance for action recognition according to modality of data perspective; action recognition models are largely divided into depth-, skeleton-, and vision-based HAR [25].

2.1. Depth-Based Human Action Recognition

Depth- and skeleton-based HARs [26,27] recognize actions using change in motion representation of the depth map acquired through the depth sensor. The depth map comprising RGB-D video has the spatiotemporal structure. Changes in the depth information over time are extracted to the spatiotemporal features of motion [28]. In addition, depth maps clearly separate people from backgrounds to represent appearance information, so they can be used for meaningful feature extraction for action recognition.

Zhanga et al. [29] propose orientation histogram features of 3D normal vectors to extend the features of histogram oriented gradient (HOG) extracted from the depth map to a spatiotemporal depth structure and to represent appearance information of a three-dimensional depth structure of the spatiotemporal depth. Authors in [30] construct super-normal feature vectors based on depth map sequences to represent motion representations for action recognition. Feature vectors are generated by applying spatial average pooling and temporal maximum pooling to time-varying depth maps, and the evaluation results on various benchmark datasets show robustness about scale change.

The HAR models using depth information have shown high action recognition performance, but have limitations, applying only to a limited range and specific environment. Generally used depth sensors include stereo cameras using triangulation techniques, time of flight (TOF) cameras, and structured-light cameras. Depth sensors using stereo cameras are inexpensive, but the process of calculating depth information is complicated, making it difficult to acquire accurate depth information. Depth sensors based on TOF cameras and structured-light cameras have limitations of difficulty to be applied outdoors, which are heavily influenced by light. In addition, the depth sensors represent the information included in the measurable distance as data, which is difficult to apply to outdoor surveillance fields. Some sensors, such as light detection and ranging, are robust to lighting and measure a relatively wide range of depth but are expensive and not suitable for use in video-based surveillance systems.

2.2. Skeleton-Based Human Action Recognition

Skeleton-based HAR [31,32] recognizes action through joint points extracted via CNN-based pose estimation algorithms from depth maps or RGB images. The location of a person's joint points represented by the time axis extracted from the video is used as the feature vector, which is connected according to the body structure of the person, and adjacent points have an important correlation with each other.

Warcho et al. [33] proposed a CNN-based HAR model that automatically learns the spatial and temporal features of data based on joint points. To reduce the redundancy of data and preserve spatiotemporal features, key frames are extracted using interframe difference methods, and joint points are generated through the open-pose [34] and then CNN is applied. Recently, more research related to skeleton-based HAR has been proposed than depth-based HAR, but since the process of extracting joint points with pose estimation algorithms is preceded, there is a disadvantage that the accuracy of joint points varies depending on sensor performance. The whole recognition performance of systems can be degraded if joint points containing noise are obtained by sensor performance, or if they are affected by external environmental factors such as lighting and occlusion [35].

2.3. Vision-Based Human Action Recognition

Before a deep learning-based CNN was applied to HAR, the conventional method recognized actions based on hand-crafted features for some human-performed actions in a simple background. Hand-crafted features include spatiotemporal interest points (STIPs) [36], 3-dimensional HOG [37], 3-Dimensional Scale Invariant Feature Transform (3D-SIFT) [38], which use various feature encoding schemes such as histograms or pyramids. In [39], the geometric properties of space-time volume according to human movement are extracted with action sketches, which stacked body outlines on the time axis according to direction, speed, and shape. These low-level features were entered into machine learning-based classification algorithms, such as SVM, decision tree, and K-nearest neighbor, and used in HARs.

However, the deep learning-based CNN is proposed and research on various CNN-based methods is being conducted to build HAR models; typically, there are methods such as 3D CNN [40], two-stream CNN [41], and ConvLSTM [42]. Recently, a study of building HAR models by fusing the above methods and a method of entering depth maps and joint points into an algorithm of vision-based HAR models have also been proposed. Karpathy et al. [43] propose a spatiotemporal LSTM (Spatial LSTM) model for 3-dimensional HARs that extend the RNN into the spatiotemporal domain to analyze the hidden features of motion representations. In addition, in [44], a study was conducted to classify an action in spatial and temporal encoding information of depth map sequences by applying 3D CNNs to encode motion patterns with spatiotemporal features in the depth map sequences. The advantages and disadvantages of these three HAR modalities are summarized in Table 1.

Table 1. Comparison of advantages and disadvantages of data modality for behavior recognition.

Modality	Advantage	Disadvantage
Depth	Clearly separate foreground and background	Vulnerable to light
Skeleton	Extract correlations from joint points connected by body structure	Data uncertainty
Vision	Includes various visual information with high resolution	Data capacity

3. Proposed Methodology

3.1. System Overview

The proposed embedded system-based HAR model uses a spatial feature map extracted using CNN through the weighted mean and the temporal feature extracted via frame change rate, and then, generates spatiotemporal features. The generated spatiotemporal features are entered into a multilayer perceptron (MLP), which is lighter than the existing networks that were used in the HAR models, outputting action classes and schematizing the block diagram of the entire system, as shown in Figure 2. Image sequences, which make up a video, entered into a model, are converted into N frame stacks (FS^N) with random intervals and extracted via CNN into feature map (FM_K^N) with spatial features (K means the length of a feature map that is flattened in one dimension). Here, CNN is used as a feature extractor to extract a spatial feature map, and the fully connected layer connected to the end of the CNN is removed. FS^N extracts frame difference (FD^N) of each frame constituting FS^N as a temporal feature, since the sequential frames within a video contain time information with sequentially stacked data. Finally, to fuse spatial and temporal features, FM_K^N is weighted and averaged to FD^N according to the interval of FS^N converted from video to generate feature vectors (FV_K) with spatiotemporal features and input them into MLP. The MLP, which is independent from CNN, outputs predefined action classes and schematized the process of recognizing action by receiving video data, which is shown in Figure 3.

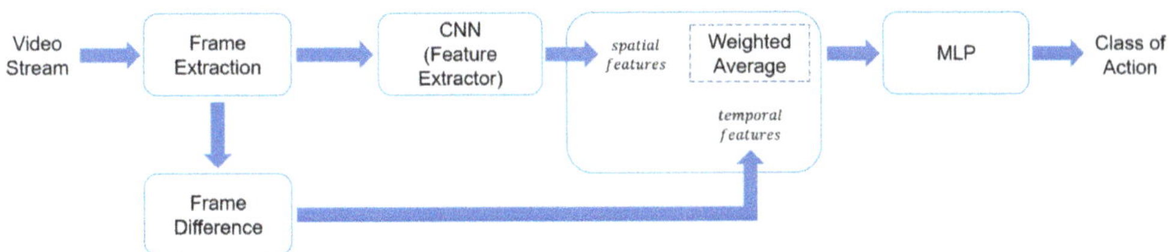

Figure 2. Block diagram of the proposed weighted mean-based human action recognition (HAR) model.

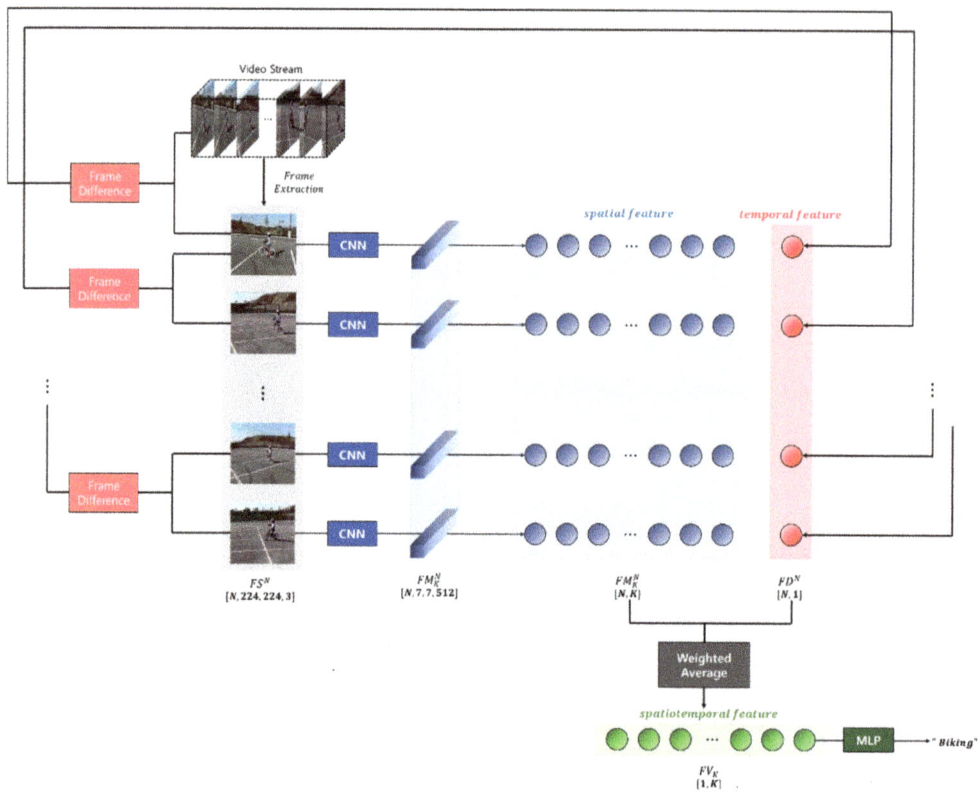

Figure 3. Action recognition process of the proposed system.

3.2. Extract Spatial Features

Videos are sequential data, in which frames representing spatial representations are listed according to time information, and the sequence of frames is the time information required to recognize the context of the action contained in the video. The video from the UCF-101 dataset processes frames at 25 fps. When all frame sequences per unit time are entered into the model, the processing time according to the computational cost increases and has a lot of redundant data. Considering processing complexity according to the data, the proposed system extracts N frames at random intervals and converts the image sequence of input video into FS^N.

FS^N is inputted to the CNN structure $VGGNet$ [45], which is pretrained with ImageNet dataset and is extracted to feature maps. Moreover, the resolution of the images in FS^N is converted to (224, 224, 3) according to the kernel size of $VGGNet$. $VGGNet$ is a CNN structure that has deeply stacked convolutional layers through (3, 3) filters, with fewer parameters than convolutional layers, which were shallowly stacked through a larger sized filter. Therefore, although $VGGNet$ is a deep layer structure, it is a CNN structure that has been proven to extract feature maps at high speed from input data and proven to extract significant features, which reduce classification errors as the layer deepens. In addition, $VGGNet$ has the size (7, 7, 512) of a feature map generated from the convolutional layer prior to the FC layer, which is smaller than other CNN structures, such as $ResNet$ and $Inception$. The proposed system is light weighted using $VGGNet$, which extracts relatively small size feature maps to build HAR models that process data at high speed. $MobileNet$, a CNN structure that could be used in embedded systems with relatively low performance

GPUs and memory, was also proposed. However, the size of the final feature map was (7, 7, 1024), which was not suitable for the purpose of the system to be built in this paper, which requires faster data processing by outputting a larger feature map than the $VGGNet$ used in this paper.

3.3. Integrated Spatial and Temporal Features Based on Weighted Mean

FS^N is a collection of image data including time information sequentially listed according to any interval, and the model extracts FD^N, which is the frame change rate of each image included in FS^N, as a temporal feature. N means the number of frames that constitute FS^N, and since they are extracted according to the same time interval, the frame change rate, FD^N, that changes according to the time interval can be used as a temporal feature of motion representation. FD^N is calculated as the average of the frame change sequentially calculated according to the image sequence of FS^N and it is transformed to one-dimensional temporal feature vectors. As compared to optical flow applied to two-stream CNNs, LSTM applied to single-stream CNNs is simpler to operate and it is extracted at high speed. The frame change rate calculated by the frame change of the images constituting FS^N additionally calculates the frame change for the whole frame of the video and the first frame of FS^N because it has the length of $N-1$. Therefore, FD^N has the same length as the data length N, indicating the time information of FS^N extracted according to any interval in the video, and FD^N in N means the frame change rate of FM_K^{N-1} and FM_K^N. FM_K^N has a size of (N, K) over the entire time because K one-dimensional vectors are extracted at each time. The data in each row (N) of FM_K^N mean spatial features extracted from images at a specific time and are sorted in a column direction over time. In the proposed system, FV_K with spatiotemporal features is generated by weighted mean of FD^N, which represents the degree of change of each frame over time in the spatial feature FM_K^N sorted according to time information, as shown in Equation (1).

$$FV_K = \left[\frac{\sum_{i=1}^{N} FM_1^i \times FD^i}{\sum_{i=1}^{N} FD^i} \; \frac{\sum_{i=1}^{N} FM_2^i \times FD^i}{\sum_{i=1}^{N} FD^i} \; \cdots \; \frac{\sum_{i=1}^{N} FM_{K-1}^i \times FD^i}{\sum_{i=1}^{N} FD^i} \; \frac{\sum_{i=1}^{N} FM_K^i \times FD^i}{\sum_{i=1}^{N} FD^i} \right] \quad (1)$$

The weighted mean is used to convert the spatial feature map, FM_K^N, listed over time into a spatiotemporal feature during the entire time, and the elements of each feature map are averaged by weighting FD^N at the corresponding time according to size K of spatial feature maps. FD^N is a temporal feature extracted from spatial information represented by the motion information of an object and background information, which changes according to the camera's point of view, so it can be matched with spatial features simultaneously when FM_K^N is extracted. According to Equation (1), the system divides the sum of the result multiplying the Kth element of each feature map by FD^N at the corresponding time and the sum of FD^N to generate spatiotemporal features FV. The elements of feature maps are generated as spatiotemporal features for action recognition through weighted mean calculations considering frame change rate during the entire time. Figure 4 is a result of schematizing the operation process when $K = 1$ and generates FV_K by repeating the operation K times according to the size of the spatial feature map. The weighted mean is weighted to the feature point of video when the motion or background information of an object changes greatly according to the time information of FD^N, thereby enabling the context of the spatial feature that changes throughout the whole time. Finally, FV_K generated through the weighted mean is inputted to the FC layer of the MLP structure, and the MLP outputs the class of the predefined videos.

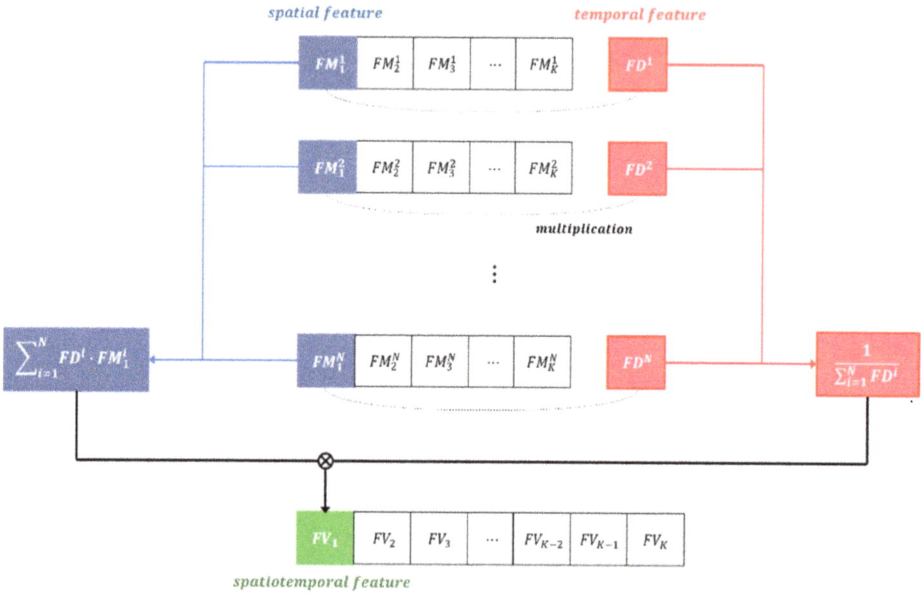

Figure 4. Process for generating weighted mean-based spatiotemporal features.

3.4. Action Recognition Using Multilayer Perceptron

Finally, the proposed HAR model recognizes an action by inputting the spatiotemporal feature vector, FV_K, generated by the weighted mean into the MLP. The MLP receiving the FV_K has three hidden layers, the input layer consists of 25,088 nodes, and the output layer consists of 101 nodes, which is the number of predefined classes. Each hidden layer consists of 2048, 1024, and 512 nodes. The batch size set in the input layer of the model is 64, and to prevent overfitting during the learning process, dropout is applied at a ratio of 0.5 in the first hidden layer. The categorical cross-entropy, which is a loss function to classify multiple classes according to the purpose of the system, was selected and adaptive moment estimation (ADAM) was applied to the model.

4. Experimental Results

4.1. Experimental Setup

The performance evaluation results of the proposed weighted mean-based action recognition model are described. The learning and testing of the model proceed with HAR benchmark dataset UCF-101.

UCF-101 consists of 13,220 video clips and 101 actionable classes taken on YouTube. The actionable categories are divided into (1) human–object interaction, (2) body-motion only, (3) human–human interaction, (4) playing musical instruments, and (5) sports. Video is a relatively challenging dataset, filmed at various illuminations, poses of people, and viewpoints of cameras. According to the train–test list of provided datasets, 9.5K datasets are divided into learning and 3.8K datasets are divided into tests, and 20% of the learning datasets are used as validation datasets. In addition, the train–test list consists of three scenarios with random order of video data, so the average of three performance evaluations was selected as the final result.

The proposed system was trained at the workstation, and performance evaluations were conducted at the workstation and NVIDIA Jetson NANO. At the workstation, tests were conducted to compare and evaluate the accuracy of the proposed method with the existing HAR models, and at Jetson NANO, the tests were performed to evaluate usability based on throughput of low-cost embedded systems. The model of the proposed system is built in the Python environment, Keras, and the system environment of the workstation consists of Intel i9-10900X CPU, NVIDIA Titan RTX (24 GB) GPU, and 128 GB of main memory. The system environment of Jetson NANO consists of Quad-core ARM A57 CPU, 128-core Maxwell GPU, and 4 GB of main memory, as shown in Figure 5.

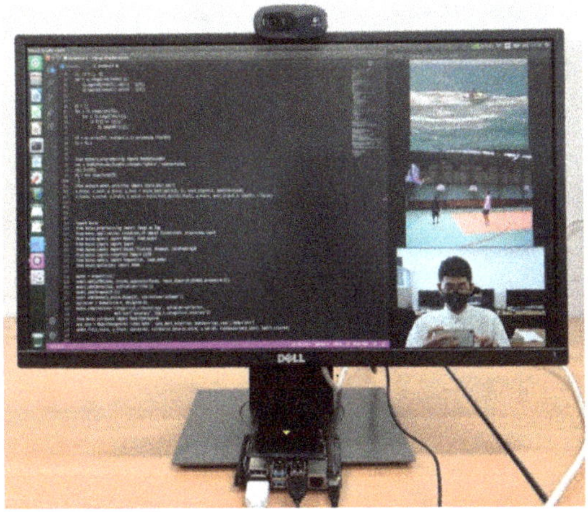

Figure 5. The proposed Jetson NANO-based HAR system appearance.

4.2. Performance Evaluation

The main idea proposed in this paper is to recognize action at a rapid rate through the spatiotemporal feature vector (FV_K) generated by weighted mean of the spatial feature map (FM_K^N) extracted from the frame stack (FS^N) consisting of sequential images with the change rate (FD^N) of each frame representing time information. The performance evaluation is conducted by comparing the action recognition accuracy and complexity of the HAR model using the proposed weighted mean and the existing HAR models. The comparison results of performance evaluation are shown in Table 2, and results are obtained using workstation. Models used for comparison of performance evaluation received RGB stream exactly as the system was built in this paper, and complexity of model and the average accuracy were compared according to the three random scenarios of the train–test split list provided by UCF-101. Equation (2) represents the complexity of each model, and the definition of each factor is defined in Table 3. If the network of the model is 3D CNN, time-axis operation is added to the convolution operation. The recognition accuracy of 3D CNN and single-stream CNN was 84.8%, 85.2%, and 88.1%, respectively, and the recognition accuracy of the model using the LSTM-based method was 90.8% and 91.21%. The LSTM-based action recognition model shows higher accuracy than 3D CNN and single-stream CNN because sequential data are received and weighted to identify connectivity. However, the proposed model based on single-stream CNN recognized action with 2.48% higher accuracy than deep LSTM. This study proved that the proposed method using the change rate of sequential frames as time information can effectively identify the connectivity of motion for action recognition.

$$
\begin{aligned}
N_{CNN} &= \sum_{conv} 2C_{k-1}C_k N_k^{(kernel)} N_k + \sum_{pool} C_k N_k \left(N_k^{(pool)} - 1\right) + 2(C_k N_k N_d + N_d N_{out}) \\
N_{LSTM} &= L \cdot \sum_{k=1}^{K} (8(N_{k-1} + N_k)N_k + 4N_k) + 2N_K N_{out} \\
N_{BILSTM} &= L \cdot \sum_{k=1}^{K} (8(N_{k-1} + N_k)N_k + 4N_k) + 2N_K N_{out}
\end{aligned} \tag{2}
$$

Table 2. Comparison of average accuracy of the proposed method for UCF-101 with other methods.

Model	Network	Accuracy (%)	Complexity
Multi-scale CNN [46]	3D CNN	84.80	N_{CNN}
C3D [47]	3D CNN	85.20	N_{CNN}
MSD [48]	3D ConvLSTM	90.80	$N_{CNN} + N_{LSTM}$
FCN [49]	Single-stream CNN	88.10	N_{CNN}
Deep LSTM [50]	2D ConvLSTM	91.21	$N_{CNN} + N_{BILSTM}$
Ours	Single-stream CNN	93.69	N_{CNN}

Table 3. Notation of equations indicating model complexity.

Symbol	Definition
K	Number of hidden layers
N_k / N_{out}	Feature vector length of the kth hidden layer/output size
$N_k^{(kernel)}$	Kernel size of the kth hidden layer
$N_k^{(pool)}$	Pooling factor in the kth hidden layer
C_k	Number of filters in the kth hidden layer
$conv/pool$	Set of indices of convolutional/pooling layers
L	Input sequence length

In addition, additional experiments for performance evaluation were conducted to verify the applicability of the action recognition model on embedded systems equipped with low-cost GPUs. Experiments were conducted by means of "Element-wise Mean," "Weighted Mean," "ConvLSTM," and "Bi-ConvLSTM." The "Element-wise Mean" means an operation that averages FM_K^N outputted from CNN according to the number of N without weighting time information.

The frame number (N) of FS^N extracted according to a random interval is increased in five intervals from 10 to 40, the action recognition accuracy was measured, and testing time times were compared according to the frame number (N) of FS^N, and then testing time was measured by the average time it takes to output the action class from each video of the test data set. The action recognition accuracy and testing time of three methods according to N number of FS^N extracted from the video data according to any interval are shown as a graph in Figure 6 and results are obtained using Jetson NANO.

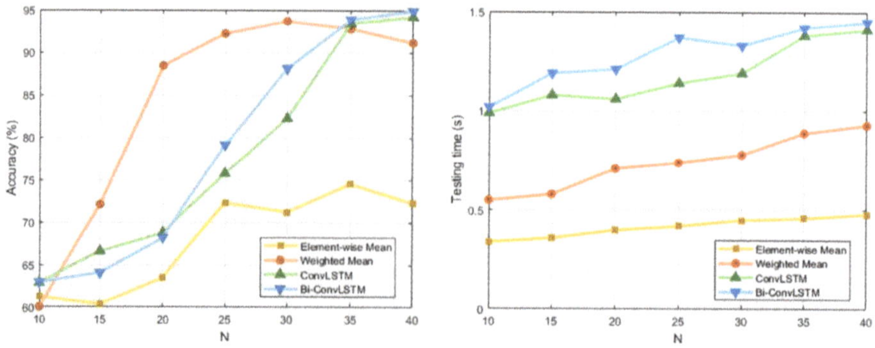

(**a**) (**b**)

Figure 6. Accuracy and testing time according to frame number of FS^N extracted at a random interval: (**a**) accuracy and (**b**) testing time.

First, the comparative evaluation was conducted with the case of applying "Element-wise Mean" to determine whether the weighted mean effectively extracts meaningful spatiotemporal features to recognize action by identifying connectivity of time-varying spatial features. Performance evaluation results show that the proposed method for weighted FD^N, in all N measures higher accuracy than "Element-wise Mean" and the weighted mean generates spatiotemporal features to recognize action efficiently. When the spatial and temporal features are fused through unweighted "Element-wise Mean" of FD^N, the spatiotemporal features are generated by relying solely on the number of N, which is the interval of FS^N. Since videos are sequential data with time-varying frames, even in the case of considering only the interval of FS^N, the connectivity of the time-varying spatial features can be identified. The motion and background information of an object have nonlinear features rather than constant changes such as the interval of FS^N, so if frames depend only on extracted intervals, changes in various environmental conditions cannot be considered. Various environmental conditions include changes in camera's viewpoints and influences of external environmental factors such as lighting and obstacles, according to data acquisition environment, and accordingly, the scale and pose of people's appearances change and motion representation changes over time. Therefore, if spatiotemporal features are generated through "Element-wise Mean," the spatiotemporal features cannot be efficiently generated because spatial representation changes according to the above conditions cannot be identified at a specific time. However, considering the change rate of frames according to the interval extracted through the weighted mean, since big changes in spatial representation are weighed with time information at a specific time and spatiotemporal features are generated, the action recognition performance was improved by considering the motion and background information of objects changing over time effectively.

In addition, as a result of the performance comparison with weighted mean and LSTM-based models, "ConvLSTM" and "Bi-ConvLSTM", when extracting more than 30 frames, LSTM-based models predict action with high recognition accuracy of 2–3% or more. However, when 20–30 frames are extracted, we can see that the proposed weighted mean-based model predicts action with high recognition accuracy of 20% or more. In particular, when 20 frames are extracted, the proposed weighted mean-based model and LSTM-based model showed the greatest performance difference, and the action prediction accuracy and testing time for videos in each method when N = 20 are shown in Table 4 where the tests were conducted in Jetson NANO.

An LSTM prediction network has been proposed to solve the problem of long input data length and disappearing influence on initial inputs in RNNs, which sequentially receive and process time series data. It controls the amount of information transmitted from hidden layers to input gates and oblivion gates. When extracting a large number of frames from video, the amount of data increases and the extracted frame interval shortens, so it is advantageous to identify time-varying spatial features. Therefore, when extracting more than 30 frames, LSTM-based models could recognize action with higher accuracy than weighted mean-based models. In addition, "Bi-ConvLSTM" utilizes information when sequentially listed FM_K^N is inputted in the reverse direction; "Bi-ConvLSTM" has higher accuracy than "ConvLSTM" when extracting frames number (N) increases. This shows that the LSTM-based model shows higher performance as the large number of frames is extracted from the videos.

Table 4. Comparison of action prediction accuracy using four methods, when $N = 20$.

	GT: ApplyEyeMakeup		GT: LongJump		GT: Kayaking	
Method	Accuracy score (%)	Testing time (s)	Accuracy score (%)	Testing time (s)	Accuracy score (%)	Testing time (s)
Element-wise Mean	65.23	0.28	68.72	0.34	67.34	0.31
Weighted Mean	88.64	0.59	89.02	0.63	88.46	0.61
ConvLSTM	69.25	0.89	71.87	1.05	68.53	1.26
Bi-ConvLSTM	71.24	0.91	72.53	1.08	67.71	1.38

	GT: Archery		GT: PlayingViolin		GT: HandStandPushups	
Method	Accuracy score (%)	Testing time (s)	Accuracy score (%)	Testing time (s)	Accuracy score (%)	Testing time (s)
Element-wise Mean	60.92	0.35	61.73	0.42	63.73	0.25
Weighted Mean	87.58	0.58	86.25	0.72	88.94	0.67
ConvLSTM	70.12	0.94	71.58	1.42	69.37	1.32
Bi-ConvLSTM	69.47	0.92	72.72	1.49	70.03	1.30

However, as the results of comparing data testing time increase, the proposed weighted mean-based HAR model showed faster data processing speed than LSTM-based models when extracting more than 30 frames. This is because the proposed model recognizes action using MLP, which is lighter than LSTM as a prediction network. LSTM receives time series data sequentially listed over time and controls the amount of information stored while repeating the operation of hidden layers by the length of data. Therefore, when extracting many frames from videos, LSTM has a structure that receives a relatively long time of data input, and the data processing speed is slowed. The data processing speeds of "ConvLSTM" and "Bi-ConvLSTM" in Figure 5 show that the data processing speed is slower as N increases. However, MLP predicts action at a faster rate than LSTM by conducting forward propagation operations regardless of data length. LSTM receives data sequentially to identify the temporal features of spatial feature maps, which are inputted, and repeats the computation of hidden layers as the frame number (N). However, the proposed model receives the spatiotemporal features generated by the weighted mean of

FM_K^N to FD^N representing time information at a time as data of the entire time. Jetson NANO, with low-cost GPUs, can recognize action at a faster speed than LSTM. In addition, the weighted mean-based model recognizes action with a high accuracy of 30% or more even when extracting ~10 frames less than LSTM-based models. This shows that the proposed method, which generates spatiotemporal features through weighted mean efficiently, extracts motion representations for action recognition. Therefore, the HAR model can be constructed by extracting relatively few frames and the instantaneous action can be recognized quickly.

5. Conclusions

In this paper, we proposed a weighted mean-based spatiotemporal feature extraction technique to build a CNN-based HAR model that recognizes action by processing video data at high speed. Previously, an action was recognized by analyzing motion patterns based on time information through models with complex structures such as 3D CNN, two-stream CNNs using optical flows, and ConvLSTMs. However, the proposed method recognizes an action by extracting frame changes, which are calculated at high speed with temporal features. The temporal feature is used to generate the spatiotemporal features during the entire time by weighing the spatial features extracted from CNNs and through the weighted mean spatial information at the point where motion changes significantly. The generated spatiotemporal features are used to recognize action by entering into MLPs with lower complexity than the prediction model used in the existing HAR, and the proposed model is verified via experiments to recognize action by processing data at a faster speed than the existing CNN-based HAR models. In addition, the efficiency of extracting spatiotemporal features was verified using the frame change rate as time information through higher action recognition performance compared to general average technique. Performance evaluation results according to the number of frames extracted from videos showed that action was recognized with high accuracy even while extracting fewer frames than the HAR model using LSTM. Finally, to assess real-time possibility in embedded systems of low-cost GPUs, the results of performance evaluation in Jetson NANO also show that data are processed at high speed. It was possible to verify the system's utilization value to recognize instantaneous action in an emergency situation.

Nevertheless, the proposed model still has a weakness in that it is vulnerable to rapid changes of background or obstacles because it recognizes the action by using the changes in image frames over time. Changes in frames caused by background or obstacles, not actions of a person to be recognized, can act as noise and deteriorate action recognition performance. In particular, it is very difficult to recognize an action in a camera that is constantly moving, not a CCTV, which is a fixed type of camera we have adopted, because the frame is continuously changing. Addressing this issue is the direction of research we are pursuing in the future.

Author Contributions: All authors took part in the discussion of the work described in this paper. Conceptualization, J.K.; Data curation, J.K.; Investigation, J.K.; Project administration, J.C.; Writing—original draft, J.C.; Writing—review & editing, J.C. All authors have read and agreed to the published version of the manuscript.

Funding: This research was funded by a National Research Foundation of Korea (NRF) grant funded by the Korean government (MOE) (No.2018R1D1A3B07041729) and the Soonchunhyang University Research Fund.

Institutional Review Board Statement: Not applicable.

Informed Consent Statement: Not applicable.

Data Availability Statement: The data presented in this study are available on request from the corresponding author.

Acknowledgments: The authors thank the editor and anonymous reviewers for their helpful comments and valuable suggestions.

Conflicts of Interest: The authors declare that they have no competing interests.

References

1. Qiu, Z.; Yao, T.; Ngo, C.W.; Tian, X.; Mei, T. Learning spatio-temporal representation with local and global diffusion. In Proceedings of the IEEE Conference on Computer Vision and Pattern Recognition (CVPR), Long Beach, CA, USA, 15–21 June 2019; pp. 12056–12065.
2. Rawat, W.; Wang, Z. Deep convolutional neural networks for image classification: A comprehensive review. *Neural Comput.* **2017**, *29*, 2352–2449. [CrossRef]
3. Zhao, Z.-Q.; Zheng, P.; Xu, S.-T.; Wu, X. Object detection with deep learning: A review. *IEEE Trans. Neural Netw. Learn. Syst.* **2019**, *30*, 3212–3232. [CrossRef] [PubMed]
4. Wu, Q.; Hu, F.; Zhu, A.; Wang, Z.; Bao, Y. Learning spatial-temporal features via a pose-flow relational model for action recognition. *AIP Adv.* **2020**, *10*, 075208. [CrossRef]
5. Liu, A.-A.; Xu, N.; Nie, W.-Z.; Su, Y.-T.; Wong, Y.; Kankanhalli, M. Benchmarking a multimodal and multiview and interactive dataset for human action recognition. *IEEE Trans. Cybern.* **2016**, *47*, 1781–1794. [CrossRef] [PubMed]
6. Gao, Z.; Zhang, Y.; Zhang, H.; Xue, Y.B.; Xu, G.P. Multi-dimensional human action recognition model based on image set and group sparsity. *Neurocomputing* **2016**, *215*, 138–149. [CrossRef]
7. Wang, L.; Xiong, Y.; Wang, Z.; Qiao, Y.; Lin, D.; Tang, X.; Van Gool, L. Temporal segment networks for action recognition in videos. *IEEE Trans. Pattern Anal. Mach. Intell.* **2019**, *41*, 2740–2755. [CrossRef] [PubMed]
8. Leong, M.C.; Prasad, D.K.; Lee, Y.T.; Lin, F. Semi-CNN architecture for effective spatio-temporal learning in action recognition. *Appl. Sci.* **2020**, *10*, 557. [CrossRef]
9. Li, S.; Zhao, Z.; Su, F. A spatio-temporal hybrid network for action recognition. In Proceedings of the IEEE Visual Communications and Image Processing (VCIP), Sydney, Australia, 1–4 December 2019; pp. 1–4.
10. Varol, G.; Laptev, I.; Schmid, C. Long-term temporal convolutions for action recognition. *IEEE Trans. Pattern Anal. Mach. Intell.* **2018**, *40*, 1510–1517. [CrossRef]
11. Ben-Ari, R.; Shpigel, M.; Azulai, O.; Barzelay, U.; Rotman, D. TAEN: Temporal aware embedding network for few-shot action recognition. *arXiv* **2020**, arXiv:2004.10141.
12. Wang, H.; Song, Z.; Li, W.; Wang, P. A hybrid network for large-scale action recognition from RGB and depth modalities. *Sensors* **2020**, *20*, 3305. [CrossRef]
13. Rodríguez-Moreno, I.; Martínez-Otzeta, J.M.; Sierra, B.; Rodriguez, I.; Jauregi, E. Video activity recognition: State-of-the-Art. *Sensors* **2019**, *19*, 3160. [CrossRef]
14. Carreira, J.; Zisserman, A. Quo Vadis, action recognition? A new model and the kinetics dataset. In Proceedings of the 2017 IEEE Conference on Computer Vision and Pattern Recognition (CVPR), Honolulu, HI, USA, 21–26 July 2017; pp. 4724–4733.
15. Diba, A.; Fayyaz, M.; Sharma, V.; Karami, A.H.; Arzani, M.; Yousefzadeh, R.; Gool, L.V. Temporal 3D ConvNets: New architecture and transfer learning for video classification. *arXiv* **2017**, arXiv:1711.08200.
16. Tran, D.; Ray, J.; Shou, Z.; Chang, S.F.; Paluri, M. ConvNet architecture search for spatiotemporal feature learning. *arXiv* **2017**, arXiv:1708.05038.
17. Simonyan, K.; Zisserman, A. Two-stream convolutional networks for action recognition in videos. *arXiv* **2014**, arXiv:1406.2199.
18. Zhao, Y.; Man, K.L.; Smith, J.; Siddique, K.; Guan, S.-U. Improved two-stream model for human action recognition. *EURASIP J. Image Video Process.* **2020**, *2020*, 1–9. [CrossRef]
19. Majd, M.; Safabakhsh, R. A motion-aware ConvLSTM network for action recognition. *Appl. Intell.* **2019**, *49*, 2515–2521. [CrossRef]
20. Lee, J.; Ahn, B. Real-time human action recognition with a low-cost RGB camera and mobile robot platform. *Sensors* **2020**, *20*, 2886. [CrossRef]
21. Shidik, G.F.; Noersasongko, E.; Nugraha, A.; Andono, P.N.; Jumanto, J.; Kusuma, E.J. A systematic review of intelligence video surveillance: Trends, techniques, frameworks, and datasets. *IEEE Access* **2019**, *7*, 170457–170473. [CrossRef]
22. Fahimeh, R.; Sareh, S.; Upcrofit, B.; Michael, M. Action recognition: From static datasets to moving robots. In Proceedings of the IEEE International Conference on Robotics and Automation (ICRA), Marina Bay Sands, Singapore, 29 May–3 June 2017; pp. 3185–3191.
23. Sreenu, G.; Durai, M.A.S. Intelligent video surveillance: A review through deep learning techniques for crowd analysis. *J. Big Data* **2019**, *6*, 48. [CrossRef]
24. Krizhevsky, A.; Sutskever, I.; Hinton, G.E.; Geoffrey, E. ImageNet classification with deep convolutional neural networks. *Adv. Neural Inf. Process. Syst.* **2012**, *25*, 1097–1105. [CrossRef]
25. Zhang, H.-B.; Zhang, Y.-X.; Zhong, B.; Lei, Q.; Yang, L.; Du, J.-X.; Chen, D.-S. A comprehensive survey of vision-based human action recognition methods. *Sensors* **2019**, *19*, 1005. [CrossRef] [PubMed]
26. Chen, C.; Liu, K.; Kehtarnavaz, N. Real-time human action recognition based on depth motion maps. *J. Real-Time Image Process.* **2016**, *12*, 155–163. [CrossRef]
27. Zhanga, J.; Lia, W.; Ogunbonaa, P.O.; Wanga, P.; Tang, C. RGB-D-based action recognition datasets: A survey. *Pattern Recognit.* **2016**, *60*, 86–105. [CrossRef]
28. Yang, X.; Tian, Y. Effective 3D action recognition using EigenJoints. *J. Vis. Commun. Image Represent.* **2014**, *25*, 2–11. [CrossRef]

29. Oreifej, O.; Liu, Z. HON4D: Histogram of oriented 4D normals for activity recognition from depth sequences. In Proceedings of the IEEE Conference on Computer Vision and Pattern Recognition (CVPR), Portland, OR, USA, 23–28 June 2013; pp. 716–723.
30. Yang, X.; Tian, Y.L. Super normal vector for activity recognition using depth sequences. In Proceedings of the IEEE Conference on Computer Vision and Pattern Recognition (CVPR), Columbus, OH, USA, 24–27 June 2014; pp. 804–811.
31. Warcho, D.; Kapuściński, T. Human action recognition using bone pair descriptor and distance descriptor. *Symmetry* **2014**, *12*, 1580. [CrossRef]
32. Muralikrishna, S.N.; Muniyal, B.; Acharya, U.D.; Holla, R. Enhanced human action recognition using fusion of skeletal joint dynamics and structural features. *J. Robot.* **2020**, *2020*, 3096858. [CrossRef]
33. Yang, Y.; Cai, Z.; Yu, Y.D.; Wu, T.; Lin, L. Human action recognition based on skeleton and convolutional neural network. In Proceedings of the Photonics & Electromagnetics Research Symposium-Fall (PIERS-Fall), Xiamen, China, 17–20 December 2019; pp. 1109–1112.
34. Cao, Z.; Hidalgo, G.; Simon, T.; Wei, S.E.; Sheikh, Y. OpenPose: Realtime multi-person 2D pose estimation using part affinity fields. *IEEE Trans. Pattern Anal. Mach. Intell.* **2019**, *43*, 172–186. [CrossRef]
35. Chaaraoui, A.A.; Padilla-Lopez, J.R.; Florez-Revuelta, F. Fusion of skeletal and silhouette-based features for human action recognition with RGB-D devices. In Proceedings of the IEEE Conference on Computer Vision and Pattern Recognition (CVPR), Portland, OR, USA, 23–28 June 2013; pp. 91–97.
36. Laptev, I. On Space-Time Interest Points. *Int. J. Comput. Vis.* **2005**, *64*, 107–123. [CrossRef]
37. Klaser, A.; Marszałek, M.; Schmid, C. A spatio-temporal descriptor based on 3D-gradients. In *BMVC 2008-19th British Machine Vision Conference 2008*; British Machine Vision Association: Durham, UK, 2008; p. 275.
38. Scovanner, P.; Ali, S. A 3-dimensional sift descriptor and its application to action recognition. In Proceedings of the 15th ACM International Conference on Multimedia: Augsburg, Germany, 24–29 September 2007; pp. 357–360.
39. Yilmaz, A.; Shah, M. Actions sketch: A novel action representation. In Proceedings of the IEEE Computer Society Conference on Computer Vision and Pattern Recognition (CVPR'05), San Diego, CA, USA, 20–25 June 2005; pp. 984–989.
40. Ji, S.; Xu, W.; Yang, M.; Yu, K. 3D Convolutional neural networks for human action recognition. *IEEE Trans. Pattern Anal. Mach. Intell.* **2013**, *35*, 221–231. [CrossRef]
41. Karpathy, A.; Toderici, G.; Shetty, S.; Leung, T.; Sukthankar, R.; Fei-Fei, L. Large-scale video classification with convolutional neural networks. In Proceedings of the IEEE Conference on Computer Vision and Pattern Recognition (CVPR), Columbus, OH, USA, 23–28 June 2014; pp. 1725–1732.
42. Si, C.; Chen, W.; Wang, W.; Wang, L.; Tan, T. An attention enhanced graph convolutional LSTM network for skeleton-based action recognition. In Proceedings of the IEEE Conference on Computer Vision and Pattern Recognition (CVPR), Long Beach, CA, USA, 15–21 June 2019; pp. 1227–1236.
43. Liu, J.; Shahroudy, A.; Xu, D.; Wan, G. Spatio-temporal LSTM with trust gates for 3D human action recognition. In Proceedings of the European Conference on Computer Vision (ECCV), Amsterdam, The Netherlands, 8–16 October 2016; pp. 816–833.
44. Sanchez-Caballero, A.; López-Diz, S.; Fuentes-Jimenez, D.; Losada-Gutiérrez, C.; Marrón-Romera, M.; Casillas-Perez, D.; Sarker, M.I. 3DFCNN: Real-time action recognition using 3D deep neural networks with raw depth information. *arXiv* **2020**, arXiv:2006.07743.
45. Simonyan, K.; Zisserman, A. Very deep convolutional networks for large-scale image recognition. *arXiv* **2014**, arXiv:1409.1556.
46. Hara, K.; Kataoka, H.; Satoh, Y. Towards good practice for action recognition with spatiotemporal 3D convolutions. In Proceedings of the 24th International Conference on Pattern Recognition (ICPR), Beijing, China, 20–24 August 2018; pp. 2516–2521.
47. Tran, D.; Bourdev, L.; Fergus, R.; Torresani, L.; Paluri, M. Learning spatiotemporal features with 3D convolutional networks. In Proceedings of the IEEE Conference on Computer Vision and Pattern Recognition (CVPR), Boston, MA, USA, 7–12 June 2015; pp. 4489–4497.
48. Li, Q.; Qiu, Z.; Yao, T.; Mei, T.; Rui, Y.; Luo, J. Action recognition by learning deep multi-granular spatio-temporal video representation. In Proceedings of the ACM on International Conference on Multimedia Retrieval, Melbourne, Australia, 6–9 June 2016; pp. 159–166.
49. Sun, L.; Jia, K.; Yeung, D.Y.; Shi, B.E. Human action recognition using factorized spatio-temporal convolutional networks. In Proceedings of the IEEE Conference on Computer Vision and Pattern Recognition (CVPR), Boston, MA, USA, 7–12 June 2015; pp. 4597–4605.
50. Ullah, A.; Ahmad, J.; Muhammad, K.; Sajjad, M.; Baik, S. Action recognition in video sequences using deep bi-directional LSTM with CNN features. *IEEE Access* **2017**, *6*, 1155–1166. [CrossRef]

applied sciences

MDPI

Article

3D Skeletal Joints-Based Hand Gesture Spotting and Classification

Ngoc-Hoang Nguyen, Tran-Dac-Thinh Phan, Soo-Hyung Kim, Hyung-Jeong Yang and Guee-Sang Lee *

Department of Artificial Intelligence Convergence, Chonnam National University, 77 Yongbong-ro, Gwangju 500-757, Korea; hoangnguyenkcv@gmail.com (N.-H.N.); phantrandacthinh2382@gmail.com (T.-D.-T.P.); shkim@jnu.ac.kr (S.-H.K.); hjyang@jnu.ac.kr (H.-J.Y.)
* Correspondence: gslee@jnu.ac.kr

Abstract: This paper presents a novel approach to continuous dynamic hand gesture recognition. Our approach contains two main modules: gesture spotting and gesture classification. Firstly, the gesture spotting module pre-segments the video sequence with continuous gestures into isolated gestures. Secondly, the gesture classification module identifies the segmented gestures. In the gesture spotting module, the motion of the hand palm and fingers are fed into the Bidirectional Long Short-Term Memory (Bi-LSTM) network for gesture spotting. In the gesture classification module, three residual 3D Convolution Neural Networks based on ResNet architectures (3D_ResNet) and one Long Short-Term Memory (LSTM) network are combined to efficiently utilize the multiple data channels such as RGB, Optical Flow, Depth, and 3D positions of key joints. The promising performance of our approach is obtained through experiments conducted on three public datasets—Chalearn LAP ConGD dataset, 20BN-Jester, and NVIDIA Dynamic Hand gesture Dataset. Our approach outperforms the state-of-the-art methods on the Chalearn LAP ConGD dataset.

Keywords: continuous hand gesture recognition; gesture spotting; gesture classification; multi-modal features; 3D skeletal; CNN

Citation: Nguyen, N.-H.; Phan, T.-D.-T.; Kim, S.-H.; Yang, H.-J.; Lee, G.-S. 3D Skeletal Joints-Based Hand Gesture Spotting and Classification. *Appl. Sci.* **2021**, *11*, 4689. https://doi.org/10.3390/app11104689

Academic Editor: Hyo-Jong Lee

Received: 14 April 2021
Accepted: 18 May 2021
Published: 20 May 2021

Publisher's Note: MDPI stays neutral with regard to jurisdictional claims in published maps and institutional affiliations.

1. Introduction

Nowadays, the role of dynamic hand gesture recognition has become crucial in vision-based applications for human-computer interaction, telecommunications, and robotics, due to its convenience and genuineness. There are many successful approaches to isolated hand gesture recognition with the recent development of neural networks, but in real-world systems, the continuous dynamic hand gesture recognition remains a challenge due to the diversity and complexity of the sequence of gestures.

Initially, most continuous hand gesture recognition approaches were based on traditional methods such as Conditional Random Fields (CRF) [1], Hidden Markov Model (HMM), Dynamic Time Warping (DTW), and Bézier curve [2]. Recently, deep learning methods based on convolution neural networks (CNN) and recurrent neural networks (RNN) [3–7] have gained popularity.

The majority of continuous dynamic hand-gesture recognition methods [3–6] include two separate procedures: gesture spotting and gesture classification. They utilized the spatial and temporal features to improve the performance mainly in gesture classification.

However, there are limitations in the performance of gesture spotting due to its inherent variability in the duration of the gesture. In existing methods, gestures are usually spotted by detecting transitional frames between two gestures. Recently, an approach [7] simultaneously performed the task of gesture spotting and gestures classification, but it turned out to be suitable only for feebly segmented videos.

Most of the recent researches [8–11] intently focus on improving the performance of the gesture classification phase, while the gesture spotting phase is often neglected on the assumption that the isolated pre-segmented gesture sequences are available for input to the gesture classification.

However, in real-world systems, spotting of the gesture segmentation plays a crucial role in the whole process of gesture recognition, hence, it greatly affects the final recognition performance. In paper [3], they segmented the videos into sets of images and used them to predict the fusion score, which means they simultaneously did the gesture spotting and gesture classification. The authors in [5] utilized the Connectionist temporal classification to detect the nucleus of the gesture and the no-gesture class to assist the gesture classification without requiring explicit pre-segmentation. In [4,6], the continuous gestures are often spotted into isolation based on the assumption that hands will always be put down at the end of each gesture which turned out to be inconvenient. It does not work well for all situations, such as in "zoom in", "zoom out" gestures, i.e., when only the fingers move while the hand stands still.

In this paper, we propose a spotting-classification algorithm for continuous dynamic hand gestures which we separate the two tasks like [4,6] but we avoid the existing problems of those methods. In the spotting module, as shown in Figure 1, the continuous gestures from the unsegmented and unbounded input stream are firstly segmented into individually isolated gestures based on 3D key joints extracted from each frame by 3D human pose and hand pose extraction algorithm. The time series of 3D key poses are fed into the Bidirectional Long Short-Term Memory (Bi-LSTM) network with connectionist temporal classification (CTC) [12] for gesture spotting.

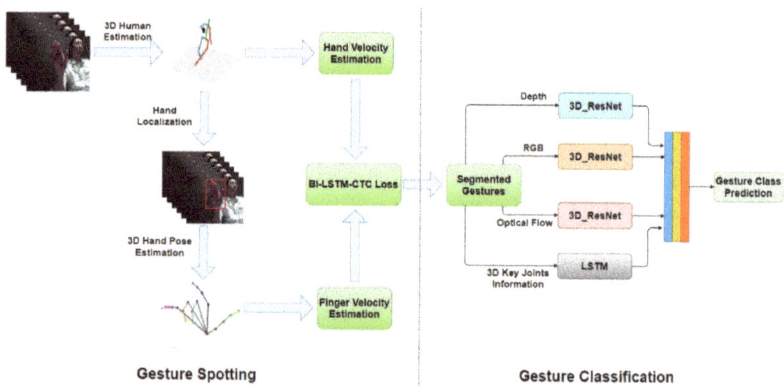

Figure 1. Gesture Spotting-Classification Module.

The isolated gestures segmented using the gesture spotting module are classified in the gesture classification module with a multi-modal M-3D network. As indicated in Figure 1, in the gesture classification module, the M-3D network is built by combining multi-modal data inputs which comprise RGB, Optical Flow, Depth, and 3D pose information data channels. Three residual 3D Convolution Neural Network based on ResNet architectures (3D_ResNet) [13] stream networks of RGB, Optical Flow and Depth channel along with an LSTM network of 3D pose channel are effectively combined using a fusion layer for gesture classification.

The preliminary version of this paper has appeared in [14]. In this paper, depth information has been considered together with 3D skeleton joints information with extensive experiments, resulting in upgraded performance.

The remainder of this paper is organized as follows. In Section 2, we review the related works. The proposed continuous dynamic hand gesture recognition algorithm is intently discussed in Section 3. In Section 4, the experiments with proposed algorithms conducted on three published datasets—Chalearn LAP ConGD dataset, 20BN-Jester, and NVIDIA

Dynamic Hand Gesture Dataset are presented with discussions. Finally, we conclude the paper in Section 5.

2. Related Works

In general, the continuous dynamic gesture recognition task is more complicated than the isolated gesture recognition task, where the sequence of gestures from an unsegmented and unbounded input stream are separated into complete individual gestures, called gesture spotting or gesture segmentation before classification. The majority of recent researchers solve the continuous dynamic gesture recognition task using two separate processes—gesture spotting and gesture recognition [1,4–6].

In the early years, the approaches for gesture spotting were commonly based on traditional machine learning techniques for the time series problems such as Conditional Random Fields (CRF) [1], Hidden Markov Model (HMM) [2], and Dynamic Time Warping (DTW) [3]. Yang et al. [1] presented a CRF threshold model that recognized gestures based on system vocabulary for labeling sequence data. Similar to the method introduced by Yang, Lee et al. [2] proposed the HMM-based method, which recognized gestures by the likelihood threshold estimation of the input pattern. Celebi et al. [3] proposed a template matching algorithm, i.e., the weighted DTW method, which used the time sequence of the weighted joint positions obtained from a Kinect sensor to compute the similarity of the two sequences. Krishnan et al. [15] presented a method using the Adaptive Boosting algorithm based on the threshold model for gesture spotting using continuous accelerometer data and the HMM model for gesture classification. The limitations of these methods are the parameter of the model has been decided through experience and the algorithm is sensitive to noise. In the recent past, with the success of deep learning applications in computer vision, deep learning approaches have been utilized for hand gesture recognition to achieve impressive performance compared to traditional methods.

The majority of the methods using recurrent neural networks (RNN) [16–18] or CNN [8,10,19–21] focus only on isolated gesture recognition, which ignores the gesture spotting phase. After the dataset for continuous gesture spotting-Chalearn LAP ConGD dataset was provided, a number of methods have been proposed to solve both phases of gesture spotting and gestures recognition [3,4,6]. Naguri et al. [6] applied 3D motion data input from infrared sensors into an algorithm based on CNN and LSTM to distinguish gestures. In this method, they segmented gestures by detecting transition frames between two isolated gestures. Similarly, Wang et al. [3] utilized transition frame detection using two streams CNN to spot gestures. In another approach proposed by Chai et al. [4], continuous gestures were spotted based on the hand position detected by Faster R-CNN and isolated gesture was classified by two parallel recurrent neural network SRNN with RGB_D data input. The multi-modal network, which combines a Gaussian-Bernoulli Deep Belief Network (DBN) with skeleton data input and a 3DCNN model with RGB_D data, was effectively utilized for gesture classification by Di et al. [7]. Tran et al. [22] presented CNN based method using a Kinect Camera for spotting and classification of hand gestures. However, the gesture spotting was done manually from a pre-specified hand shape or finger-tip pattern. And classification of hand gestures used only fundamental 3DCNN networks without employing the LSTM network. The system is based on the Kinect system and the comparison using a commonly used public dataset is almost impossible.

Recently, Molchanov et al. [5] proposed a method for joint gesture spotting and gesture recognition using a zero or negative lag procedure through a recurrent three-dimensional convolution neural network (R3DCNN). This network is highly effective in recognizing weakly segmented gestures from multi-modal data.

In this paper, we propose an effective algorithm for both spotting and classification tasks by utilizing extracted 3D human and hand skeletal features.

3. Proposed Algorithm

In this section, we intently focus on the proposed method using two main modules: gesture spotting and gesture classification. For entire frames of continuous gesture video, the speed of hand and finger estimated from the extracted 3D human pose and 3D hand pose are utilized to segment continuous gesture. The isolated gesture segmented by gesture spotting module is classified using the proposed M-3D network with RGB, Optical flow, Depth, and 3D key joints information.

3.1. Gesture Spotting

The gesture spotting module is shown on the left of Figure 1. All frames of continuous gesture sequence are utilized to extract 3D human pose using the algorithm proposed in [23]. Through RGB hand ROI localized from 3D hand palm position $J_h(x,y,z)$ when the hand palm stands still and over spine base joint, we use a 3D hand pose estimation algorithm to effectively extract the 3D position of the finger joints. From the extracted 3D human pose, the hand speed v_{hand} is estimated using the movement distance of the hand joint between two consecutive frames.

- **3D human pose extraction:** From each RGB frame, we obtain a 3D human pose by using one of the state-of-the-art methods for 2D/3D human pose estimation in the wild-pose-hgreg-3d network. This network has been proposed by Zhou et al. [23] which provides the pre-trained model on the Human3.6M dataset [24]. This is the largest dataset providing both 2D, 3D annotations of human poses in 3.6 million RGB images. This network is a fast, simple, and accurate neural network based on 3D geometric constraints for weakly-supervised learning of 3D pose with 2D joint annotations extracted through the state-of-the-art of 2D pose estimation method, i.e., stacked hourglass network of Newell et al. [25]. In our proposed approach, we use this 3D human pose estimation network to extract the exact 3D hand joint information, which is effectively utilized for both gesture spotting and gesture recognition task.

Let $J_h(x_{hk}, y_{hk}, z_{hk})$, $J_h(x_{hk-1}, y_{hk-1}, z_{hk-1})$ be the 3D position of the hand joint at the kth frame, and $(k-1)$th frame, respectively. The hand speed is estimated as

$$v_{hand} = \alpha \cdot \sqrt{(x_{hk} - x_{hk-1})^2 + (y_{hk} - y_{hk-1})^2 + (z_{hk} - z_{hk-1})^2} \quad (1)$$

where α is the frame rate.

The finger speed is estimated by the change in distance between the 3D position of fingertips of the thumb and the index finger in sequence frames. Let denote $J_{ft}(x_{ftk}, y_{ftk}, z_{ftk})$, $J_{fi}(x_{ink}, y_{ink}, z_{ink})$ the 3D position fingertips of the thumb and the index finger at the kth frame, respectively. The distance between the two fingertips at the kth frame is given as

$$d_{fk} = \sqrt{\left(x_{ftk} - x_{ink}\right)^2 + \left(y_{ftk} - y_{ink}\right)^2 + \left(z_{ftk} - z_{ink}\right)^2} \quad (2)$$

where d_{fk} and d_{fk-1} represent the distances of the kth frame and previous frame, respectively, the finger speed v_{finger} is estimated as

$$v_{finger} = \alpha \cdot \left(d_{fk} - d_{fk-1}\right) \quad (3)$$

The function utilizes v_{hand} and v_{finger} extracted from each frame:

$$v_k = v_{hand} + v_{finger} \quad (4)$$

and is used as the input of the Bi-LSTM network to spot gestures from video streams, as shown in Figure 2. In our network, the Connectionist temporal classification [12] CTC loss is used to identify whether the sequence frames are in gesture frames or transition frames.

- **3D hand pose extraction:** Using the hand palm location detected by the 3D human pose estimation network, we extract hand ROI and use it for 3D hand pose extraction when the hand palm stands still over the spine base joint. We also estimate the 3D hand pose by using the real-time 3D hand joints tracking network of OccludedHands proposed by Mueller et al. [26] and further additionally fine-tuned it with the hand pose dataset of Stereo Hand Pose Tracking Benchmark [27]. In this method, they utilized both RGB and Depth information to robustly and accurately localize the hand center position and regress the 3D joint from the 2D hand position heat-map. Firstly, they used a CNN network called HALNet to estimate the heat-map of the hand center and then crop the hand region. Secondly, they applied another CNN network called JORNet for a hand cropped frame to generate a heat-map of 2D hand joints and regress 3D hand joint positions from it. The Stereo Hand Pose Tracking Benchmark is a large dataset for 2D and 3D hand pose estimation with 21 joint points for 18,000 images. Due to the robustness and accuracy of its performance, the 3D position of the thumb and index fingertips detected by the network are used for finger speed calculation and other recognition features. In the case where the predicted joint becomes invisible with very low confidence, we estimate this joint position based on its last known position.
- **LSTM:** An LSTM network is a recurrent neural network of a special kind, in which current network output is influenced by previously memorized inputs. The network can learn the contextual information of a temporal sequence. In an LSTM network, the gates and memory cells at time t are given as follows:

$$\begin{cases} i_t = \sigma(W_i[x_t, h_{t-1}] + b_i \\ f_t = \sigma\Big(W_f[x_t, h_{t-1}] + b_f \\ o_t = \sigma(W_o[x_t, h_{t-1}] + b_o \\ \widetilde{c_t} = \tanh(W_c[x_t, h_{t-1}] + b_c), \\ c_t = f_t * c_{t-1} + i_t * \widetilde{c_t}, \\ h_t = \tanh(c_t) * o_t \end{cases} \quad (5)$$

where i, f, and o are the vectors of input, forget and output gate, respectively. $\widetilde{c_t}$ and c_t are called the "candidate" hidden state and internal memory of the unit. h_t represents the output hidden state. $\sigma(.)$ is a sigmoid function while W and b are connected weights matrix and bias vectors, respectively.

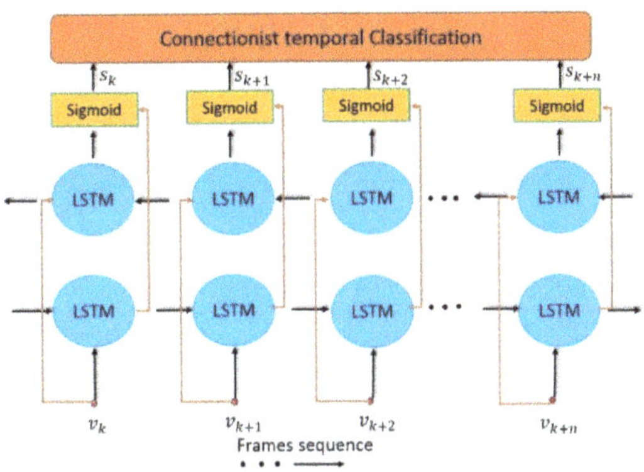

Figure 2. Gesture segmentation with Bi_LSTM and CTC loss.

- **Bi-LSTM network**: While the output of a single forward LSTM network depends only on previous input features, the Bi-LSTM network is known as an effective method for sequence labeling tasks, which is beneficial to both previous and future input features. Bi-LSTM can be considered as a stack of two LSTM layers, in which, a forward LSTM layer utilizes the previous input features while the backward LSTM layer captures the future input features. The benefit of the fact that the Bi-LSTM network considers both previous and future input features is its effectiveness to classify the frame in sequence frame, gesture frame, or transition frame. The prediction error can be reduced by using Bi-LSTM instead of LSTM.

- **Connectionist temporal classification:** The Connectionist temporal classification CTC is known as the loss function which is highly effective in sequential label prediction problems. The proposed algorithm utilizes CTC to detect whether the sequence frames are in gesture frames or transition frames with input from a sequence of Soft-Max layer outputs.

3.2. Gesture Classification

The isolated gestures segmented by the present gesture spotting module are classified into individual gesture classes in the gesture recognition module. The proposed gesture recognition module is a multi-model network called the M-3D network. This model is based on a multi-channel network with three different data modalities, as shown on the right of Figure 1.

In our approach, from each frame of a video, we extract optical flow, 3D pose (hand joint, thumb tip, and index fingertip joint) information of multi-channel features input to the model. Optical flow is determined by two adjacent frames. There are some existing methods of optical flow extraction such as Farneback [28], MPEG flow [29], and Brox flow [30]. The quality motion information of optical flow clearly affects the performance of the gesture recognition model. Therefore, the Brox flow technique is applied to our approach as it has better quality performance compared to other optical flow extraction techniques.

While the key hand and finger joints positions are extracted by the 3D human pose and 3D hand pose extraction network presented in Section 3.1, we only focus on the two most important joints of thumb tip and index fingertip which can describe all gesture types. Our gesture classification algorithm is based on the combination of three 3D_ResNet stream networks of RGB, Optical Flow, Depth channels with an LSTM network of 3D key joint features.

- **Three stream RGB, Optical Flow, and Depth 3D_ResNet networks:** The 3D_CNN framework is regarded as one of the best frameworks for spatiotemporal feature learning. The 3D_ResNet network is an improved version of the residual 3D_CNN framework based on ResNet [31] architecture. The effectiveness of 3D_ResNet has been proved by remarkable performance in action video classification.

The single 3D_ResNet is described in Figure 3. The 3D_ResNet consists of a 3D convolutional layer and is followed by a batch normalization layer and rectified-linear unit layer. Each RGB and Optical Flow stream model is pre-trained on the largest action video classification dataset of the Sports-1M dataset [32]. Input videos are resampled into 16 frames-clips before being fed into the network. Let a resampled sequence of 16 frames RGB frames be $V_c = \{x_{c1}, x_{c2}, \ldots, x_{c16}\}$, Optical Flow frames be $V_{of} = \{x_{of1}, x_{of2}, \ldots, x_{of16}\}$ and Depth frames be $V_d = \{x_{d1}, x_{d2}, \ldots, x_{d16}\}$ and operation function 3D_ResNet network of RGB, Optical Flow and Depth modalities be $\Theta_c(.)$, $\Theta_{of}(.)$ and $\Theta_d(.)$, respectively. Hence, the prediction probability of two single networks for i classes is

$$P_c\{p_1, p_2, \ldots, p_{16}|V_c\} = \Theta_c(V_c) \tag{6}$$

$$P_{of}\left\{p_1, p_2, \ldots, p_{16}\middle|V_{of}\right\} = \Theta_{of}\left(V_{of}\right) \tag{7}$$

$$P_D\{p_1, p_2, \ldots, p_{16}|V_D\} = \Theta_D(V_D) \tag{8}$$

where p_i is the prediction probability of video belonging to the ith class.

- **LSTM network with 3D pose information:** In dynamic gesture recognition, temporal information learning plays a critical role in the performance of the model. In our approach, we utilize the temporal features by tracking the trajectory of the hand palm together with the specific thumb tip and index fingertip joint. LSTM framework is suitably proposed to learn the features for the gesture classification task. The parameters of our LSTM refer to the approach [33]. Input vectors from a sequence of the LSTM network frames is defined as: $V_j = \{v_{j1}, v_{j2}, \ldots, v_{j16}\}$ where: $v_{jk} = \{J_h(x_{hk}, y_{hk}, z_{hk}), J_{ft}(x_{ftk}, y_{ftk}, z_{ftk}), J_{fi}(x_{ink}, y_{ink}, z_{ink})\}$ is a 9×1 vector which contains 3D position information of key joints at kth frame. The input of the LSTM network corresponds to the dimension of a single frame of sequences of 16 sampled frames in a gesture video that is a tensor for 1×9 numbers. The prediction probability output using LSTM with input V_j is

$$P_L\{p_1, p_2, \ldots, p_{16}|V_j\} = \Theta_L(V_j) \tag{9}$$

where $\Theta_L(.)$ denotes the operation function of the LSTM network.

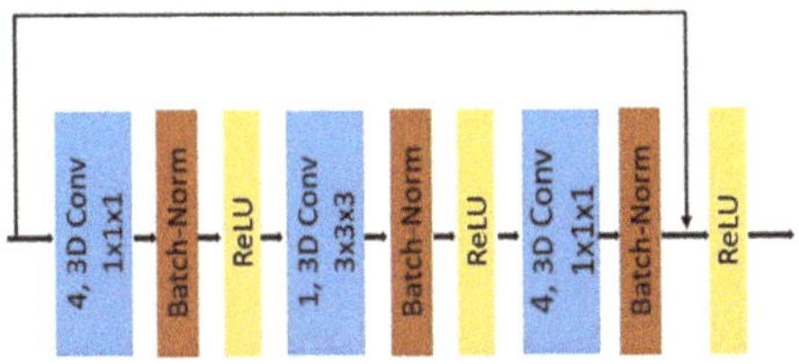

Figure 3. The overview of 3D_ResNet architecture. This figure showed the number of feature map, kernel size of the 3D convolutional layer (3D Conv), batch normalization layer (Batch-Norm), and Rectified-Linear unit layer (ReLU).

- **Multi-modality fusion:** The results of the multiple different channel networks are fused in the final fusion layer to predict a gesture class. It is a fully connected layer where the number of output units is equal to the number of classes on the dataset. The output probability of each class is estimated by pre-trained last fusion layer with $\Theta_{fusion}(.)$ operation function:

$$P\{p_1, p_2, \ldots, p_{16}|V_c\} = \Theta_{fusion}\left\{\begin{array}{ll} P_c\{p_1, p_2, \ldots, p_{16}|V_c\}, & P_{of}\left\{p_1, p_2, \ldots, p_{16}\middle|V_{of}\right\}, \\ P_D\{p_1, p_2, \ldots, p_{16}|V_D\}, & P_L\{p_1, p_2, \ldots, p_{16}|V_j\} \end{array}\right\} \tag{10}$$

The performance of the gesture recognition task is improved by combining the temporal information learning by LSTM network with spatiotemporal features learning by 3D_ResNet that is proved through experimental results.

4. Experiments and Results

In this section, we describe the experiments that evaluate the performance of the proposed approach on three public datasets: 20BN_Jester dataset [34], NVIDIA Dynamic Hand Gesture dataset [5], and Chalearn LAP ConGD dataset [35].

4.1. Datasets

- **20BN_Jester dataset:** is a large dataset collected from 148,092 densely-labeled RGB video clips for hand gesture recognition tasks from 27 gestures classes. The dataset is divided into three subsets: the training set having 118,562 videos, 14,787 videos for the validation set, and 14,743 videos (without labels) for the test set. This dataset has only been used for the gesture classification module.
- **NVIDIA Dynamic Hand Gesture dataset** is a collection of 1532 feebly segmented dynamic hand gesture RGB-Depth videos captured using SoftKinetic DS325 sensor with a frame rate of 30 fps of 20 subjects for 25 gesture classes. The continuous data streams are captured in an indoor car with both dim and bright lighting conditions.

This weakly segmented gesture video includes the preparation, nucleus, and transition frames of gesture.

- **Chalearn LAP ConGD dataset:** is a large dataset containing 47,933 gesture instances with 22,535 RGB-Depth videos for both continuous gesture spotting and gesture recognition task. The dataset includes 249 gestures performed by 21 different individuals. This dataset is further divided into three subsets: training set (14,314 videos), validation set (4179 videos), and test set (4042 videos).

The summary of the three datasets is shown in Table 1.

Table 1. Ablation studies on the ISBI 2016 and ISBI 2017 datasets.

Dataset	Number of Classes	Number of Videos	umber of Videos for Train, Validation, Test Set	Gesture Segmentation Task Provided
20BN_Jester	27	148,092	118,562 | 14,787 | 14,743	No
NVIDIA Hand Gesture	25	1532	1050 | – | 428	Yes
Chalearn LAP ConGD	249	22,535 (47,933 instances)	14,314 | 4179 | 4042	Yes

4.2. Training Process

- **Network training for hand gesture spotting:** To train the Bi-LSTM network for segmentation of continuous gestures, we firstly use a pre-trained 3D human pose extraction network (on Human3.6M dataset) and a pre-trained 3D hand pose extraction network (on Stereo Hand Pose Tracking dataset) to extract the 3D position of key poses. The quality between human and hand pose extraction algorithms are demonstrated in Figure 4. Using those extracted input features for the network, we train the Bi_LSTM network with the provided gesture segmentation labels by a training set of Chalearn LAP ConGD dataset.

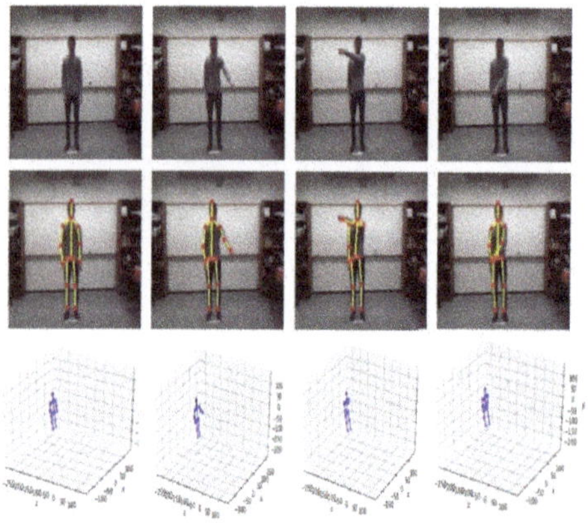

(a)

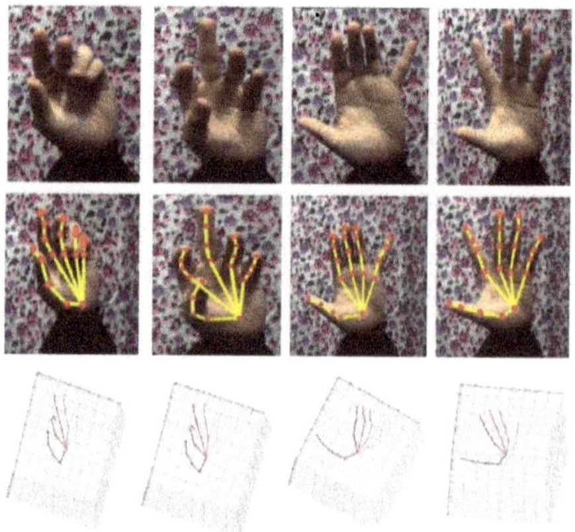

(b)

Figure 4. (**a**) The 2D and 3D human pose estimation examples and (**b**) The 2D and 3D hand pose estimation examples.

Bi-LSTM network is trained with CTC loss for predicting the sequence of binary output values to classify whether the frame belongs to gesture frame or transition frame. In Bi_LSTM, the input layer has 20 time-steps, the hidden layer has 50 memory units, and the last fully connected layer output has one binary value per time-step with a sigmoid active function. The efficient ADAM optimization algorithm [36] is applied to find the optimal weight of the network. The spotting output of the Bi-LSTM network by a given speed input is displayed as in Figure 5.

- **Network training for hand gesture classification:** The single-stream network (pre-trained on Sports-1M dataset) is separately fine-tuned on the huge dataset Chalearn LAP ConGD dataset. Each fine-tuned stream 3D_CNN network weights is learned using ADAM optimization, learning rate with an initial value of 0.0001 reducing by half for every 10 epochs on 200 epochs. Ensemble modeling with 5 3D_ResNet models is applied to increase the classification accuracy. The LSTM network parameters are selected through the observations of experimental results. The optimal LSTM model parameters are 3 memory blocks, and 256 LSTM Cells per memory block. The pre-trained LSTM network is trained with a learning rate of 0.0001 on 1000 epochs. After pre-training of each streaming network, we retrain these networks with a specific dataset. Finally, we concatenate the prediction probability outputs of these trained models to train the weights of the last fusion fully connected layer for gesture classification. Besides training with the 3D_ResNet framework, we also train with the 3D_CNN framework to prove the effectiveness of the proposed algorithm.

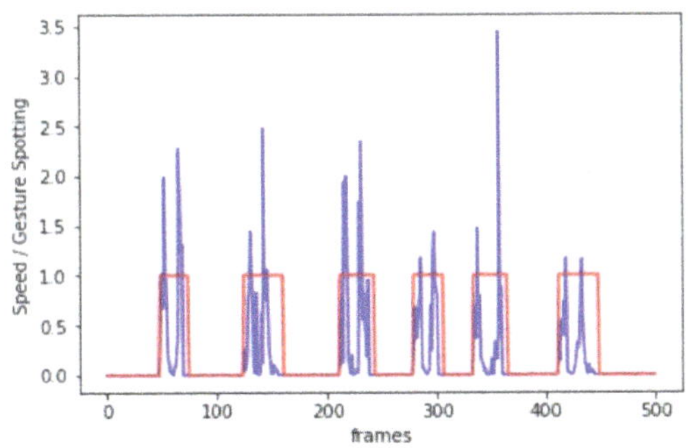

Figure 5. Example of sequence frames segmentation by the Bi_LSTM network. The blue line is the given speed input, and the red line is gesture spotting output (a value of 1.0 indicates the gesture frames).

4.3. Results and Analysis

- **Hand gesture spotting:** To prove the performance of the proposed hand gesture spotting module, we evaluated the model on two datasets—the NVIDIA Dynamic Hand Gesture dataset and Chalearn LAP ConGD dataset. The frame-wise accuracy metric and edit distance score [37] are used to measure the gesture segmentation performance. The results and comparison with other methods are shown in Tables 2 and 3. From the results shown in these tables, our proposed approach achieved the best performance as compared to other methods in both datasets. Our approach gets higher frame-wise accuracy and edits distance score on NVIDIA Dynamic Hand Gesture dataset and Chalearn LAP ConGD dataset than existing works. The significantly improved experimental results proved the effectiveness of the proposed approach.
- **Hand gesture classification:** The performance of our gesture classification module is evaluated by experiments conducted on the 20BN_Jester dataset (without Depth modality) and NVIDIA Dynamic Hand Gesture dataset. The accuracy comparison with other approaches for isolated dynamic hand gesture classification is shown in Table 3.

Table 3 shows that our gesture recognition module obtained a positive result. The recognition performance is improved by using 3D data information of key joints. Moreover, the recognition performance of our method is among the top performers of existing approaches, with an accuracy of 95.6% on the 20BN-Jester Dataset and an accuracy of 82.4% on the NVIDIA Hand Gesture dataset.

- **Continuous hand gesture spotting classification:** To entirely evaluate our approach on continuous dynamic hand gesture spotting recognition, we apply the Jaccard index [3] for measuring the performance. For a given gesture video, the Jaccard index estimates the average relative overlap between the ground truth and the predicted sequences of frames. A sequence S is given by *i*th class gesture label and binary vector ground truth $G_{s,i}$, while the binary vector prediction for the *i*th class is denoted as $P_{s,i}$. The binary vector $G_{s,i}$ and $P_{s,i}$ are vectors with 1-values indicating the corresponding frames in which the *i*th gesture class is being performed. So, the Jaccard Index for the given sequence S is computed by the following formula of

$$J_{s,i} = \frac{G_{s,i} \cap P_{s,i}}{G_{s,i} \cup P_{s,i}} \tag{11}$$

When $G_{s,i}$ and $P_{s,i}$ are empty vectors, the Jaccard Index $J_{s,i}$ is set as 0. For a given sequence S containing L number of true class labels l_s, the Jaccard Index is estimated by the function:

$$J_s = \frac{1}{l_s} \sum_{i=1}^{l} J_{s,i} \tag{12}$$

For all testing sequence of n gestures: $s = \{s_1, s_2, \ldots, s_n\}$ the mean Jaccard Index $\overline{J_s}$ $J \rightarrow_s J \rightarrow_s$ is applied to evaluation as follows:

$$\overline{J_s} = \frac{1}{n} \sum_{j=1}^{n} J_{s,j} \tag{13}$$

The spotting-recognition performance comparison of our proposed approach to the existing methods by evaluation experiment on the test set of Chalearn LAP ConGD dataset is shown in Table 4.

Table 2. Gestures spotting performance comparison with different methods on NVIDIA Hand Gesture dataset. Bold values are highest indices.

Method	NVIDIA Hand Gesture		Chalearn LAP ConGD	
	Frame-Wise Accuracy	Edit Distance Score	Frame-Wise Accuracy	Edit Distance Score
2S-RNN [4]	80.3	74.8	87.5	86.3
Proposed in [3]	84.6	79.6	90.8	88.4
R-3DCNN [5]	90.1	88.4	90.4	90.1
Our proposed	**91.2**	**89.6**	**93.1**	**93.8**

Table 3. Gesture classification performance comparison of different methods on the 20BN_Jester dataset and NVIDIA Hand Gesture dataset. Bold values are highest indices.

Method	Accuracy on 20BN-Jester Dataset	Accuracy on NVIDIA Hand Dataset
iDT-HOG [2]	-	59.1
C3D [8]	91.6	69.3
R-3DCNN [5]	95.0	79.3
MFFs [9]	**96.2**	**84.7**
3D_ResNet (RGB)	92.8	75.5
3D_ResNet (RGB + Optical flow)	93.3	77.8
Ours M3D	95.6	82.4

Table 4. The spotting-recognition performance comparison of our proposed approach to existing methods on the test set of the Chalearn LAP ConGD dataset. Bold values are highest indices.

Method	Mean Jaccard Index
2S-RNN [4]	0.5162
Proposed in [3] (RGB + Depth)	0.5950
R-3DCNN [5]	0.5154
Our proposed (3D_CNN)	0.5982
Our proposed	**0.6159**

From the results illustrated in Table 4, the mean Jaccard Index on the test set of the Chalearn LAP ConGD dataset shows that the proposed method achieves satisfactory performance on the dataset. By using 3D key joint features and multiples, the recognition performance is significantly enhanced.

5. Discussions

In Section 4.3, we have shown the effectiveness of our method on the three datasets. In terms of hand gesture spotting, we get the best results of both indexes on the NVIDIA Dynamic Hand Gesture dataset and Chalearn LAP ConGD dataset. The extraction of human pose and hand pose helps us track the hand movement more accurately and detect the beginning and the end of the sequence, avoiding the minor motion that could contaminate the following classification task. In the task of hand gesture classification, Table 3 presents the efficiency of the addition of modalities into our model on both the 20BN_Jester dataset and the NVIDIA Dynamic Hand Gesture dataset. Different views of data are crucial to the performance of the hand gesture classification. Continuous gesture classification is more difficult when there are several kinds of gestures in one video, which means the capability of gesture spotting greatly influences the performance of gesture classification. In Table 4, we get the best results when doing both tasks on the Chalearn LAP ConGD dataset.

6. Conclusions

In this paper, we presented an effective approach for continuous dynamic hand gesture spotting recognition for RGB input data. The continuous gesture sequences are firstly segmented into separate gestures by utilizing the motion speed of key 3D poses as the input of the Bi-LSTM network. After that, each segmented gesture is defined in the gesture classification module using a multi-modal M-3D network. In this network, three 3D_ResNet stream networks of RGB, Optical Flow, Depth data channel, and LSTM networks of 3D key pose features channel are effectively combined for gesture classification purposes. The results of the experiments conducted on the ChaLearn LAP ConGD Dataset, NVIDIA Hand Gesture dataset, and 20_BN Jester dataset proved the effectiveness of our proposed method. In the future, we will try to include other different modalities to improve the performance. The tasks of gesture spotting and classification in this paper are performed separately into 2 steps. The upcoming plan is to do both tasks by one end-to-end model so that it is more practical in real-world problems.

Author Contributions: Conceptualization, G.-S.L. and N.-H.N.; methodology, N.-H.N.; writing—review and editing, N.-H.N., T.-D.-T.P., and G.-S.L.; supervision, G.-S.L., S.-H.K., and H.-J.Y.; project administration, G.-S.L., S.-H.K., and H.-J.Y.; funding acquisition, G.-S.L., S.-H.K., and H.-J.Y. All authors have read and agreed to the published version of the manuscript.

Funding: This work was supported by the National Research Foundation of Korea (NRF) grant funded by the Korean government (MSIT). (NRF-2020R1A4A1019191) and also by the Bio & Medical Technology Development Program of the National Research Foundation (NRF)& funded by the Korean government (MSIT) (NRF-2019M3E5D1A02067961).

Institutional Review Board Statement: Not Applicable.

Informed Consent Statement: Not Applicable.

Data Availability Statement: Not Applicable.

Conflicts of Interest: The authors declare no conflict of interest.

References

1. Yang, H.-D.; Sclaroff, S.; Lee, S.-W. Sign Language Spotting with a Threshold Model Based on Conditional Random Fields. *IEEE Trans. Pattern Anal. Mach. Intell.* **2008**, *31*, 1264–1277. [CrossRef] [PubMed]
2. Słapek, M.; Paszkiel, S. Detection of gestures without begin and end markers by fitting into Bézier curves with least squares method. *Pattern Recognit. Lett.* **2017**, *100*, 83–88. [CrossRef]
3. Wang, H.; Wang, P.; Song, Z.; Li, W. Large-scale multimodal gesture recognition using heterogeneous networks. In Proceedings of the 2017 IEEE International Conference on Computer Vision Workshops (ICCVW), Venice, Italy, 22–29 October 2017; pp. 3129–3137.
4. Chai, X.; Liu, Z.; Yin, F.; Liu, Z.; Chen, X. two streams recurrent neural networks for large-scale continuous gesture recognition. In Proceedings of the 2016 23rd International Conference on Pattern Recognition (ICPR), Cancun, Mexico, 4–8 December 2016; pp. 31–36.
5. Molchanov, P.; Yang, X.; Gupta, S.; Kim, K.; Tyree, S.; Kautz, J. Online detection and classification of dynamic hand gestures with recurrent 3D convolutional neural networks. In Proceedings of the 2016 IEEE Conference on Computer Vision and Pattern Recognition (CVPR), Las Vegas, NV, USA, 27–30 June 2016; pp. 4207–4215.
6. Naguri, C.R.; Bunescu, R.C. Recognition of Dynamic Hand Gestures From 3D Motion Data Using LSTM and CNN Architectures. In Proceedings of the 2017 16th IEEE International Conference on Machine Learning and Applications (ICMLA), Cancun, Mexico, 18–21 December 2017; pp. 1130–1133.
7. Wu, D.; Pigou, L.; Kindermans, P.-J.; Le, N.D.-H.; Shao, L.; Dambre, J.; Odobez, J.-M. Deep Dynamic Neural Networks for Multimodal Gesture Segmentation and Recognition. *IEEE Trans. Pattern Anal. Mach. Intell.* **2016**, *38*, 1583–1597. [CrossRef] [PubMed]
8. Tran, D.; Bourdev, L.; Fergus, R.; Torresani, L.; Paluri, M. Learning Spatiotemporal Features with 3D Convolutional Networks. In Proceedings of the 2015 IEEE International Conference on Computer Vision (ICCV), Santiago, Chile, 7–13 December 2015; pp. 4489–4497.
9. Kopuklu, O.; Kose, N.; Rigoll, G. Motion Fused Frames: Data Level Fusion Strategy for Hand Gesture Recognition. In Proceedings of the 2018 IEEE/CVF Conference on Computer Vision and Pattern Recognition Workshops (CVPRW), Salt Lake City, UT, USA, 18–22 June 2018; pp. 2184–21848.
10. Narayana, P.; Beveridge, J.R.; Draper, B.A. Gesture Recognition: Focus on the Hands. In Proceedings of the 2018 IEEE/CVF Conference on Computer Vision and Pattern Recognition, Salt Lake City, UT, USA, 18–22 June 2018; pp. 5235–5244.
11. Zhu, G.; Zhang, L.; Mei, L.; Shao, J.; Song, J.; Shen, P. Large-scale Isolated Gesture Recognition Using Pyramidal 3D Convolutional Networks. In Proceedings of the 2016 23rd International Conference on Pattern Recognition (ICPR), Cancun, Mexico, 4–8 December 2016; pp. 19–24.
12. Graves, A.; Fernández, S.; Gomez, F.; Schmidhuber, J. Connectionist Temporal Classification: Labelling Unsegmented Sequence Data with Recurrent Neural Networks. In Proceedings of the 23rd International Conference on Machine Learning, Pittsburgh, PA, USA, 25–29 June 2006; pp. 369–376.
13. Hara, K.; Kataoka, H.; Satoh, Y. Can Spatiotemporal 3D CNNs Retrace the History of 2D CNNs and ImageNet? In Proceedings of the 2018 IEEE/CVF Conference on Computer Vision and Pattern Recognition, Salt Lake City, UT, USA, 18–22 June 2018; pp. 6546–6555.
14. Hoang, N.N.; Lee, G.-S.; Kim, S.-H.; Yang, H.-J. Continuous Hand Gesture Spotting and Classification Using 3D Finger Joints Information. In Proceedings of the 2019 IEEE International Conference on Image Processing (ICIP), Taipei, Taiwan, 22–25 September 2019; pp. 539–543.
15. Krishnan, N.C.; Lade, P.; Panchanathan, S. Activity gesture spotting using a threshold model based on Adaptive Boosting. In Proceedings of the 2010 IEEE International Conference on Multimedia and Expo, Singapore, 19–23 July 2010; pp. 155–160.
16. Ullah, A.; Ahmad, J.; Muhammad, K.; Sajjad, M.; Baik, S.W. Action Recognition in Video Sequences using Deep Bi-Directional LSTM With CNN Features. *IEEE Access* **2018**, *6*, 1155–1166. [CrossRef]
17. Donahue, J.; Hendricks, L.A.; Guadarrama, S.; Rohrbach, M.; Venugopalan, S.; Saenko, K.; Darrell, T. Long-term Recurrent Convolutional Networks for Visual Recognition and Description. In Proceedings of the IEEE Conference on Computer Vision and Pattern Recognition, Boston, MA, USA, 7–12 June 2015; pp. 2625–2634.
18. Du, Y.; Wang, W.; Wang, L. Hierarchical recurrent neural network for skeleton based action recognition. In Proceedings of the 2015 IEEE Conference on Computer Vision and Pattern Recognition, Boston, MA, USA, 7–12 June 2015; pp. 1110–1118.
19. Zhu, T.; Zhou, Y.; Xia, Z.; Dong, J.; Zhao, Q. Progressive Filtering Approach for Early Human Action Recognition. *Int. J. Control Autom. Syst.* **2018**, *16*, 2393–2404. [CrossRef]
20. Ding, Z.; Chen, Y.; Chen, Y.L.; Wu, X. Similar Hand Gesture Recognition by Automatically Extracting Distinctive Features. *Int. J. Control Autom. Syst.* **2017**, *15*, 1770–1778. [CrossRef]

21. Zhu, T.; Xia, Z.; Dong, J.; Zhao, Q. A Sociable Human-robot Interaction Scheme Based on Body Emotion Analysis. *Int. J. Control Autom. Syst.* **2019**, *17*, 474–485. [CrossRef]
22. Tran, D.-S.; Ho, N.-H.; Yang, H.-J.; Baek, E.-T.; Kim, S.-H.; Lee, G. Real-Time Hand Gesture Spotting and Recognition Using RGB-D Camera and 3D Convolutional Neural Network. *Appl. Sci.* **2020**, *10*, 722. [CrossRef]
23. Zhou, X.; Huang, Q.; Sun, X.; Xue, X.; Wei, Y. Towards 3D Human Pose Estimation in The Wild: A Weakly-Supervised Approach. In Proceedings of the IEEE International Conference on Computer Vision, Venice, Italy, 22–29 October 2017; pp. 398–407.
24. Ionescu, C.; Papava, D.; Olaru, V.; Sminchisescu, C. Human3. 6m: Large Scale Datasets and Predictive Methods for 3D Human Sensing in Natural Environments. *IEEE Trans. Pattern Anal. Mach. Intell.* **2013**, *36*, 1325–1339. [CrossRef] [PubMed]
25. Newell, A.; Yang, K.; Deng, J. Stacked Hourglass Networks for Human Pose Estimation. In Proceedings of the European Conference on Computer Vision, Amsterdam, The Netherlands, 8–16 October 2016; pp. 483–499.
26. Müeller, F.; Mehta, D.; Sotnychenko, O.; Sridhar, S.; Casas, D.; Theobalt, C. Real-Time Hand Tracking Under Occlusion from an Egocentric RGB-D Sensor. In Proceedings of the 2017 IEEE International Conference on Computer Vision Workshops (ICCVW), Venice, Italy, 22–29 October 2017; pp. 1284–1293.
27. Zhang, J.; Jiao, J.; Chen, M.; Qu, L.; Xu, X.; Yang, Q. 3D Hand Pose Tracking and Estimation Using Stereo Matching. *arXiv* **2016**, arXiv:1610.07214.
28. Farnebäck, G. Two-Frame Motion Estimation Based on Polynomial Expansion. In Proceedings of the Scandinavian Conference on Image Analysis, Halmstad, Sweden, 29 June–2 July 2003; Springer: Berlin/Heidelberg, Germany, 2003; pp. 363–370.
29. Kantorov, V.; Laptev, I. Efficient Feature Extraction, Encoding and Classification for Action Recognition. In Proceedings of the IEEE Conference on Computer Vision and Pattern Recognition, Columbus, OH, USA, 23–28 June 2014; pp. 2593–2600.
30. Brox, T.; Bruhn, A.; Papenberg, N.; Weickert, J. High Accuracy Optical Flow Estimation Based on a Theory for Warping. In Proceedings of the European Conference on Computer Vision, Prague, Czech Republic, 11–14 May 2004; Springer: Berlin/Heidelberg, Germany, 2004; pp. 25–36.
31. He, K.; Zhang, X.; Ren, S.; Sun, J. Deep residual learning for image recognition. In Proceedings of the IEEE Conference on Computer Vision and Pattern Recognition, Las Vegas, NV, USA, 27–30 June 2016; pp. 770–778.
32. Karpathy, A.; Toderici, G.; Shetty, S.; Leung, T.; Sukthankar, R.; Fei-Fei, L. Large-Scale Video Classification with Convolutional Neural Networks. In Proceedings of the IEEE conference on Computer Vision and Pattern Recognition, Columbus, OH, USA, 23–28 June 2014; pp. 1725–1732.
33. Sarkar, A.; Gepperth, A.; Handmann, U.; Kopinski, T. Dynamic Hand Gesture Recognition for Mobile Systems Using Deep LSTM. In Proceedings of the 9th International Conference on Intelligent Human Computer Interaction, Evry, France, 11–13 December 2017; pp. 19–31.
34. Materzynska, J.; Berger, G.; Bax, I.; Memisevic, R. The Jester Dataset: A Large-Scale Video Dataset of Human Gestures. In Proceedings of the IEEE/CVF International Conference on Computer Vision Workshops, Seoul, Korea, 28–29 October 2019.
35. Wan, J.; Zhao, Y.; Zhou, S.; Guyon, I.; Escalera, S.; Li, S.Z. ChaLearn Looking at People RGB-D Isolated and Continuous Datasets for Gesture Recognition. In Proceedings of the IEEE Conference on Computer Vision and Pattern Recognition Workshops, Las Vegas, NV, USA, 26 June–1 July 2016; pp. 56–64.
36. Kingma, D.P.; Ba, J. Adam: A method for stochastic optimization. In Proceedings of the 3rd International Conference for Learning Representation (ICLR), San Diego, CA, USA, 5–8 May 2015.
37. Lea, C.; Vidal, R.; Hager, G.D. Learning convolutional action primitives for fine-grained action recognition. In Proceedings of the 2016 IEEE International Conference on Robotics and Automation (ICRA), Stockholm, Sweden, 16–21 May 2016; pp. 1642–1649.

applied sciences

MDPI

Article

Gesture Recognition Based on 3D Human Pose Estimation and Body Part Segmentation for RGB Data Input

Ngoc-Hoang Nguyen, Tran-Dac-Thinh Phan, Guee-Sang Lee *, Soo-Hyung Kim and Hyung-Jeong Yang

Department of Electronics and Computer Engineering, Chonnam National University, 77 Yongbong-ro, Gwangju 500-757, Korea; hoangnguyenkcv@gmail.com (N.-H.N.); phantrandacthinh2382@gmail.com (T.-D.-T.P.); shkim@jnu.ac.kr (S.-H.K.); hjyang@jnu.ac.kr (H.-J.Y.)

* Correspondence: gslee@jnu.ac.kr

Received: 31 July 2020; Accepted: 4 September 2020; Published: 6 September 2020

Abstract: This paper presents a novel approach for dynamic gesture recognition using multi-features extracted from RGB data input. Most of the challenges in gesture recognition revolve around the axis of the presence of multiple actors in the scene, occlusions, and viewpoint variations. In this paper, we develop a gesture recognition approach by hybrid deep learning where RGB frames, 3D skeleton joint information, and body part segmentation are used to overcome such problems. Extracted from the RGB images are the multimodal input observations, which are combined by multi-modal stream networks suited to different input modalities: residual 3D convolutional neural networks based on ResNet architecture (3DCNN_ResNet) for RGB images and color body part segmentation modalities; long short-term memory network (LSTM) for 3D skeleton joint modality. We evaluated the proposed model on four public datasets: UTD multimodal human action dataset, gaming 3D dataset, NTU RGB+D dataset, and MSRDailyActivity3D dataset and the experimental results on these datasets proves the effectiveness of our approach.

Keywords: dynamic gesture recognition; human action recognition; multi-modalities network

1. Introduction

Gesture recognition has recently attracted much attention because of its wide applications such as the human–computer interaction, telecommunications, and robotics, but it still remains as one of the major challenges because of the inherent complexity of human motions. In early times, gesture recognition based on conventional techniques of classification with handcrafted features, such as support vector machine (SVM), bag-of-features and multiclass SVM, and hidden Markov model (HMM), have been proposed [1–3]. Recently, deep learning-based methods are increasingly employed due to their advantages of end-to-end learning by automatic extraction of spatiotemporal features from raw data. The development of deep learning methods based on a convolution neural network (CNN) and recurrent neural network (RNN) or long short-term memory network (LSTM) have achieved positive results in handling gesture recognition tasks [4–8]. However, there are limitations in the performance of gesture classification due to the complexity of the scene, e.g., the presence of multiple actors in the background, occlusions, illumination changes, or viewpoint variations.

In existing methods, to overcome the challenges caused by the issue of background or viewpoint variations, gesture recognition is usually developed by combining multiple modalities of data inputs (such as skeleton joints information, human body shape, RGB, optical flow, and depth frames) with newly developed deep learning models [9–11]. By utilizing skeleton joints information or depth information, gesture recognition performance has been significantly improved because they are helpful in the representation of gestures and played an important role in gesture identification. Although human skeleton joints and depth data can be collected directly from time-of-flight (ToF) cameras, gesture recognition on RGB video input is a substantial challenge because the human pose should be estimated with high accuracy. Skeleton joints convey vital information to represent gesture from the human pose, but it is not enough to identify complicated motions when it does not match to the shape of body parts correctly. For instance, if there are other actors present in the scene together with the main target, the action of other objects from the background can cause confusion to the correct extraction of the target person's skeleton. Figure 1b shows possible errors in skeleton joints extraction of the main person due to the presence of other moving objects. However, when the body parts are segmented beforehand, the skeleton of a target can be obtained with much higher accuracy because the layout of the skeleton is restricted by the body part. Moreover, skeleton joints of the frame sequence of a video can be temporally incoherent due to independent errors in each frame as we illustrate in Figure 2; and this can cause incorrect classification.

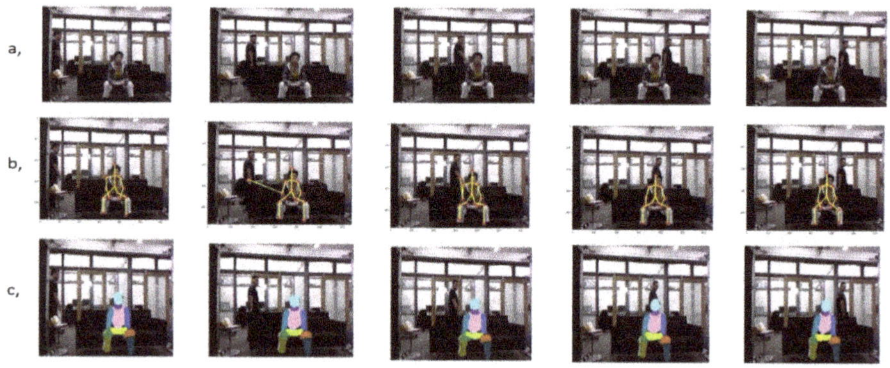

Figure 1. Illustration of the case in which many actors are present in the scene. The first row (**a**) RGB input images, the second row (**b**) extracted skeleton joints images, and the last row (**c**) color-encoded body part segmentation images. In this case, the skeleton joints extraction can be distracted by other objects in the background, but the color-encoded body part segmentation can help avoiding such a case.

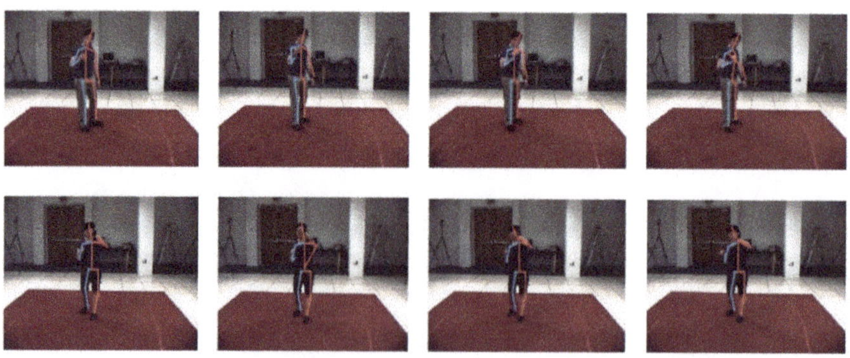

Figure 2. Illustration of skeleton joints of sequence sequential frames can be incoherent due to independent error in each frame.

In this paper, we propose a multi-modal gesture recognition method for RGB data input with a multi-modal algorithm. The algorithm consists of three submodels: two residual 3D convolutional neural networks based on ResNet architecture (3DCNN_ResNet) [12] and a long short-term memory network (LSTM) to perform on three data modalities: RGB images, color body part segmentations, and 3D human skeleton joints, respectively. We extract color body part segmentations and 3D human skeleton joints from RGB input.

The color body part segmentations with 12 semantic parts are obtained by the segmentation CNN network RefineNet [13]. By utilizing body part segmentation, the gesture recognition network can focus on only the target person in the presence of multiple objects in the scene. The 3D human skeleton joints information of the target person is extracted by the 3D human pose estimation network called Pose3D [14]. Pose3D is a deep learning network that obtains the sequence of 3D poses based on the temporal information across the sequence of 2D joint locations in order to prevent temporally incoherent estimation. The gesture classes are predicted by a combination of these three submodels that are effectively fused by an integrated stacking module as the late fusion layers.

The contributions of the paper are:

- The reason for the distraction of skeleton joints extraction has been addressed, which hinders from the proper functioning of gesture recognition methods. In other words, the existence of noise or extra persons in the background can cause such distractions.
- The solution to the problem of multiple actors in the scene, which caused the distraction of skeleton joints extraction, has been presented with target person extraction. The target person is segmented, and used for eliminating distraction in skeleton joints extraction.

This idea of using target person extraction has never been addressed before in the literature as far as we know.

The remainder of this paper is organized as follows: related works are given in Section 2. In Section 3, the proposed algorithm for gesture recognition is presented. The experiments on four public datasets to evaluate the effectiveness of our approach are given in Section 4. Finally, a conclusion is given in Section 5.

2. Related Works

In this section, we briefly described the previous methods relevant to our work for gesture recognition. Although significant improvement on gesture recognition has been reported, but new challenges appear with different input modalities and restrictions.

In early years, gesture or action recognition problem has been dealt with classical machine learning methods such as support vector machine (SVM) [1], bag-of-features [2], and hidden Markov model (HMM) [3]. In these methods, gestures are classified based on the features extracted by a hand-engineered extractor. Hussein et al. [1] presented a discriminative descriptor for action classification based on the covariance matrix for 3D skeleton joint locations. Dardas [2] similarly used the SVM classifier to identify gesture classes, but via a bag-of-words vector mapped from key points extracted by scale-invariant feature transform (SIFT). Lee [3] presented a method using HMM based on the likelihood threshold estimation of the input pattern. The gesture recognition task has been tackled by conventional machine learning methods, but there are significant limitations in these approaches. For instance, the parameters of the model depend on experience, and the system is sensitive to noise.

Recently, due to the revolution of deep learning, gesture recognition approaches have been presented with impressive performances compared to the traditional methods. The deep learning-based methods became popular due to its capability to extract spatiotemporal features from the raw video input automatically. A convolutional neural network (CNN) was initially used for extracting spatial features for static images, but it has been extended to deal with different input types or to extract different types of features. Various approaches [4,5,15,16] have utilized CNN to treat sequential frames of a video as multi-channel inputs for the purpose of video classification. Feichtenhofer [4] incorporated a two-stream network with separate ConvNets for RGB images and optical flow images to extract motion information for gesture classification. Kopuklu [15] proposed data level fusion to combine RGB and optical flow modalities with static images to extract action features. CNN can be incorporated with RNN or LSTM to learn both spatial and temporal features of a video for action classification.

Donahue [8] deployed long-term recurrent convolutional network (LRCN), in which CNN is used to extract spatial features of images and LSTM is applied to capture temporal dependencies in the sequence of such features.

For dynamic gesture recognition, [7,17,18] applied a 3D convolutional neural network to capture discriminative features along both spatial and temporal dimensions due to 3D convolutions and 3D pooling. Tran [17] used spatiotemporal features extracted by 3DCNN to classify gesture with SVM classifier. Molchanov [16] proposed recurrent 3D convolutional neural networks (R3DCNN) to recognize gestures online, where 3DCNN is used as a feature extractor. The gesture recognition based on a human pose has also achieved impressive results. Utilizing RNN or LSTM to capture temporal features from human skeleton joints for gesture classification is gaining popularity [19–21], in which skeleton joints are extracted by depth information with a ToF camera, but it becomes more challenging when only RGB data input is used. In other works, multiple data modalities or multiple deep learning models are combined to achieve better performance in gesture classification. Duan [9] presented a convolutional two-stream consensus voting network based on 3DCNN for RGB and depth channels to identify the gesture classes in a video input. Chai [10] also proposed a multi-stream model based on RNN with hand location information extracted from RGB-D input. The summary of the related works is given in Table 1.

Table 1. The summary of related works.

Author	Approach/Features	Details	Comments
Hussen et al. [1]	Support Vector Machine (SVM)	Discriminative classification from 3D skeleton joints	Classical machine learning methods.
Dardas et al. [2]	Bag of features	Classification via bag-of-words vector	
Lee et al. [3]	Hidden Markov Model	HMM based on the likelihood threshold estimation	
Feichtenhofer et al. [4]	Two-stream network with separate ConvNets	Extract motion from RGB images and optical flow images	Early deep learning-based methods
Kopuklu et al. [15]	Data level fusion	Combine RGB and optical flow modalities with static images	
Donahue et al. [8]	Long-term recurrent convolutional network	CNN and LSTM for spatio-temporal features	
Tran et al. [17]	3DCNN, SVM classifier	Spatio-temporal features extracted by 3DCNN	3DCNN to capture spatial and temporal features
Molchanov et al. [16]	Recurrent 3D Convolutional Neural Networks	Online gesture recognition by 3DCNN	
Yan et al. [19], Li et al. [20], and Omran et al. [21]	RNN or LSTM, human skeleton joints	Capture temporal features from skeleton joints and depth	Deep learning + human pose
Duan et al. [9]	Convolutional two-stream consensus voting network	Combine the results from RGB and depth in a video input	Multi-modal or multiple deep learning models
Chai et al. [10]	Multi-stream model based on RNN	Extract hand location information from RGB-D input	

Most of the gesture recognition methods exploit the human pose estimation, however often the pose estimation in the video can often be inconsistent because of independent errors in the sequence of frames. Additionally, the extraction of skeleton joints may not be successful when the background contains multiple objects or human beings. In this paper, we try to solve these problems through the use of segmentation of the target person. We propose a novel method for gesture recognition with multi-modalities: RGB images, color body part segmentation images, and 3D skeleton joints information.

3. Proposed Method

In this section, we present in detail our proposed approach for gesture recognition. The proposed method consisted of three submodels: two 3D_ ResNet networks and an LSTM network to deal with RGB frames, color body part segmentation images, and 3D skeleton joints information. These three submodels were effectively fused by integrated stacking module as a late fusion layer in order to decide the gesture class. We show an overview of the proposed algorithm in Figure 3.

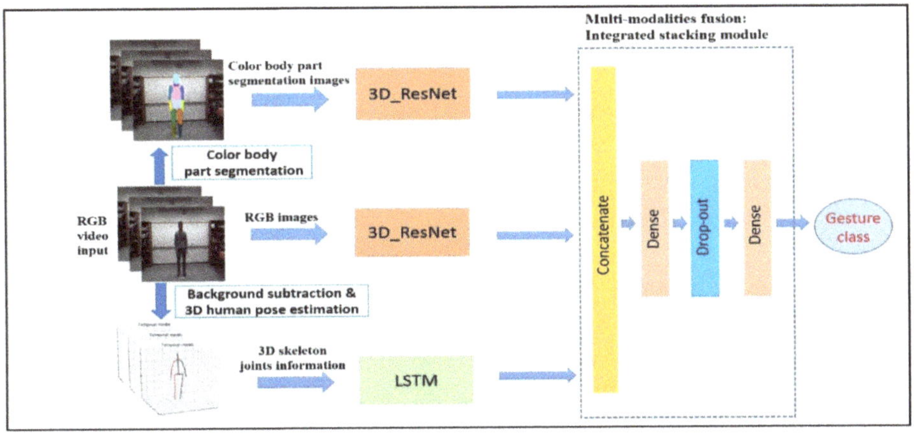

Figure 3. The proposed approach. This algorithm consists of three subnetworks and the fusion module.

In our algorithm, given the RGB image sequence of a video, the color body part segmentation images were generated by the segmentation network—RefineNet [22]. Some of the datasets, e.g., MSRDailyActivity3D, contained sometimes more than one person on the scene. The target person was a person whose position was nearest to the center of the screen and closest to the camera in most of the frames of the video. Therefore, a person closer to the center area was selected first. If more than one person were around the center of the screen, the person closer to the camera was selected from the depth information. From the color body part segmentation, the background of the input image was excluded from the scene, leaving only the target person, which was used for extracting 3D human skeleton joints by temporal 3D human pose estimation network—Pose3D. Two-stream 3D_ResNet networks were used to learn the features from the RGB images and color body part segmentation images for gesture classification. In the other subnetwork, the extracted 3D human skeleton joints were utilized by the LSTM network.

- Color body part segmentation: The RefineNet model, given the RGB input frames, produced the color-coded 12 body parts segmentation. The RefineNet is a multi-path refinement network for semantic segmentation via multi-level features and potentially long-range connections. The RefineNet model typically consists of four blocks: adaptive convolution, multi-resolution fusion, chained residual pooling, and output convolution. The batch-normalization layers were simplified from the convolution block but it still contained the remaining convolution units of the original ResNet. Multi-resolution fusion performed feature map fusion by convolutions and a summation. Multiple resolution feature maps extracted from varied input paths were fused into a high-resolution feature map. Additionally, the output feature map was fused through multiple pooling blocks of chained residual pooling blocks. The final prediction was given by the output convolution block, which had another remaining convolution unit and soft-max layer. The RefineNet network was based on ResNeXt-101 [23] with trained weights by the UP-3D dataset [24] for color-coded 12 body parts segmentation. Figure 4 shows the example of color body parts segmentation on sampled frames on a video.

Figure 4. Example of color-encoded 12 body parts segmentation from RGB images sequence by the RefineNet model.

- Temporal video 3D human pose estimation: We extracted 3D joints skeleton information for gesture recognition from temporal video 3D human pose estimation called Pose_3D [14] from background subtracted RGB images. The Pose_3D network is a sequence-to-sequence network that predicts a sequence of temporally consistent 3D human pose from the sequence of 2D human poses. The 2D human pose was obtained by the state-of-art 2D human pose estimation framework—the stacked-hourglass network [25] trained on the Human3.6M dataset [26]. The decoder of the Pose_3D consists of LSTM units and residual connection to predict temporally consistent 3D poses of the current frame using the 3D poses of previous frames and 2D joints information of all frames, which were taken from the final state of the encoder. The temporal smoothness constraint was imposed on the 3D pose extraction of a video. Since the stacked-hourglass network was used for 2D pose estimation on individual frames, this constraint made the predicted 3D poses more stable and reliable even with 2D pose estimation failure in a few frames within the temporal window. Figure 5 shows the example of temporal 3D skeleton joints of video frames extracted by the Pose_3D network.

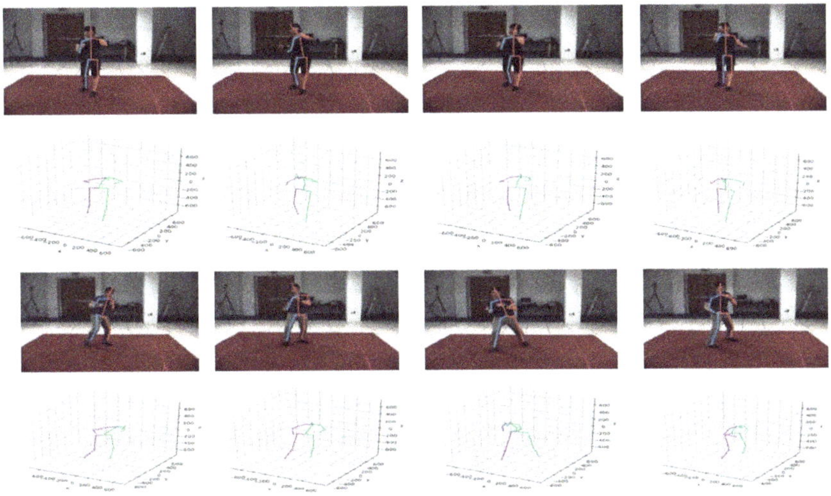

Figure 5. Example of skeleton joints extraction from RGB images sequence by temporal video 3D human pose estimation-Pose_3D.

- Two-stream 3D_ResNet networks for RGB and color body part segmentation modalities: The residual 3D_CNN network based on ResNet architecture [23] was applied to benefit from two data modalities: the background subtracted RGB images and color-coded 12 body parts segmentation images for gesture classification. For the input of 3D_CNN of the RGB image branch, we subtracted the background of the RGB image by color body part segmentation and fed it into the network. Due to spatiotemporal feature learning by 3D convolution and 3D pooling, the 3D_CNN network is known as one of the essential frameworks for video classification. Residual 3D_CNN could significantly improve the classification performance of basic 3D_CNN framework. The 3D_ResNet is one of the current residual 3D_CNN versions. Various ResNet-based architectures with 3D convolutions were studied, but the 3D_ResNet network based on ResNeXt-101 was employed because of the quality performance for the proposed method.

Different from other original bottleneck blocks with a standard convolutional layer, the ResNeXt block employs a group convolution layer with its capacity to divide the feature maps into small groups. The single 3D_ResNet stream modality network in our proposed method included five ResNeXt blocks. The structure of each ResNeXt block consisted of convolution layers (group convolution layer), a batch normalization (BatchNorm) layer, and a rectified-linear unit (ReLu) layer, as shown in Figure 6. The input of each modality stream network was a fixed number of T sampled frames of a video: $V_c = \{x_{c1}, x_{c2}, \dots, x_{cT}\}$ for RGB modality and $V_{ps} = \{x_{ps1}, x_{ps2}, \dots, x_{psT}\}$ for color body part segmentation modality. The operation function of these stream-networks was $\Theta_c(.)$ and $\Theta_{ps}(.)$ respectively. Then, the prediction probabilities of the stream networks for RGB input and color body part segmentation for i classes can be described as:

$$P_c\{p_1|V_c, p_2|V_c, \dots, p_i|V_c\} = \Theta_c(V_c) \tag{1}$$

$$P_{ps}\{p_1|V_{ps}, p_2|V_{ps}, \dots, p_i|V_{ps}\} = \Theta_{ps}(V_{ps}) \tag{2}$$

where p_i is the prediction probability of the video belonging to the class ith, and $\mathrm{P_c}$ and $\mathrm{P_{ps}}$ denote the network outputs, which are the vectors or class prediction probabilities.

- The LSTM network for 3D skeleton joints modality: The LSTM network was proposed as a submodel for gesture recognition to benefit from the extracted 3D skeleton joints data. The 3D skeleton data provides useful information about the temporal features such as the localization of the relevant body joints over a time series to recognize the performed action. The LSTM networks are utilized to capture the contextual information of a temporal sequence for long periods by the gates and memory cells. In an LSTM network, the operation of gates and memory cells by time can be described as follows:

$$\begin{cases} i_t = \sigma(W_i[x_t, h_{t-1}] + b_i), \\ f_t = \sigma\left(W_f[x_t, h_{t-1}] + b_f\right), \\ o_t = \sigma(W_o[x_t, h_{t-1}] + b_o), \\ \widetilde{c_t} = \tan h(W_c[x_t, h_{t-1}] + b_c), \\ c_t = f_t * c_{t-1} + i_t * \widetilde{c_t}, \\ h_t = \tan h(c_t) * o_t \end{cases} \quad (3)$$

where $i, f,$ and o are the vectors of the input gate, the forget gate and the output gate, respectively. $\widetilde{c_t}$ and c_t denote the "candidate" hidden state and internal memory of the unit respectively. ht is the output hidden state and $\sigma(.)$ is a sigmoid function while W and b represent connected weights matrix and bias vectors, respectively.

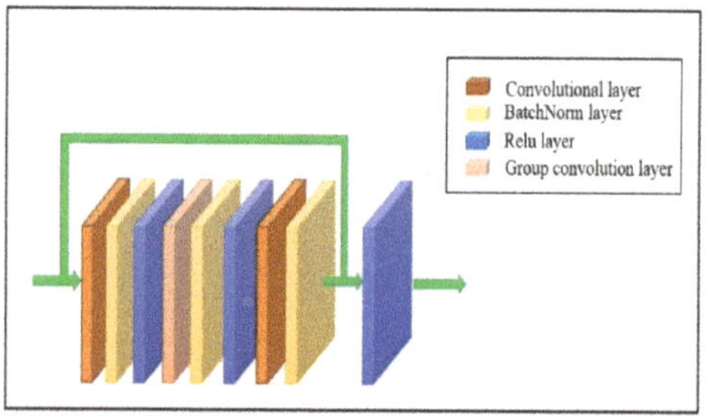

Figure 6. The overview of ResNet blocks. It consists of two convolution layers (1 × 1 × 1 kernel size), one group convolution layer (3 × 3 × 3 kernel size), and three batch normalization (BatchNorm) layers and rectified-linear unit (ReLu) layers.

In our approach, we used the relevant 17 human skeleton joints extracted by the Pose_3D network to perform the gesture classification. The input of the LSTM submodel was $T \times 51$ vector, which contains 3D location of 17 skeleton joints in a sequence of T frames of a video. This input vector is defined as $V_s = \{v_{s1}, v_{s2}, \ldots, v_{sT}\}$, where v_{sk} is a 51×1 vector of 3D position information of all skeleton joints at the kth frame. The prediction probability of the LSTM modality network for i gesture classes by a given input V_j is:

$$P_j\{p_1|V_j, p_2|V_j, \ldots, p_i|V_j\} = \Theta_j(V_j) \quad (4)$$

where $\Theta_j(.)$ represents the operation function of the LSTM network.

- Multi-modality fusion by an integrated stacking module: The final gesture class was predicted by the fusion score of three submodels. The multi-modalities fusion block combined the prediction score of submodels in the integrated stacking module, as shown in Figure 3. This module is a more extensive multi-headed neural network, which consists of a concatenation layer, two fully connected layers and a dropout layer to avoid overfitting. We used the outputs of each subnetworks as separate input-heads to the module. The fusion score computed by the fusion block with non-linear operation function $\Theta_{\text{fusion}}(.)$ to predict gesture classes as follows:

$$P\{p_1|V_c, p_2|V_c, \ldots, p_i|V_c\} = \Theta_{fusion}\left\{\begin{array}{l} P_c\{p_1|V_c, p_2|V_c, \ldots, p_i|V_c\}, P_{ps}\{p_1|V_{ps}, p_2|V_{ps}, \ldots, p_i|V_{ps}\}, \\ P_j\{p_1|V_j, p_2|V_j, \ldots, p_i|V_j\} \end{array}\right\} \quad (5)$$

4. Experimental Results

To evaluate the performance of the proposed method, we did experiments on four public datasets: UTD multimodal human action dataset [27], gaming 3D dataset [28], NTU RGB+D dataset [29] and MSRDailyActivity3D dataset [30]. In these datasets, we used only RGB modality for our work.

4.1. Datasets

- UTD Multimodal Human Action dataset [27]: is a multimodal human action dataset that was collected by using a Microsoft Kinect sensor and a wearable inertial sensor. This data includes 27 different types of actions performed by eight gestures, where each actor repeats each gesture four times. This dataset contains a total of 861 gesture videos after removing corrupted sequences.
- Gaming 3D dataset [28]: is a collection of 600 videos for action recognition in the gaming scenario. This dataset consisted of 20 action classes performed by ten subjects, and each subject repeated each action three times. Besides providing the action class label for each video, this dataset also had the ground truth for the peak frame of each action.
- NTU RGB+D dataset [29]: is a large dataset that contains 56,880 RGB-D videos captured by three Kinect V2 sensors concurrently. This dataset consisted of 60 human activities related to daily actions, mutual actions, and medical conditions. In our work, we only focused on the daily actions category with the RGB video data.
- MSRDailyActivity3D dataset [30]: is a dataset captured for daily human activities in a living room. This collection contained 16 regular activity classes performed by ten different individuals. There are 320 activity videos. This dataset was made with a noisy background with other activities from untargeted people.

For the MSRDailyActivity3D dataset and Gaming 3D dataset, we stratified a random split of each dataset into 5 folds with a train/valid/test set by ratio 8:1:1, respectively. For the UTD-MHAD dataset, we applied the 8-fold cross-validation method. Seven sets were used for training and the remaining set was used for testing. For the NTU-RGB+D dataset, there are two kinds of benchmarks recommended by the creator of the NTU-RGB+D dataset, which are cross-subject and cross-view benchmarks. We chose the cross-view benchmark, which included 37,462 clips for training and 18,817 clips for testing without validation.

4.2. Implementation

Three subnetworks were trained with the corresponding data modalities extracted from the datasets described above. The color body parts segmentation images and 3D skeleton joints were extracted first from the dataset to separately train corresponding subnetworks. The color body parts segmentation images were obtained by the RefineNet network, pretrained with UP-3D dataset as described in Section 3. We used the Pytorch framework of the RefineNet network, which was provided publicly. The 3D skeleton joints information was estimated by the Pose_3D network with the pretrained weight on the large dataset Human3.6M.

For two-stream networks for RGB and color-encoded body parts segmentation modalities, they were pretrained with UCF101—action recognition dataset [31]. The optimal settings for the LSTM network comprise of three memory blocks, and 256 LSTM Cells per memory blocks as in [32].

We combined the output scores of three subnetworks and fed it into the integrated stacking module to generate the final gesture class prediction. This multi-modality fusion is trained after subnetworks generate their own outputs.

The adaptive moment estimation [33] was used to optimize the parameters during the training process due to its effectiveness with a large number of parameters. We utilized the adaptive learning rate method to get the optimal parameters of the model.

All of the above codes were run with Nvidia GTX 1080 Ti GPU. The experiment was conducted with PC of CPU i7-7700 and 32GB of memory. It took 2–4 days per dataset for the experiment but the primary concern was the performance, which we mainly focused on.

4.3. Results

The performance of the proposed approach for gesture recognition was evaluated by the experiments conducted on four presented datasets: UTD multimodal human action dataset, gaming 3D dataset, NTU RGB+D dataset, and MSRDailyActivity3D dataset. Comparisons of the results with previous approaches are reported in Table 1.

Table 2 shows the comparison of the proposed network with existing networks on different datasets [7,8,19,34–40]. The comparison shows that the proposed network outperformed the existing works. Note that the combination of the three subnetworks was better than each single data modality network as shown in Table 3. The color body part segmentation modality produced better performance than the RGB modality due to the removal of background noise. Classification of gestures by important change detection with the spatiotemporal interest point (STIP) [36] also exhibited the effectiveness of the proposed method. The proposed method was outperformed by only [37] in which additional sensor devices like an accelerometer or gyroscope were utilized in addition to the RGB input. These sensors were available only on special circumstances.

Table 2. Gesture classification performance with different datasets ('-': not available).

Datasets/Methods	UTD Multimodal Human Action	Gaming 3D	NTU RGB+D	MSRDaily Activity3D
C3D [7]	85.3	89.1	83.3	87.5
LRCN [8]	83.0	-	-	-
ST-GCN [19]	-	-	88.3	-
I3D [34]	90.7	93.8	85.8	88.4
T_C3D [35]	89.5	90.3	85.7	88.9
STIP [36]	70.3	-	-	-
[37] (RGB, Accelerometer, Gyroscope)	97.6	-	-	-
TSSI + GLAN + SSAN [38]	-	-	89.1	-
Structure Preserving Projection (RGB+ Depth) [39]	-	-	-	89.8
ScTPM + CS-Mltp(RGB+ Depth) [40]	-	-	-	90.6
Proposed method	96.7	95.3	90.4	90.3

Table 3. Gesture classification performance in each modality and the fusion result on the UTD multimodal human action dataset.

Method	Accuracy (%)
3D_ResNet (RGB)	92.1
3D_ResNet (color body part segmentation)	94.6
LSTM (3D skeleton joints)	95.4
Fusion Result	96.7

For the NTU RGB+D dataset, to compare with the related model using only the RGB data input, our proposed method achieved outstanding results. While the ST-GCN network [19], which used a spatial–temporal graph convolutional network for a sequence of skeleton graphs, and the TSSI + GLAN + SSAN network [38], which utilized a two-branch attention architecture on skeleton images, obtained the highest accuracy in the literature, our method outperformed as shown in Table 1.

In the MSRDailyActivity3D dataset, our method outperformed the quality networks for video classification with only RGB video data input such as C3D, T_C3D, or I3D. Comparing with two state-of-the-art models, the structure preserving projection method [39] and ScTPM + CS-Mltp [40], our method had the same or even a close performance.

5. Conclusions

In this paper, we presented a novel approach for dynamic gesture recognition by using RGB images. This approach is a combination of two 3D_ResNet networks and one LSTM network to benefit from multi-modalities of RGB frames, 3D skeleton joint information, and color body part segmentation. The output scores of the submodels were fused at the integrated stacking module to obtain the final gesture class prediction. The effectiveness of the proposed method was shown by the experimental results on four public datasets.

Author Contributions: Conceptualization, G.-S.L. and N.-H.N.; methodology, N.-H.N.; writing—review and editing, N.-H.N., T.-D.-T.P. and G.-S.L.; supervision, G.-S.L., S.-H.K. and H.-J.Y.; project administration, G.-S.L., S.-H.K. and H.-J.Y.; funding acquisition, G.-S.L., S.-H.K. and H.-J.Y. All authors have read and agreed to the published version of the manuscript.

Funding: This research was supported by the Basic Science Research Program through the National Research Foundation of Korea (NRF) funded by the Ministry of Education (NRF-2018R1D1A3B05049058) and NRF funded by MSIT(NRF-2020R1A4A1019191).

Conflicts of Interest: The authors declare no conflict of interest.

References

1. Hussein, M.E.; Torki, M.; Gowayyed, M.A.; El-Saban, M. Human action recognition using a temporal hierarchy of covariance descriptors on 3d joint locations. In Proceedings of the Twenty-Third International Joint Conference on Artificial Intelligence, Beijing, China, 3–19 August 2013; Volume 13, pp. 2466–2472.
2. Dardas, N.H.; Georganas, N.D. Real-time hand gesture detection and recognition using bag-of-features and support vector machine techniques. *IEEE Trans. Instrum. Meas.* **2011**, *60*, 3592–3607. [CrossRef]
3. Lee, H.K.; Kim, J.H. An HMM-based threshold model approach for gesture recognition. *IEEE Trans. Pattern Anal. Mach. Intell.* **1999**, *21*, 961–973.
4. Feichtenhofer, C.; Pinz, A.; Zisserman, A. Convolutional two-stream network fusion for video action recognition. In Proceedings of the IEEE Conference on Computer Vision and Pattern Recognition (CVPR), Las Vegas, NV, USA, 26 June–1 July 2016; pp. 1933–1941.
5. Karpathy, A.; Toderici, G.; Shetty, S.; Leung, T.; Sukthankar, R.; Fei-Fei, L. Large-scale video classification with convolutional neural networks. In Proceedings of the IEEE conference on Computer Vision and Pattern Recognition, Columbus, OH, USA, 23–28 June 2014; pp. 1725–1732.
6. Ohn-Bar, E.; Trivedi, M.M. Hand gesture recognition in real time for automotive interfaces: A multimodal vision-based approach and evaluations. *IEEE Trans. Intell. Transp. Syst.* **2014**, *15*, 2368–2377. [CrossRef]
7. Tran, D.; Bourdev, L.; Fergus, R.; Torresani, L.; Paluri, M. Learning spatiotemporal features with 3d convolutional networks. In Proceedings of the IEEE International Conference on Computer Vision, Santiago, Chile, 7–13 December 2015; pp. 4489–4497.
8. Donahue, J.; Anne Hendricks, L.; Guadarrama, S.; Rohrbach, M.; Venugopalan, S.; Saenko, K.; Darrell, T. Long-term recurrent convolutional networks for visual recognition and description. In Proceedings of the IEEE Conference on Computer Vision and Pattern Recognition, Boston, MA, USA, 7–12 June 2015; pp. 2625–2634.
9. Duan, J.; Zhou, S.; Wan, J.; Guo, X.; Li, S.Z. Multi-Modality Fusion based on consensus-voting and 3D convolution for isolated gesture recognition. *arXiv* **2016**, arXiv:1611.06689.
10. Chai, X.; Liu, Z.; Yin, F.; Liu, Z.; Chen, X. Two streams recurrent neural networks for large-scale continuous gesture recognition. In Proceedings of the 23rd International Conference on Pattern Recognition (ICPR), Cancun, Mexico, 4–8 December 2016; pp. 31–36.
11. Wu, D.; Pigou, L.; Kindermans, P.J.; Le, N.D.H.; Shao, L.; Dambre, J.; Odobez, J.M. Deep dynamic neural networks for multimodal gesture segmentation and recognition. *IEEE Trans. Pattern Anal. Mach. Intell.* **2016**, *38*, 1583–1597. [CrossRef] [PubMed]

12. Hara, K.; Kataoka, H.; Satoh, Y. Can spatiotemporal 3D CNNs retrace the history of 2D CNNs and ImageNet? In Proceedings of the IEEE Conference on Computer Vision and Pattern Recognition, Salt Lake City, UT, USA, 18–22 June 2018; pp. 6546–6555.
13. Lin, G.; Milan, A.; Shen, C.; Reid, I. RefineNet: Multi-path refinement networks for high-resolution semantic segmentation. In Proceedings of the IEEE Conference on Computer Vision and Pattern Recognition, Honolulu, HI, USA, 21–26 July 2017; pp. 1925–1934.
14. Rayat Imtiaz Hossain, M.; Little, J.J. Exploiting temporal information for 3D human pose estimation. In Proceedings of the European Conference on Computer Vision, Munich, Germany, 8–14 September 2018; pp. 68–84.
15. Kopuklu, O.; Kose, N.; Rigoll, G. Motion fused frames: Data level fusion strategy for hand gesture recognition. In Proceedings of the IEEE Conference on Computer Vision and Pattern Recognition Workshops, Salt Lake City, UT, USA, 18–22 June 2018; pp. 2103–2111.
16. Molchanov, P.; Yang, X.; Gupta, S.; Kim, K.; Tyree, S.; Kautz, J. Online detection and classification of dynamic hand gestures with recurrent 3D convolutional neural networks. In Proceedings of the IEEE Conference on Computer Vision and Pattern Recognition (CVPR), Las Vegas, NV, USA, 27–30 June 2016; pp. 4207–4215.
17. Tran, D.; Ray, J.; Shou, Z.; Chang, S.F.; Paluri, M. ConvNet architecture search for spatiotemporal feature learning. *arXiv* **2017**, arXiv:1708.05038, preprint.
18. Luvizon, D.C.; Picard, D.; Tabia, H. 2D/3D Pose estimation and action recognition using multitask deep learning. In Proceedings of the IEEE Conference on Computer Vision and Pattern Recognition, Salt Lake City, UT, USA, 18–22 June 2018; pp. 5137–5146.
19. Yan, S.; Xiong, Y.; Lin, D. Spatial temporal graph convolutional networks for skeleton-based action recognition. In Proceedings of the Thirty-second AAAI Conference on Artificial Intelligence, New Orleans, LA, USA, 2–7 February 2018.
20. Li, M.; Chen, S.; Chen, X.; Zhang, Y.; Wang, Y.; Tian, Q. Actional-structural graph convolutional networks for skeleton-based action recognition. In Proceedings of the IEEE Conference on Computer Vision and Pattern Recognition, Long Beach, CA, USA, 16–20 June 2019; pp. 3595–3603.
21. Omran, M.; Lassner, C.; Pons-Moll, G.; Gehler, P.; Schiele, B. Neural body fitting: Unifying deep learning and model based human pose and shape estimation. In Proceedings of the International Conference on 3D Vision, Verona, Italy, 5–8 September 2018; pp. 484–494.
22. Fu, K.; Zhao, Q.; Gu, I.Y.H. Refinet: A deep segmentation assisted refinement network for salient object detection. *IEEE Trans. Multimed.* **2018**, *21*, 457–469. [CrossRef]
23. He, K.; Zhang, X.; Ren, S.; Sun, J. Deep residual learning for image recognition. In Proceedings of the IEEE Conference on Computer Vision and Pattern Recognition, Las Vegas, NV, USA, 27–30 June 2016; pp. 770–778.
24. Lassner, C.; Romero, J.; Kiefel, M.; Bogo, F.; Black, M.J.; Gehler, P.V. Unite the people: Closing the loop between 3d and 2d human representations. In Proceedings of the IEEE Conference on Computer Vision and Pattern Recognition, Honolulu, HI, USA, 21–26 July 2017; pp. 6050–6059.
25. Newell, A.; Yang, K.; Deng, J. Stacked hourglass networks for human pose estimation. In Proceedings of the European Conference on Computer Vision, Amsterdam, The Netherlands, 11–14 October 2016; pp. 483–499.
26. Ionescu, C.; Papava, D.; Olaru, V.; Sminchisescu, C. Human3.6m: Large scale datasets and predictive methods for 3d human sensing in natural environments. *IEEE Trans. Pattern Anal. Mach. Intell.* **2013**, *36*, 1325–1339. [CrossRef] [PubMed]
27. Chen, C.; Jafari, R.; Kehtarnavaz, N. UTD-MHAD: A multimodal dataset for human action recognition utilizing a depth camera and a wearable inertial sensor. In Proceedings of the IEEE International Conference on Image Processing, Quebec City, QC, Canada, 27–30 September 2015; pp. 168–172.
28. Bloom, V.; Makris, D.; Argyriou, V. G3D: A gaming action dataset and real time action recognition evaluation framework. In Proceedings of the IEEE Computer Society Conference on Computer Vision and Pattern Recognition Workshops, Providence, RI, USA, 16–21 June 2012; pp. 7–12.
29. Shahroudy, A.; Liu, J.; Ng, T.T.; Wang, G. NTU RGB+D: A large scale dataset for 3D human activity analysis. In Proceedings of the IEEE Conference on Computer Vision and Pattern Recognition, Las Vegas, NV, USA, 27–30 June 2016; pp. 1010–1019.
30. Wang, J.; Liu, Z.; Wu, Y.; Yuan, J. Mining actionlet ensemble for action recognition with depth cameras. In Proceedings of the IEEE Conference on Computer Vision and Pattern Recognition, Providence, RI, USA, 16–21 June 2012; pp. 1290–1297.

31. Soomro, K.; Zamir, A.R.; Shah, M. UCF101: A dataset of 101 human action classes from video in the wild. *arXiv* **2012**, arXiv:1212.0402, preprint.
32. Sarkar, A.; Gepperth, A.; Handmann, U.; Kopinski, T. Dynamic hand gesture recognition for mobile systems using deep LSTM. In Proceedings of the International Conference on Intelligent Human Computer Interaction, Evry, France, 11–13 December 2017; pp. 19–31.
33. Kingma, D.P.; Ba, J. Adam: A method for stochastic optimization. In Proceedings of the 3rd International Conference for Learning Representations, San Diego, CA, USA, 7–9 May 2015.
34. Carreira, J.; Zisserman, A. Quo vadis, action recognition? In A new model and the kinetics dataset. In Proceedings of the IEEE Conference on Computer Vision and Pattern Recognition, Honolulu, HI, USA, 21–26 July 2017; pp. 6299–6308.
35. Liu, K.; Liu, W.; Gan, C.; Tan, M.; Ma, H. T-C3D: Temporal convolutional 3D network for real-time action recognition. In Proceedings of the Thirty-second AAAI Conference on Artificial Intelligence, New Orleans, LA, USA, 2–7 February 2018.
36. Mahjoub, A.B.; Atri, M. Human action recognition using RGB data. In Proceedings of the 11th International Design & Test Symposium, Hammamet, Tunisia, 18–20 December 2016; pp. 83–87.
37. Ehatisham-Ul-Haq, M.; Javed, A.; Azam, M.A.; Malik, H.M.; Irtaza, A.; Lee, I.H.; Mahmood, M.T. Robust human activity recognition using multimodal feature-level fusion. *IEEE Access* **2019**, *7*, 60736–60751. [CrossRef]
38. Yang, Z.; Li, Y.; Yang, J.; Luo, J. Action recognition with spatio-temporal visual attention on skeleton image sequences. *IEEE Trans. Circuits Syst. Video Technol.* **2018**, *29*, 2405–2415. [CrossRef]
39. Yu, M.; Liu, L.; Shao, L. Structure-preserving binary representations for RGB-D action recognition. *IEEE Trans. Pattern Anal. Mach. Intell.* **2015**, *38*, 1651–1664. [CrossRef] [PubMed]
40. Luo, J.; Wang, W.; Qi, H. Spatio-temporal feature extraction and representation for RGB-D human action recognition. *Pattern Recognit. Lett.* **2014**, *50*, 139–148. [CrossRef]

applied sciences

MDPI

Article

Robust Hand Shape Features for Dynamic Hand Gesture Recognition Using Multi-Level Feature LSTM

Nhu-Tai Do, Soo-Hyung Kim *, Hyung-Jeong Yang and Guee-Sang Lee

Department of Artificial Intelligence Convergence, Chonnam National University, 77 Yongbong-ro, Gwangju 500-757, Korea; 176680@jnu.ac.kr (N.-T.D.); hjyang@jnu.ac.kr (H.-J.Y.); gslee@jnu.ac.kr (G.-S.L.)

* Correspondence: shkim@jnu.ac.kr

Received: 1 August 2020; Accepted: 7 September 2020; Published: 10 September 2020

Abstract: This study builds robust hand shape features from the two modalities of depth and skeletal data for the dynamic hand gesture recognition problem. For the hand skeleton shape approach, we use the movement, the rotations of the hand joints with respect to their neighbors, and the skeletal point-cloud to learn the 3D geometric transformation. For the hand depth shape approach, we use the feature representation from the hand component segmentation model. Finally, we propose a multi-level feature LSTM with Conv1D, the Conv2D pyramid, and the LSTM block to deal with the diversity of hand features. Therefore, we propose a novel method by exploiting robust skeletal point-cloud features from skeletal data, as well as depth shape features from the hand component segmentation model in order for the multi-level feature LSTM model to benefit from both. Our proposed method achieves the best result on the Dynamic Hand Gesture Recognition (DHG) dataset with 14 and 28 classes for both depth and skeletal data with accuracies of 96.07% and 94.40%, respectively.

Keywords: Dynamic Hand Gesture Recognition; human-computer interaction; hand shape features

1. Introduction

Besides the common language modalities, hand gestures are also often used in our daily lives to communicate with each other. For example, close friends can greet each other with a wave of their hands instead of words. Furthermore, hand gestures are the language of communication for deaf and mute people. In addition, hand gesture recognition is also one of the ways that computers can interact with humans by translating the human hand gestures into commands. Recently, hand gesture recognition research has developed rapidly, which is an essential element in the development of new technologies in the computer vision and pattern recognition fields. Especially, real-time 3D hand pose estimation combined with depth cameras has contributed to the successful launch of virtual reality and augmented reality applications such as sign language recognition [1], virtual reality [2], robotics [3], interaction systems [4], and interactive gaming [5]. Nevertheless, there exist various challenges hindering the achievement of accurate results due to the complex topology of the hand skeleton with high similarity among fingers and a small size. In addition, the cultural factors or personal habits of humans such as position, speed, and style can lead to variations in the hand gesture. Due to these special features, the hands can have various shapes describing the same pose. Feix et al. [6] found 17 different hand shapes that humans commonly use in everyday tasks to preform grasping.

Recent studies have suggested a number of solutions for challenges such as using reliable tools to capture 3D hand gestures and motion or using color gloves with attached sensors to capture real-time measurements of the hand [7,8]. However, their calibration setup process is complex and expensive. In 2013, Shotton et al. [9] presented a concept called the "body skeleton" to accurately predict the 3D positions of 20 body joints from depth images. The author demonstrated that the position, movement, and orientation of the joints can be a great description of human action. As such, the hand skeleton can also process accurate information about the hand shape, and later, Potter et al. [10] proposed research on Australian Sign Language by using a reliable dataset of the labeled 3D hand skeleton corresponding to the 22 joints, which was provided by the Leap Motion Controller (LMC) device. Even so, the result was still inaccurate when the hand was near or perpendicular to the camera or when the person performed a quick gesture. In 2016, some 3D hand gesture datasets were proposed by [11], and Smedt et al. [12] gave a promising solution for performing gesture recognition tasks.

In addition to the dataset, the algorithm method also must meet the optimization needs for gesture recognition. Previously, traditional methods produced the feature descriptors in the spatial and temporal dimension to encode the statuses of hand motion and hand shape [13,14]. Currently, methods based on deep learning are considered solutions to recognize and classify images efficiently and reliably. Specifically, dynamic gesture recognition also applies deep learning such as [15,16]; however, they are limited in real-time execution.

As shown in Figure 1, the diversity of hand features improves the dynamic hand gesture recognition under the challenges of the complex topology of the hand and hand pose variations due to cultural factors and personal habits. The inputs of a gesture are the depth data and the hand skeleton perceived by a depth camera, as shown in the first row of Figure 1. We can focus on the global hand movement as the hand bounding box in red color and the hand posture in Row 2. We refer to the hand posture as the hand shape to highlight the local movements among hand components.

There are two kinds of hand shapes based on the input data skeleton or depth data, as in Row 3 and Row 4 in Figure 1, respectively. As the brightness constrains optical flow problems, the hand shapes are robust features because they not only focus on the local movements of the hand components, but also track the displacement of every element in the data such as the depth value in the depth data or the hand joint in the skeletal data between two consecutive times.

With our hypothesis that robust hand shape features impact the learning of local movements directly and global movements indirectly, our work explores the hand shape approach with the derivatives from the depth data and skeletal data. For the hand depth feature approach, we exploited hand component features from the hand segmentation model as shown in Row 5 in Figure 1. This can capture the shape changes of a gesture based on the hand component labels referring to the local movement in a hand gesture. For the hand skeleton features, we could exploit the point-cloud data from the depth data by 3D hand reconstruction; however, due to the time constraints in real-time hand applications, we focused on the 3D point-cloud under the hand joint coordinates in the 3D world space from the skeletal data. Our model will be able to learn 3D geometric transformation features. Simultaneously, we also use the displacement and rotation of the hand joints with respect to their neighbors for the hand skeletal shape features. Therefore, our model of dynamic hand gesture recognition addresses the diversity problem in hand features from the two modalities of the hand depth and hand skeletal data.

To benefit from the various features, we propose multi-level feature LSTM using Conv1D, the Conv2D pyramid, and the LSTM block to exploit the diversity of the hand features from handcrafted data to automatically generate data from the deep learning model for the time series data and depth data. Our proposed method achieves the best accuracy on Dynamic Hand Gesture Recognition (DHG) [12] on 14 and 28 classes, respectively, with the skeletal data. It is good for real-time applications requiring low processing costs.

Our paper's contribution consists of three parts. Firstly, we identify various hand features from two modalities with depth and skeletal data. We then propose the best features for exploiting skeletal data and depth data to achieve the best results. Secondly, we build the multi-level feature LSTM with Conv1D, the Conv2D pyramid, and the LSTM block to use the effective hand features. Finally, we experimented on DHG 14 and 28 and obtained the best results.

The rest of this paper is composed of the following sections: Related Work, Proposed Idea, Hand Posture Feature Extraction, Network Architecture, Experiment, and Discussion and Conclusion. In Related Work, we discuss the datasets and approaches of 3D hand gestures. In the next two parts, our proposed method is described in detail. We then analyze the strengths of our method and make comparisons with the state-of-the-art in the experiments and discussion. The conclusions are given in the final part.

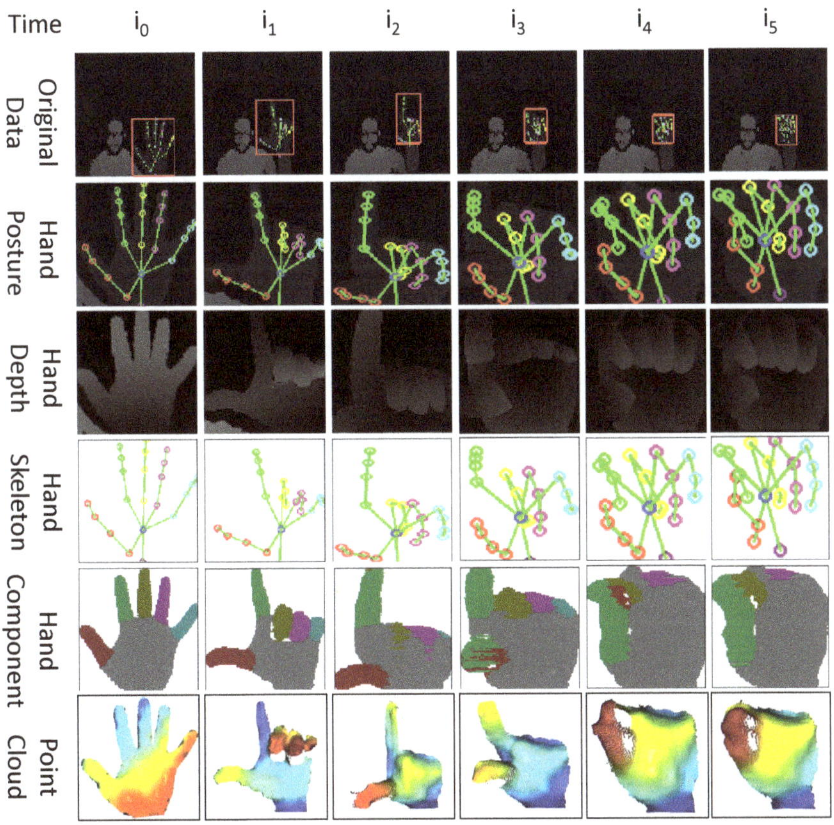

Figure 1. Overview of the features of a dynamic hand gesture. Left to right shows the time axis of the gesture, and top to bottom shows the types of hand data features, consisting of the original data, hand posture, hand depth, hand skeleton, hand component, and hand point-cloud.

2. Related Works

Hand gesture recognition research has robustly developed with a variety of approaches in recent years. The advancement in 3D depth sensors with a low cost has been one of the key elements that has increased the research into 3D hand gestures. With this technology, light variations, background clutter, and occlusions are major concerns in the detection and segmentation of hands. Furthermore, the depth sensors can capture 3D information in the scene context, which helps give faster estimation of the hand skeleton from the hand pose. Hence, there is much information to recognize hand gestures such as the hand skeleton, depth, and color images [17]. Below are the main categories of the approaches to 3D hand gesture recognition: static and dynamic hand gesture recognition or hand gesture recognition using deep images and/or hand skeletal data.

2.1. Dynamic Hand Gesture Recognition

The first approach is to identify static hand gestures. The 3D depth information and various traditional methods are utilized to detect hand shadows and hand regions used to extract features. Namely, Kuznetsova et al. [18] used the hand point cloud to compute invariant features, then they applied a multi-layered random forest for training to recognize hand signs. In the same way, Pugeault et al. [19] combined the Gabor filter and random forest for gesture classification to detect hand shape for the American Sign Language (ASL) finger-spelling system. Dong et al. [20] suggested a hierarchical mode-seeking method to localize hand joint positions under kinematic constraints and applied random forest to classify ASL signs based on joint angles. Ohn-Bar et al. [14] leveraged a modified HOG algorithm, named the spatial-temporal HOG2 descriptor, to extract useful information from spatial-temporal descriptors.

Ren et al. [21] expressed the hand shape as a time-series curve to facilitate the classification and clustering of shapes without using HOG, SIFT, and random forest. Furthermore, they proposed a new distance metric named the finger-earth mover's distance to discriminate hand gestures. Recently, Zhang obtained remarkable results using his proposed histogram of 3D facets to encode 3D hand shape information from the depth map [13]. Furthermore, Oreifej et al. [22] built a histogram of the normal orientations' distribution by integrating the time, depth, and spatial coordinates into 4D space to recognize the activity from depth sequences.

If the static approach handles the hand region and extracts hand features from a single image, the dynamic methods deem hand gesture recognition as recognizing a sequence of hand shapes by exploiting the temporal features of motion. Zhang et al. solved the gesture recognition problems by linking the histogram of 3D facets, the N-gram model, and dynamic programming on depth maps [13]. On the other hand, Monnier et al. [23] used a boosted classifier cascade to detect the gesture. They also leveraged body skeletal data and the histogram of oriented gradients to obtain the features.

Recently, the significant progress of Convolutional Neural Networks (CNNs) has led to various groundbreaking studies in the computer vision field, and hand gesture recognition in particular, such as image classification [24], object detection [25], and image segmentation [26]. Aghbolaghi et al. [27] performed a survey to demonstrate the effectiveness of deep learning approaches in action and gesture recognition. In [28], a factorized spatial-temporal convolutional network, which is a cascaded deep architecture, learned spatio-temporal features and transferred learning from the pre-trained ImageNet on a 2D CNN, while Varol et al. [29] used a neural network having long-term temporal convolutions to compute motion features for temporal information. In order to study real-time hand gesture recognition, Neverova et al. [30] proposed a method that combines both video data and articulated pose with multi-scale and multi-modal deep learning. Similarly, Molchanov et al. [16] applied multi-scale 3D-CNN models with depth, color, and stereo-IR sensor data.

2.2. Depth and 3D Skeleton Dynamic Hand Gesture Recognition

In recent works, along with the advances in hand pose estimation and the technology of depth-based cameras, skeleton-based recognition has gained more traction. In [11], they extracted the features of the distances, angles, elevations, and adjacent angles of fingertips by employing the data of direction, normal position, the central location of the palm, and fingertip position. Garcia et al. built the mo-capsystem to make hand pose annotations and gather 6D object poses from RGB-D video data for hand recognition [31]. De Smedt et al. [12] published an approach with results better than the results of depth-based methods. They calculated the shape of connected joints descriptor from the connected joints of the hand skeleton and used a Fisher vector to encode it. A temporary pyramid was used to model Fisher vectors and skeleton-based geometric features before extracting the final feature.

There are various methods that are based on deep learning, for dynamic hand gesture recognition using skeleton-like information. Chen et al. [32] performed training on a bidirectional Recurrent Neural Network (RNN) using the movement features of fingers and hand and skeleton sequences. Devineau et al. proposed parallel convolutions to handle sequences of the hand skeleton.

De Smedt [33] proposed a new way based on CNN and LSTM by fusing two streams of the hand shape and skeleton model.

Collectively, all the above methods have problems in recognizing gestures at some distance from the camera, with variable illumination.

3. Proposed Idea

3.1. Problem Definition

A dynamic hand gesture $\boldsymbol{G} = (\boldsymbol{S}, \boldsymbol{D})$ in this study can be described as a time-series stream of the 3D hand skeletal data $\boldsymbol{S}$ and hand depth data $\boldsymbol{D}$ with the length N_t frames. The goal of our method is based on $\boldsymbol{S}$ and $\boldsymbol{D}$ to classify whether a dynamic hand gesture belongs to one of the given gesture types $C = \{c_1, c_2, ..., c_{N_c}\}$. The gesture types are determined based on the specific dataset.

Let $\boldsymbol{x}_j^t \triangleq \left(x_j^t, y_j^t, z_j^t\right)$ be the world space coordinate of the 3D hand joint j at time t; the hand skeleton posture $S_t \in \mathbb{R}^{N_j \times 3}$ at time t is the set J of hand joints with N_j coordinates in 3D space defined as follows:

$$\boldsymbol{S}_t = \left\{\boldsymbol{x}_j^t\right\}_{j=1..N_j}^{t} \tag{1}$$

The 3D hand skeletal data $S \in \mathbb{R}^{N_t \times N_j \times 3}$ are expressed as the set of hand skeleton postures S_t as below:

$$\boldsymbol{S} = \{\boldsymbol{S}_t\}^{t=1..N_t} \tag{2}$$

In the case of the hand depth data, we let $D_t \in \mathbb{R}^{w \times h}$ be the depth image at time t with the width w and height h. The 3D hand depth data $D \in \mathbb{R}^{N_t \times w \times h}$ are represented by the set of hand depth postures D_t as follows:

$$\boldsymbol{D} = \left\{\boldsymbol{D}_t \in \mathbb{R}^{w \times h}\right\}^{t=1..N_t} \tag{3}$$

3.2. Problem Overview

The overview of our proposed pipeline is as shown in Figure 2. We first apply the temporal frame sub-sampling for every dynamic gesture G to the specific length N_t. Then, we split the dynamic gesture into hand skeletal data S and hand depth data D as the input to the feature extraction step.

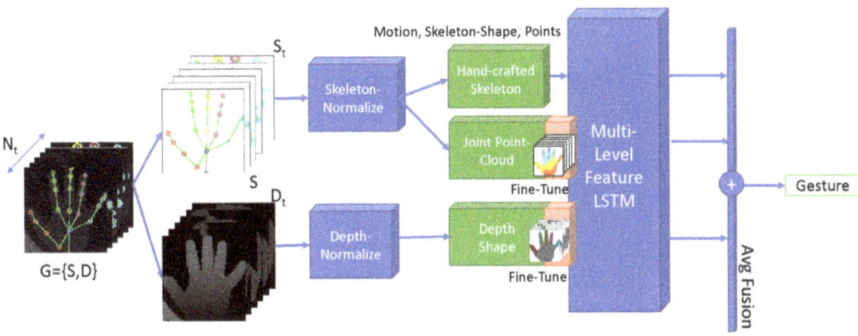

Figure 2. Overview of the proposed system for dynamic hand gesture recognition.

In the normalize phase, with hand skeletal data, to deal with the variants in the size and pose (translation and rotation) of the gestures due to the changes in the performer and the camera pose, we need to normalize the 3D hand joint coordinates of all frames by the same transformation of rotation and translation and uniformly scale from the first hand posture in the gesture to the given reference hand posture in front of the camera with the palm joint at the root coordinate. With the hand depth data, we apply image processing techniques to eliminate the isolated depth pixels in the hand region and convert the depth value to the range $[0, 1]$.

In our proposed framework, the feature extraction phase consists of five feature types to exploit the robust features for the dynamic hand gesture as follows:

Three first three feature types are the handcrafted skeleton features including the motion, skeleton shape, and normalized hand skeletal coordinates. The motion feature captures the changes of the translation and rotation of the overall hand. A new contribution of this study is using the pairwise joint distance instead of using the Shape of Connected Joints (SoCJ) descriptor as in [34] to reduce the complex feature space. Moreover, we add more than one feature with the oriented angle of the three joint tuples to represent the hand skeletal shape with the oriented values among the hand joints.

The next two feature types are the joint point-cloud feature and the hand depth shape feature automatically extracted by the deep learning models in an end-to-end fashion. With the recent success of the hand pose regression problem in using 3D reconstruction from depth data to the point-cloud and voxel [35,36], we opted to use Point-Net [37] to learn the 3D geometric features. Instead of using all 3D points in the hand depth data reconstructed from the point-cloud, we propose to only use the 3D world space hand joints as the key points for Point-Net. Regarding the hand depth shape feature, we propose to use the middle layer of the encoder-decoder hand segmentation model as the hand depth shape feature.

Finally, we propose the multi-level feature LSTM model to train on every hand gesture feature. Our architecture firstly uses the LSTM filter layer to exploit the long-term dependencies between the frames and reduce the complexity of the input feature. After the first LSTM layer, we use the self-attention mechanism and three kinds of blocks, namely, Con1D, Conv2D, and LSTM, to exploit the spatial and temporal coherency in the feature spaces. The LSTM filter layer will send the encode states to the feature LSTM layer to help the second LSTM learn better.

Note that for hand gesture feature extracting from deep learning models, they will be integrated into the hand gesture recognition model to fine-tune again during the training phase of the dynamic gesture recognition model.

Finally, we use the average pooling layer to integrate the classification probability for all separate models for every hand gesture feature. Our result will classify the type of gestures.

3.3. Hand Skeleton Normalization

The hand skeletal data are received from various camera sensors, the pose of the camera, as well as the performer. Therefore, we need to normalize the data to the reference pose and size to prevent over-fitting of the method with respect to the environmental elements.

First, we choose the reference hand $\boldsymbol{S}_{ref} = [x_1, x_2, ..., x_{N_j}]^T$ in the dataset with the status of open and in front of the camera, as in Figure 3. Then, we transform the reference hand so that the palm joint is at the root coordinate and scale the hand size to fit in the unit sphere as follows:

$$\boldsymbol{S}_{norm} = \boldsymbol{S}_{ref} - \boldsymbol{x}_{palm} \tag{4}$$

$$scale = \max \|\boldsymbol{x}_i\|_2, \text{ where } \boldsymbol{x}_i \in \boldsymbol{S}_{norm} \tag{5}$$

$$\boldsymbol{S}_{norm} = \frac{\boldsymbol{S}_{norm}}{scale} \tag{6}$$

For every skeletal data sequence, we find the optimal rotation $\boldsymbol{R}$ and translation $\boldsymbol{t}$ using [38] between the hand skeletal data at the first frame $\boldsymbol{S}_{t_1}$ and the reference hand $\boldsymbol{S}_{norm}$ based on the seven hand joints corresponding to the 3D points (palm, wrist, from the thumb to pinky base joints) as the least squares error minimization problem:

$$E = \sum_{i=1}^{N_b} \left\| \boldsymbol{R}\boldsymbol{x}_i^{t_1} + \boldsymbol{t} - \boldsymbol{x}_i^{norm} \right\|_2 \longrightarrow 0 \tag{7}$$

where $\boldsymbol{x}_i^{t_1} \in$ base joints $(\boldsymbol{S}_{t_1})$, $\boldsymbol{x}_i^{norm} \in$ base joints $(\boldsymbol{S}_{norm})$, and $N_b = 7$ is the number of base joints (palm, wrist, from the thumb to pinky base joints).

To solve Equation (7), we find the center of base joints (S_{t_1}), base joints (S_{norm}), respectively:

$$x_{center}^{t_1} = \frac{1}{n_{t_1}} \sum_{i=1}^{n_{t_1}} x_i^{t_1} \tag{8}$$

$$x_{center}^{norm} = \frac{1}{n_{norm}} \sum_{i=1}^{n_{norm}} x_i^{norm} \tag{9}$$

where $n_{t_1} = |\text{base joints}(S_{t_1})|$, and $x_i^{t_1} \in \text{base joints}(S_{t_1})$; similarly, $n_{norm} = |\text{base joints}(S_{norm})|$, and $x_i^{norm} \in \text{base joints}(S_{norm})$. Then, we calculate the Singular Value Decomposition (SVD) of the co-variance matrix $\boldsymbol{H}$ to find the rotation transformation $\boldsymbol{R}$ as follows:

$$\boldsymbol{H} = \left(\text{base joints}(S_{t_1}) - x_{center}^{t_1}\right)\left(\text{base joints}(S_{norm}) - x_{center}^{norm}\right)^T \tag{10}$$

$$\boldsymbol{U}, \boldsymbol{S}, \boldsymbol{V} = SVD(\boldsymbol{H}) \tag{11}$$

$$\boldsymbol{R} = \boldsymbol{V}\boldsymbol{U}^T \tag{12}$$

We address the reflection case of the SVD results by checking the determinant $|R| < 0$ and fixing again as the equation below:

$$\boldsymbol{U}, \boldsymbol{S}, \boldsymbol{V} = SVD(\boldsymbol{R}) \tag{13}$$

$$\boldsymbol{V}[column3] = -1 * \boldsymbol{V}[column3] \tag{14}$$

$$\boldsymbol{R} = \boldsymbol{V}\boldsymbol{U}^T \tag{15}$$

Finally, the translation is calculated as below:

$$\boldsymbol{t} = x_{center}^{norm} - \boldsymbol{R} \times x_{center}^{t_1} \tag{16}$$

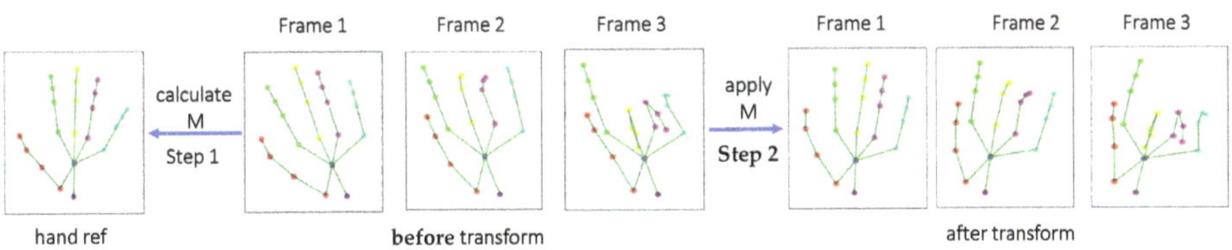

Figure 3. Hand skeleton normalization. We will calculate the rotation, translation, and uniform scale of the first hand skeleton to the reference hand skeleton. Then, the matrix transform will be applied to the remaining hand skeletons.

3.4. Hand Depth Normalization

Given that $\boldsymbol{H}_t = (x_t, y_t, w_t, h_t)$ is the bounding box of the hand region at time t, the hand skeleton data $\boldsymbol{D}_t$ are extracted from the depth data $\boldsymbol{I}_t$ by $\boldsymbol{I}_t(\boldsymbol{H}_t)$. There are many background and noisy pixels, which are often the isolated depth pixels in the hand depth data $\boldsymbol{D}_t$. The morphology operator in image processing is used to eliminate them.

For the background pixel elimination problem, the depth values of pixels in the hand region gather around the centroid $M_{centroid}$ of the hand depth values. Therefore, the depth threshold t_{depth} is used to remove the background pixels as follows:

$$\boldsymbol{D}_t(x,y) = \begin{cases} \boldsymbol{D}_t(x,y) & |\boldsymbol{D}_t(x,y) - M_{centroid}| < t_{depth} \\ 0 & otherwise \end{cases} \tag{17}$$

where the centroid $M_{centroid} = Mode(\boldsymbol{D}_t)$. All depth values after removing isolated and background pixels are normalized to [0, 1].

4. Hand Posture Feature Extraction

A dynamic hand gesture often consists of two components: motion and shape. Motion components contain the changing information with respect to time of the overall hand for global motions and the fingertip positions for local motions. The shape components represent the hand shape at the specific time by the hand joint positions and hand components (palm, wrist, fingers at the base, middle or tip regions) in the corresponding domains (skeleton or depth).

In this study, we only calculate the global motions of the overall hand. The local motions can be exploited from the hand shape changes with respect to time. Therefore, the hand shape feature models sometimes archive better performance than the motion models due to capturing the local motions from the shape changes.

Furthermore, there are three ways to divide the types of hand posture features. The first group is based on skeleton and depth data. The second group is based on the means of feature extraction, such as the handcrafted features and deep learning features. The last group is the components of the gesture, such as motion and shape.

In this section, we mention three main groups: handcrafted skeleton features (motion, skeleton shape, and normalized points), joint point-cloud feature (input from normalized points to exploit 3D geometric characteristic), and depth shape feature (input from the depth data to determine the hand components).

4.1. Handcrafted Skeleton Features

4.1.1. Motion Feature Extraction

We represent the global motion S_{motion} of the specific hand by the changes of the palm coordinate S_{dir}, the angle between the palm and wrist joint S_{rot}, and the major axis of all hand joints S_{maj}:

$$S_{motion} = \{S_{dir}, S_{rot}, S_{maj}\} \tag{18}$$

The translations of the hand that determine S_{dir} by the direction of the two palm joints at two consecutive times t_i and t_{i+1} are calculated as below:

$$S^t_{dir} = \frac{x^t_{palm} - x^{t-1}_{palm}}{\left\| x^t_{palm} - x^{t-1}_{palm} \right\|_2} \tag{19}$$

$$S_{dir} = \{S^t_{dir}\} \tag{20}$$

The rotation of the hand represents the sign of the angles between the wrist and palm joints using the dot product operator as below:

$$S^t_{rot} = \frac{x^t_{wrist} \cdot x^t_{palm}}{\left\| x^t_{wrist} \right\| \left\| x^t_{palm} \right\|} \tag{21}$$

$$S_{rot} = \{S^t_{rot}\} \tag{22}$$

Furthermore, we propose to use the changes in the major axis of all hand joints to more precisely express the orientation of all hand joints. The major and minor axes of the hand joints correspond to the eigenvectors of the covariance matrix of the set of hand joints:

$$\boldsymbol{Cov}\left(S^t\right) = \frac{1}{N_j} \sum_{i=1}^{N_j} \left(x^t_i - \bar{x}^t_i\right) \left(x^t_i - \bar{x}^t_i\right)^T \tag{23}$$

$$U^t, S^t, V^t = SVD\left(\boldsymbol{Cov}\left(S^t\right)\right) \tag{24}$$

$$U^t = \left[v^t_1, v^t_2, v^t_3\right] \tag{25}$$

where $\bar{x}_i^t$ is the center of the 3D hand joint coordinates S^t and $\{v_i\}$ is the eigenvectors forming the orthogonal basis, as well as the major axes. Hence, the major axis feature S_{maj} is expressed as:

$$S_{maj}^t = \{v_i^t\}_{i=1}^3 \quad (26)$$

$$S_{maj} = \left\{S_{maj}^t\right\} \quad (27)$$

4.1.2. Hand Skeleton Shape Extraction

The role of hand shape in the dynamic hand gesture recognition determines the movement of the joints and components of the local motion. Therefore, we represent the hand skeleton shape S_{kshape} with two components: the movement of the joints with respect to their neighbors S_{kmov} and the angle of the joints with respect to two neighboring joints S_{krot}, as shown in Figure 4:

$$S_{kshape} = \{S_{kmov}, S_{krot}\} \quad (28)$$

Regarding the shape descriptor of the movement of a hand joint with respect to its neighbors, we use all displacements at a point with respect to the remaining points with no overlap between the two points as follows:

$$S_{kmov}^t = \left\{x_j^t - x_i^t \mid i < j \text{ and } i, j \in [1, N_j]\right\} \quad (29)$$

$$S_{kmov} = \{S_{kmov}^t\} \quad (30)$$

In this study, with $N_j = 22$, there are in total $C_{N_j}^2 = 231$ elements in S_{kmov}^t. Besides the local movement features, we also suggest the angles between one joint and two distinct joints as the features to exploit the local rotation in the dynamic gesture. This is calculated as:

$$S_{krot}^t = \left\{\frac{\left(x_j^t - x_i^t\right) \cdot \left(x_k^t - x_j^t\right)}{\left\|\left(x_j^t - x_i^t\right)\right\| \left\|\left(x_k^t - x_j^t\right)\right\|} \mid i < j < k \text{ and } i, j, k \in [1, N_j]\right\} \quad (31)$$

$$S_{krot} = \{S_{krot}^t\} \quad (32)$$

Similarly, the number of angle features at time t is $C_{N_j=22}^3 = 1540$.

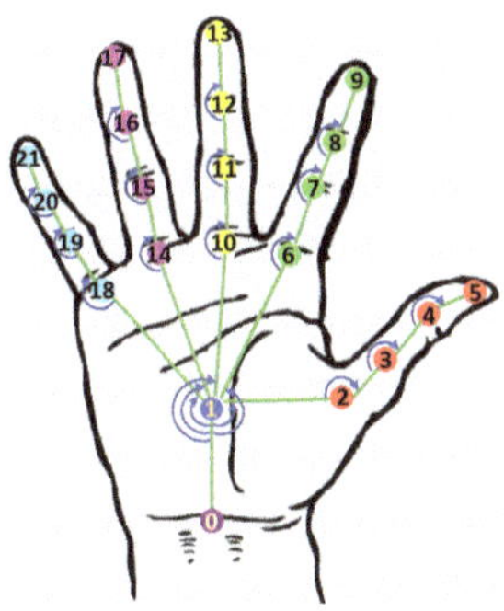

Figure 4. Hand skeleton shape calculated by the movement and rotation of joints with respect to their neighbors. There are 22 joints in the hand skeleton data numbered from 0–21: 0 (wrist), 1 (palm), 2–5 (thumb), 6–9 (index), 10–13 (middle), 14–17 (ring), and 18–21 (pinky).

4.1.3. Normalized Points

Finally, we directly use the hand joint coordinates at time t normalized S^t_{points} to the classification model for gesture recognition as below:

$$S^t_{points} = \{x^t_i\} \tag{33}$$

$$S_{points} = \left\{S^t_{points}\right\} \tag{34}$$

4.2. Joint Point-Cloud Feature Model

With the success of deep learning networks, various computer vision tasks can extract the features and automatically classify the object in an end-to-end fashion. Therefore, we aimed to build our feature models to exploit the 3D geometric transformation features from the point-cloud and the visual features from the depth data.

The joint point-cloud feature model facilitates the learning of the 3D geometric transformation features from the 3D point-cloud. In the hand gesture data, there are two ways to construct the point-cloud. Firstly, we can reconstruct the hand depth data into a set of 3D points. This approach is hindered by the unordered attributes of the point-cloud, the alignment between the points at different times, and the processing cost to convert and process. Secondly, the hand joint points of the gesture can represent the point-cloud. This has the advantages of the order of the set and the alignment of the joints by time.

Due to the low resolution of the skeletal point-cloud, we chose PointNet [37], as shown in Figure 5, to learn deep features in the 3D geometric transform in an end-to-end fashion.

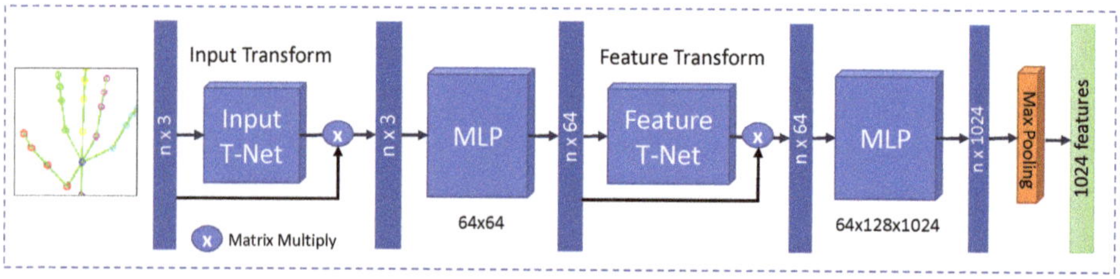

Figure 5. Point-Netarchitecture [37].

Therefore, the joint point-cloud feature model f_{point_cloud} is expressed as:

$$S^t_{point\_cloud} = f_{point\_cloud}\left(S^t_i\right) \tag{35}$$

$$S_{point\_cloud} = \left\{S^t_{point\_cloud}\right\} \tag{36}$$

where $S^t_{point\_cloud}$ is the point-cloud feature at time t from f_{point_cloud}.

Point-Net is comprised of the transform blocks, MLP blocks, and one max-pooling layer. The transform blocks can be represented as a function f to map a point set (input T-Net) or a feature point set (feature T-Net) to a feature vector with permutation invariance and geometric transformations as follow:

$$f^{trans}_{point\_cloud}\left(S^t_i\right) = \gamma\left(\max\left\{h\left(S^t_i\right)\right\}\right) \tag{37}$$

where x^t_i is the point-cloud, h is the MLP block to capture the feature of the point set, γ is the symmetric function as an appropriate rigid or affine transformation with a 3×3 matrix transform to achieve the normalization, and the operator max selects highly activated values from the point features. After every transform block, there are MLP blocks to learn and extend the feature size. Finally, the max-pooling layer will return the feature values of the point-cloud.

In this study, the joint point-cloud model was integrated as a child block of the dynamic hand gesture recognition model to learn point features from scratch while training the gesture classification.

4.3. Depth-Shape Feature Model

The depth shape feature model f_{depth_shape} plays the role of the feature extraction block to obtain the feature vector that presents the hand shape in the depth data. The depth shape feature model in our study is based on the U-Net architecture [39] to learn the hand components from segmenting hand regions from depth data. It can be expressed as:

$$\boldsymbol{D}^t_{depth\_shape} = f^e_{depth\_shape}\left(\boldsymbol{D}^t\right) \tag{38}$$

$$\boldsymbol{H}^t_{mask} = f^d_{depth\_shape}\left(\boldsymbol{D}^t_{depth\_shape}\right) \tag{39}$$

$$\boldsymbol{D}_{depth\_shape} = \left\{\boldsymbol{D}^t_{depth\_shape}\right\} \tag{40}$$

or:

$$\boldsymbol{H}^t_{mask} = f^d_{depth\_shape}\left(f^e_{depth\_shape}\left(\boldsymbol{D}^t\right)\right) \tag{41}$$

where $f^e_{depth\_shape}$ is the encoder function to encode the depth data as the depth shape feature $\boldsymbol{D}^t_{depth\_shape}$ at time t while $f^d_{depth\_shape}$ is the decoder function to map the encoder feature to the hand component masks $\boldsymbol{H}^t_{mask}$ by one-hot encoding.

U-Net, as shown in Figure 6, consists of the encoder and decoder blocks and the skip connections between them. The structure of the encoder can be based on the common visual image classification models such as VGG16 [40], Resnet50 [41], etc. The backbone of the encoder using VGG16, as shown in Figure 6, has five blocks of two convolution layers, batch-normalization, and max-pooling. The encoder block converts the depth data input into the encoded features, then the decoder blocks convert the encoded features into the semantic representation of the input data with the hand component masks in the background, palm, thumb, index, middle, ring, and finger regions. The role of the skip connections helps our model to be trained stably and achieve a better result by passing the features from the encoder to the decoder at the same levels concurrently with the features from the decoders below. Through the skip connection, the model can keep learning even when the deeper encoder and layer cannot learn by the dying ReLU problem or the vanishing problem.

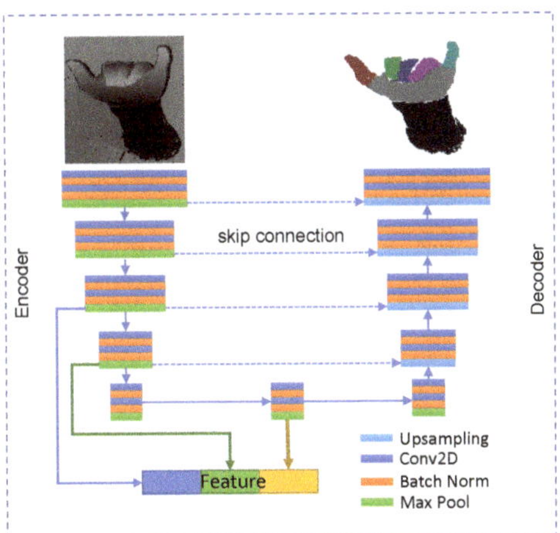

Figure 6. Hand component segmentation model to extract depth shape features.

The middle block between the encoder and decoder blocks is the feature block to return the feature vector. To enhance the feature vector, the hierarchical depth shape feature combines Blocks 3 and 4 of the encoder, as shown in Figure 6.

For the dataset for training the depth shape, it should focus on the hand components instead of only the hand pose due to the complex structure of the hand such as the small size and self-occlusions. In this study, Finger-Paint [42] is a suitable dataset with all the requirements.

For the loss function, the soft Dice loss [43] is a good choice in the unbalanced cases among the segmentation regions. Since the palm region often has a larger size compared to finger regions in the hand components, the loss function measures the overlap between the two regions the object region and the non-object region, in the binary classification. In the multi-class classification problem, the final score will be averaged with the Dice loss of each class expressed as:

$$L\left(y_{true}, y_{pred}\right) = 1 - \frac{1}{N_c}\sum_{c=1}^{N_c} \frac{2\sum_{pixels} y^c_{true} y^c_{pred} + \epsilon}{\sum_{pixels} \left(y^c_{true}\right)^2 + \sum_{pixels}\left(y^c_{pred}\right)^2 + \epsilon} \tag{42}$$

where c is the region of the hand components including the N_c regions (background, palm, thumb, index, middle, ring, pinky); y^c_{true} and y^c_{pred} are the ground-truth and the prediction of the hand component masks, respectively, in region c.

5. Our Network Architecture

From the hand feature extraction step, the system receives handcrafted skeleton features ($\boldsymbol{S}_{motion}$, $\boldsymbol{S}_{skeleton\_shape}$, and $\boldsymbol{S}_{points}$), the joint point-cloud feature $\boldsymbol{S}_{point\_cloud}$, and the depth shape feature $\boldsymbol{D}_{depth\_shape}$. In general, let $\chi\left\{\boldsymbol{G}^t\right\}$ be a feature transform of a gesture $\boldsymbol{G}^t$ at time t using one of the feature extractions mentioned. Our proposed model shown in Figure 7 uses the first LSTM layer [44] for a time series of features extracted from gesture data to exploit the long-term dependencies and encode them into a sequence the same as the length of the input gesture with the specific feature size. The encoder LSTM layer can be expressed as follows:

$$\boldsymbol{h}^t_1, \boldsymbol{c}^t_1 = LSTMCell\left(\boldsymbol{G}^t, \boldsymbol{h}^{t-1}_1, \boldsymbol{c}^{t-1}_1\right) \tag{43}$$

$$\boldsymbol{h}_1 = \left\{\boldsymbol{h}^t_1\right\}, \boldsymbol{c}_1 = \left\{\boldsymbol{c}^t_1\right\} = LSTM\left(\boldsymbol{G}\right) \tag{44}$$

where $\boldsymbol{h}_1$ and $\boldsymbol{c}_1$ are the set of hidden state $\boldsymbol{h}^t_1$ and cell state $\boldsymbol{c}^t_1$ at time t. If the feature transform χ is the deep learning feature model such as $\boldsymbol{S}_{point\_cloud}$ and $\boldsymbol{D}_{depth\_shape}$, the proposed model can be straightforward to fine-tune again the feature model integrated in it.

$\boldsymbol{h}_1$ plays the role of the normalized encoding features containing long-term dependencies, and $\boldsymbol{c}_1$ is the cell state in the LSTM used to transfer to the next layer the states for the other LSTM or as an attention vector.

The encoding feature vector $\boldsymbol{h}_1$ gives the Con1D pyramid block, the Conv2D pyramid block, as well as the LSTM block to exploit the temporal-spatial coherency, as shown in Figure 8.

The Conv1D pyramid block contains the Conv1D blocks with every block consisting of the Conv1D layers and the dropout layer with different filter sizes. Global average pooling is in the last position to capture the feature vector by exploiting $\boldsymbol{h}_1$ on its feature axis.

The Conv2D pyramid block consists of the same Conv2D blocks as the VGG16 block, which contain the Conv2D layers, dropouts, and max-pooling at the end of the block. It will exploit features in the time-step and feature axis of the input, select the high values by max-pooling, and down-sample the input on the time-step axis. Finally, the global average pooling layer will compress and return the feature vector.

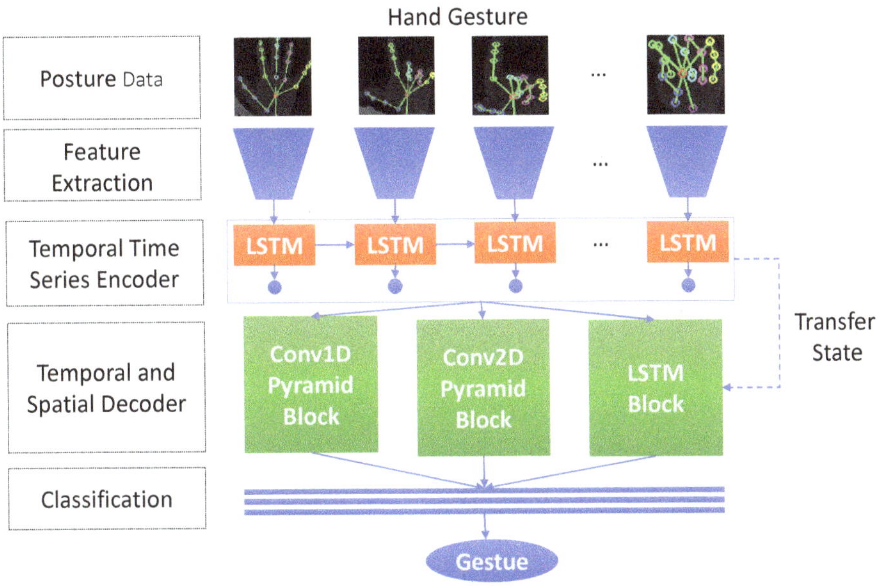

Figure 7. Multi-level feature LSTM architecture.

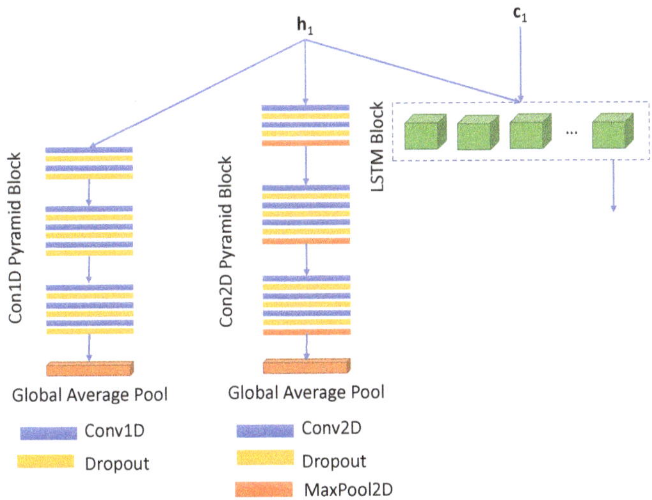

Figure 8. The structure of the LSTM block, Conv1D, and the Conv2D pyramid block.

Unlike Conv1D and the Conv2pyramid block, the input of the LSTM model from h_1 and c_1 of the previous LSTM layer c_1 will help the model learn h_1 better by inheriting the cell state from the previous LSTM layer.

Finally, all features are concatenated from the building blocks and added to the dense layers to classify the gestures.

For the loss function, we use the Category Cross-Entropy (CCE) for classification expressed as:

$$CCE\left(y_{true}, y_{pred}\right) = -\sum_{i=1}^{C} y_{true} \log y_{pred} \tag{45}$$

6. Experiments and Discussion

6.1. Training Datasets

6.1.1. Depth-Shape Model

In this work, we selected a suitable dataset for training the depth shape model. The depth shape model needs to focus on the hand components clearly for the recognition of the various hand poses including the hard cases with small-sized hands and self-occlusion hand pose. For this reason, we chose the FingerPaintdataset, as shown in Figure 9, by Sharp et al. [42]. There were five performers, A, B, C, D, and E, with three hand pose subjects to record in the dataset. Regarding the hand pose subjects, the "global" subject focused on the large global movements while the hand pose was almost static; the "pose" subject consisted of gestures with only moving fingers and no hand movement; finally, the "combined" subject was attributed to two subjects, "pose" and "global". There is also the special topic of "penalization", which was based on the performer.

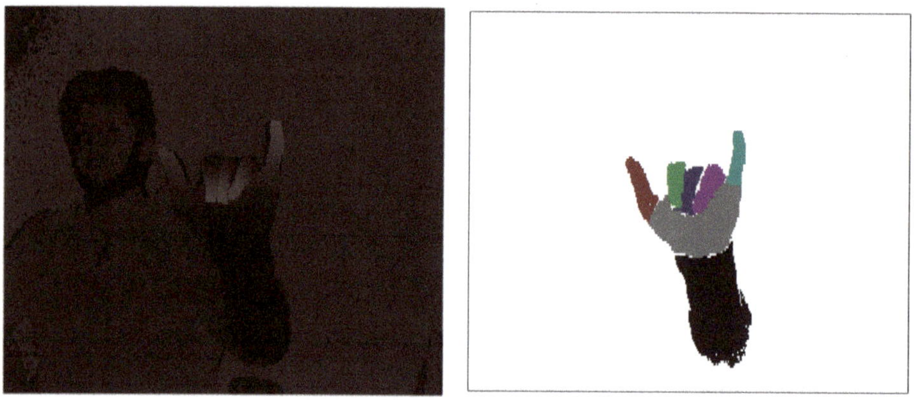

(a) (b)

Figure 9. Hand depth data (**a**) and hand component label (**b**) in the FingerPaint dataset.

There was a total 56,500 hand poses with the statistical number of hand poses based on the performer per subject shown in Table 1.

Table 1. Number of hand poses of every performer per subject.

	Subject A	Subject B	Subject C	Subject D	Subject E
Global	3500	3500	3500	3500	3500
Pose	3500	3500	3500	3500	3500
Combined	3500	3500	3500	3500	3500
Personalization	800	800	800	800	800

To enhance the segmentation performance, a survey was conducted on the number of pixels between hand components only focusing on the hand region. Figure 10a shows the statistic result of all regions and Figure 10b illustrates on the other regions except the background region.

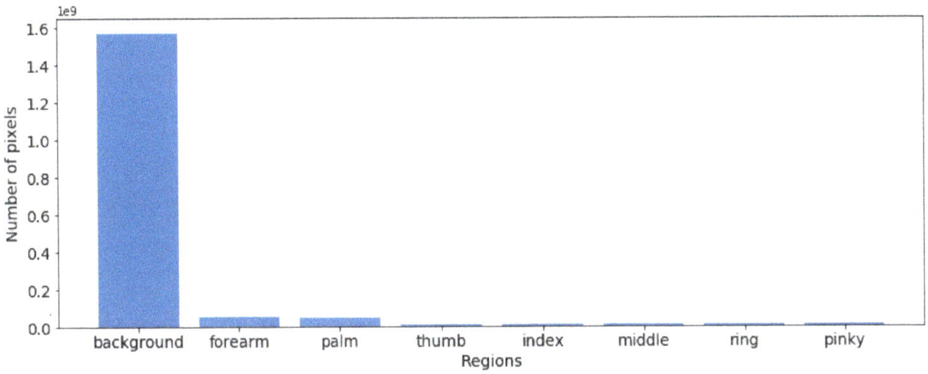

(a)

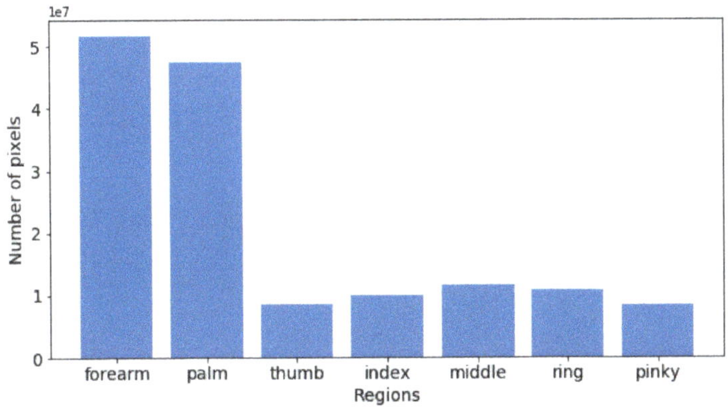

(b)

Figure 10. Number of pixels of the regions in the FingerPaint dataset on all regions (**a**) and regions without the background (**b**).

There was an unbalance between the background and the remaining regions. Without the background, the number of pixels in the forearm and palm regions was larger than the finger regions. For training, we did a split of 70%/30% of every subject and performer for training/validation. Additionally, we used rotation, translation, and scaling by 10% for data augmentation during the training process.

6.1.2. Dynamic Gesture Recognition

For the dynamic hand gesture recognition, we chose the Dynamic Hand Gesture (DHG) dataset containing the hand skeleton and hand depth data suitable for our method. There were 20 performers making up the dataset. Every person performed five gestures in two different ways: using one finger or the whole hand. The dataset had a total of 2800 sequences with 1960 for training and 840 for validation. Every sequence was labeled with 14 or 28 gestures depending on one finger or the overall hand for the ground-truth, as shown in Table 2.

A dynamic hand gesture had a length from 20 to 150 frames. Every frame consisted of a depth image of size 640 × 480, the skeleton information in the image coordinate, and the world coordinate of 22 hand joints captured by the Intel RealSense camera. Figure 11 shows samples of the gestures in the DHG dataset.

Table 2. List of gestures in the Dynamic Hand Gesture (DHG) dataset.

14 Classes	28 Classes	Gesture	Label
1	1, 2	Grab	Fine
2	3, 4	Expand	Fine
3	5, 6	Pinch	Fine
4	7, 8	Rotation CW	Fine
5	9, 10	Rotation CCW	Fine
6	11, 12	Tap	Coarse
7	13, 14	Swipe Right	Coarse
8	15, 16	Swipe Left	Coarse
9	17, 18	Swipe Up	Coarse
10	19, 20	Swipe Down	Coarse
11	21, 22	Swipe X	Coarse
12	23, 24	Swipe V	Coarse
13	25, 26	Swipe +	Coarse
14	27, 28	Shake	Coarse

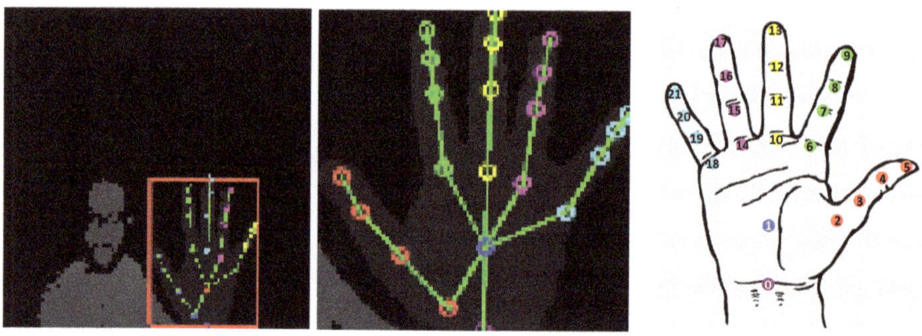

(a) (b) (c)

Figure 11. A sample hand posture in the DHG dataset: (**a**) hand depth data with drawing the hand bounding box and the 22 hand joints; (**b**) hand regions in zoom mode; (**c**) 22 hand joints with 0 (wrist), 1 (palm), 2–5 (thumb), 6–9 (index), 10–13 (middle), 14–17 (ring), and 18–21 (pinky).

6.2. Setup Environments, Metrics, and Training Parameters

Environments: Our program was developed with Python 3.5 using the TensorFlow Keras framework to build the deep learning models. Our program ran on a desktop PC with Intel Corei7 8700k with 32 GB of RAM and one graphic card GeForce GTX 1080 Ti.

Metrics: For the depth shape model in the hand component segmentation, we used the metric mean IoU to evaluate our segmentation results. This is the intersection over union between the ground-truth and prediction on every hand region. We used eight hand regions: background, palm, forearm, thumb, index, middle, ring, and pinky.

$$MeanIoU = \sum_{i=1}^{C} \frac{TP_i}{(TP_i + FP_i + FN_i)} 100 \tag{46}$$

where C is the number of hand regions and $TP_i/FP_i/FN_i$ the true positive/false positive/false negative for region i.

For the dynamic hand gesture recognition, we quantified them based on the accuracy between the prediction and ground-truth and the time cost for predicting a gesture.

Parameters in training: We performed a temporal augmentation on the sequence length of a hand gesture by randomizing the position of the first frame and getting 32 frames with equal step sizes. For the spatial augmentation of the depth data, we used basic transforms, such as random rotation by 45 degrees, translation by 5%, and scaling by 10%, based on the frame size.

For training the model, we used Adam [45] with a learning rate of 0.001 and for the first time training reducing the learning rate on the plateau. For the fine-tuning step in the previous training, we used SGD [46] to train with a learning rate ranging from 0.004 to 0.0001 using the cosine annealing learning rate schedule.

Experimenting with the features and models: We conducted the experiments based on the list of features shown in Table 3.

Table 3. List of hand features.

No.	Name	Features	Input	Description
1	Motion	Motion	Skeleton	Movement, rotation, and major axes of the hand
2	Skeleton Shape	Hand Shape	Skeleton	Movement, rotation of joints with neighbors
3	Points	Raw data	Skeleton	Normalize hand joint points
4	Joint point-cloud	3D geometric transformation	Skeleton	Point-Net model
5	Depth shape	Hand components	Depth	Palm, thumb, index, etc., regions

There was a total of five hand features from the two input types, skeleton and depth. We divided the groups of features as the motion group (learning the global motion of hand gestures) with feature motion, hand shape (capturing the changes of hand components) with feature skeleton shape, joint point-cloud, and depth shape, and the others with the feature input by the normalizing points.

For the experiments on our proposed models, we divided our proposed model into with/without Con1D-2D pyramid blocks, as in Table 4.

Table 4. List of the proposed models.

No.	Model Name
1	Multi-Level Feature LSTM without Conv1+2D (MLF LSTM)
2	Multi-Level Feature LSTM with Conv1+2D (MLF LSTM Conv1-2D)

6.3. Results on Hand Component Segmentation

We trained our depth shape models with three types of backbones: VGG16 [40], MobinetV2 [47], and Seresnext [48]. Our results are shown in Table 5.

Table 5. Results of hand component segmentation.

Backbone	Mean IoU
VGG16	82.30%
MobilenetV2	84.00%
Seresnext	86.40%

We achieved the highest mean IoU with the backbone Seresnext and, therefore, chose this backbone for our depth shape model. Figure 12 shows the quality of the depth shape model with the backbone Seresnext compared to the ground-truth.

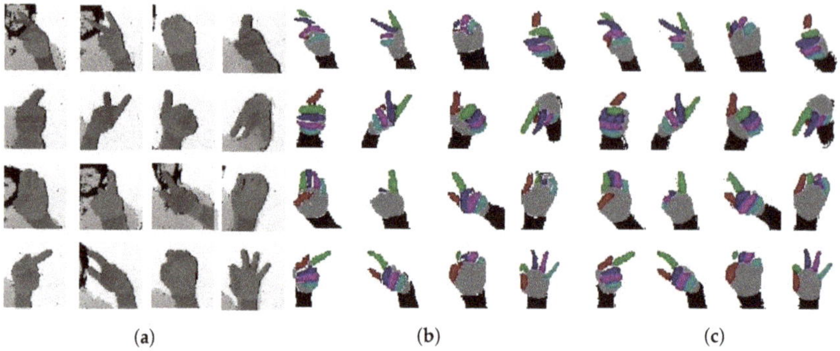

Figure 12. Results of hand component segmentation using the Seresnext backbone with the ground-truth depth (a), ground-truth labels (b), and Seresnext (c).

6.4. Results Using the Single Hand Features

We conducted the experiments on two models, MLF LSTM and MLF LSTM with Conv1D and the Conv2D pyramid block, to analyze the influence of single hand features on the DHG dataset with 14 and 28 classes. The results are shown in Table 6.

Table 6. Performance results of the two models using the separate hand features.

		14 Classes		28 Classes	
No.	Features	MLF LSTM	MLF LSTM Conv1-2D	MLF LSTM	MLF LSTM Conv1-2D
1	Motion	80.23	82.5	72.85	70.23
2	Skeleton shape	74.76	74.16	70.83	69.4
3	Joint point-cloud	68.92	85.11	56.07	70.23
4	Points	88.33	88.09	82.97	83.09
5	Depth shape	92.26	90.71	87.61	88.33

Motion features are better with the gestures focusing on global movement. However, when performing the complex gestures using from one finger (14 classes) to all fingers (28 classes), the motion features decrease significantly from 82.5% to 70.23%.

The performance of the models using the depth shape feature only was reduced slightly from 92.26% and 90.71% down to 87.61% and 88.33%. Depth shape features also give the best accuracy of all the features, because they help the model recognize the local motion and also capture the changes of the depth values between two consecutive frames, enabling the model to learn the optical flow features; therefore, the model can recognize global motion.

Upon comparison of the performance between the two models, MLF LSTM Conv1-2D gives better results when using joint point-cloud features with 85.11%/68.92% on 14 classes and 70.23%/56.07% on 28 classes. In contrast, MLF LSTM shows better results of 92.26% and 87.61% on 14 and 28 classes as compared to MLF LSTM Conv1-2D with 90.71% and 88.3%. The different between the two models is the training from scratch and the training from the pre-trained weight.

Figure 13 shows the false cases using the depth shape on DHG 14 and 28. For the 14 classes, the gestures grab and pinch are greatly confused. From 14 to 28 classes, the confusion occurred between grab and pinch in both the one finger gesture and all-finger gesture. Rotation CW(1) and (2) are nearly confused the same between the one finger gesture and the all-finger gesture.

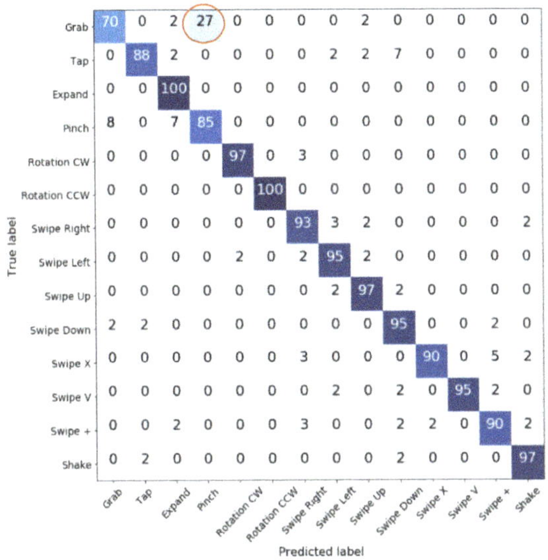

(**a**)

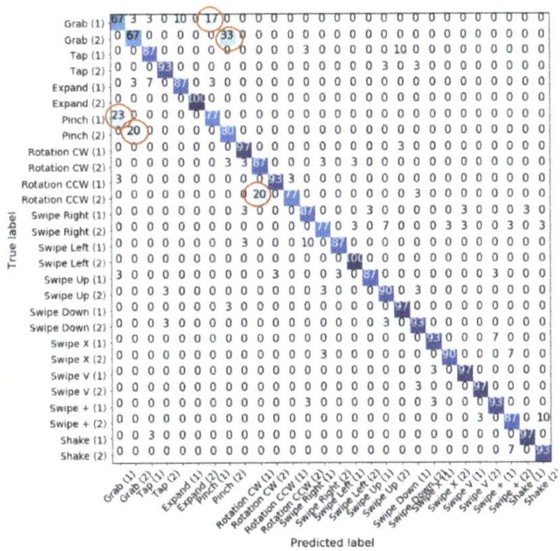

(**b**)

Figure 13. Confusion matrix of the best models using the singleshape feature on DHG 14 (accuracy of 92.26%) (**a**) and DHG 28 (accuracy of 88.33%) (**b**). The red circles are the false cases causing the confusion in recognition.

6.5. Experiment 2: Effects of Hand Features in the Skeleton Data

This experiment shows the influences among hand features from skeletons, as shown in Table 7. It has an important role in real-time applications with high requirements for the processing time.

The model using motion features achieved 82.50% on the 14 classes and 72.85% on the 28 classes. When we added hand shape skeletons, our model could capture the local motion between the fingers, increasing our accuracy from 82.50% to 84.52% on the 14 classes and from 72.85% to 82.02% on the 28 classes.

Moreover, the joint point-cloud model could learn good features in 3D geometry and improve the performance of our model by 5.07% (14 classes) and 2.98% (28 classes). Finally, when we integrated normalized point features, we achieved good accuracy on the skeleton data, 93.45% (14 classed) and 90.11% (28 classes).

In Figure 14, the red circle on the left confusion matrix points out the weakness of the model using the motion + skeleton shape feature. The model confused the grab and pinch gestures with the false cases being 27%. In contrast, the joint point-cloud was better with the false cases being 13%. Therefore, the combined results of the two models increased the accuracy from 84.52% to 90.59% (6.07% increase).

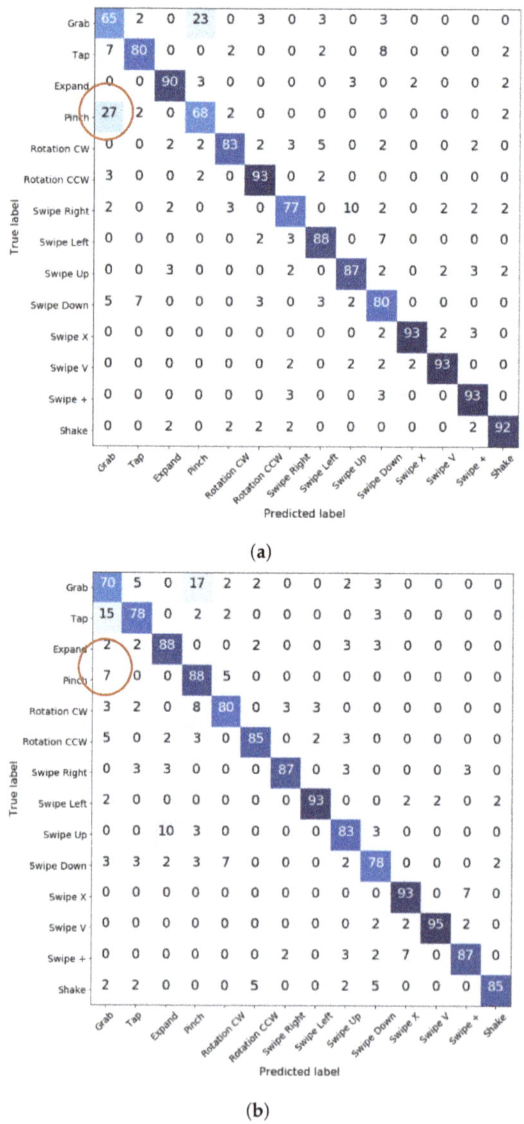

Figure 14. Confusion matrix of prediction results using the MLF LSTM Conv1-2D model with the features motion + skeleton shape (accuracy of 84.52%) (**a**) and joint point-cloud (accuracy of 85.11%) (**b**) on DHG-14.

Table 7. Performance results on the skeleton data.

		14 Classes		28 Classes	
No.	Features	MLF LSTM	MLF LSTM Conv1-2D	MLF LSTM	MLF LSTM Conv1-2D
1	Motion	80.23	82.50	72.85	70.23
2	Motion + Skeleton-Shape	86.07	84.52	82.02	81.19
3	Motion + skeleton shape + joint point-cloud	89.52	90.59	85.47	86.78
4	Motion + skeleton shape + joint point-cloud + points	93.45	93.69	89.40	90.11

6.6. Experiment 3: Comparison of Input Data

Table 8 shows that our model achieved the best result on the 14 and 24 classes with 96% and 94.4% combining the skeleton and depth data. Our confusion matrices are as shown in Figures 15 and 16.

Table 8. Overall performance results.

		14 Classes		28 Classes	
No.	Input	MLF LSTM	MLF LSTM Conv1-2D	MLF LSTM	MLF LSTM Conv1-2D
1	Skeleton	93.45	93.69	89.4	90.11
2	Depth2D	92.26	90.71	87.61	88.33
3	All	96.07	94.28	94.4	92.38

Moreover, the model with skeleton input performed better than that with depth data. Because it uses much GPU resource, the model using depth data addresses the performance problem in real-time applications.

6.7. Experiment 4: Comparison with Related Works

We made a comparative survey of the previous works on the dynamic hand gesture recognition, as shown in Table 9. Smedt et al. [12] built the DHG dataset and conducted their experiments based on the Fisher vector to extract features from the Shape of Connected Joints (SoCJ), built temporal pyramid features, and classified by SVM. Their works achieved state-of-the-art performances of 86.86% and 84.22% on DHG 14 and 28, respectively, with the traditional approach. Using deep learning with the dynamic graph and attention mechanism, Chen at al. successfully achieved the highest accuracy of 91.9% and 88% on DHG 14 and 28 by the deep learning approach.

Due to the multi-modal features between enhancing traditional features and integrating the joint point-cloud model for exploiting the 3D geometric transform, our method gave better results than the other two. The proposed method gave 93.69% and 90.11% on DHG 14 and 28 using skeleton data, as well as 96.07% as 90.11% using both depth and skeleton data.

Table 9. Comparison with the related works. SoCJ, Shape of Connected Joints.

Method	Input	Year	DHG 14	DHG 28
HOG2 [14]	Depth	2013	81.85	76.53
HON4D [22]	Depth	2013	75.53	74.03
MotionManifold [49]	Skeleton	2015	76.61	62
SkeletalQuads [50]	Skeleton	2014	84.5	79.43
Fea-SVM [51]	Skeleton	2014	50.32	30.85
3D Key Frame [52]	Depth	2017	82.9	71.9
MotionFeature+RNN [32]	Skeleton	2017	84.68	80.32
CNN+LSTM [53]	Skeleton	2017	85.6	81.1
STA-Res-TCN [54]	Skeleton	2018	89.2	85
Parallel CNN [55]	Skeleton	2018	91.28	84.35
NIUKF-LSTM [56]	Skeleton	2018	84.92	80.44
ST-GCN [57]	Skeleton	2018	91.2	81.7
SoCJ+HoHD+HoWR [34]	Skeleton	2019	86.86	84.22
DG-STA [58]	Skeleton	2019	91.9	88
GREN [59]	Skeleton	2020	82.29	82.03
Our proposed method	Skeleton		93.69	90.11
Our proposed method	Depth		92.26	88.33
Our proposed method	Overall		96.07	94.4

Our confusion matrices for the proposed methods are as shown in Figures 15 and 16.

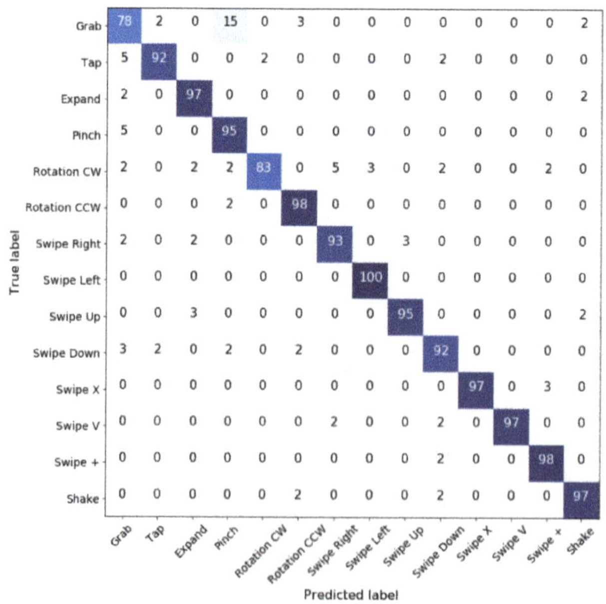

(a)

Figure 15. *Cont.*

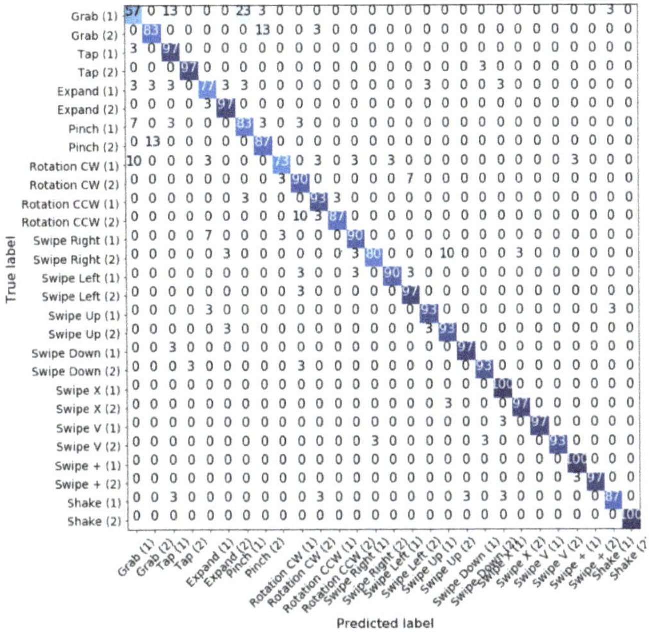

(b)

Figure 15. Confusion matrix of the best prediction results on the skeleton data: MLF LSTM Con1-2D with skeleton data (accuracy of 93.69%) (**a**) on DHG-14 and MLF LSTM Conv1-2D with skeleton data (accuracy of 90.11%) (**b**) on DHG-28.

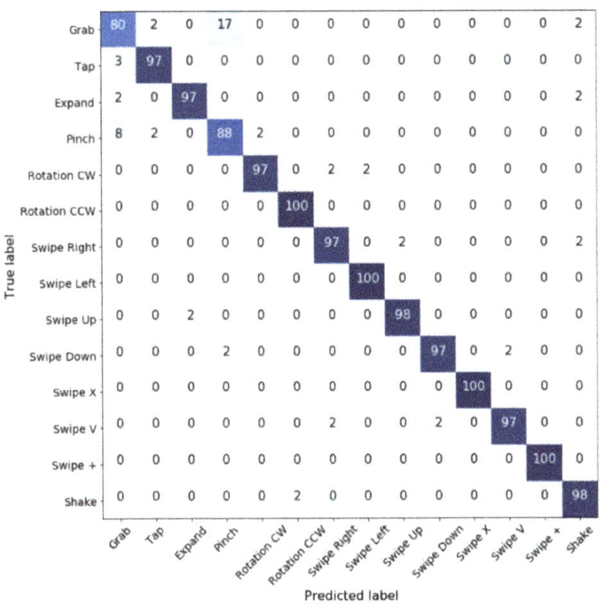

(a)

Figure 16. *Cont.*

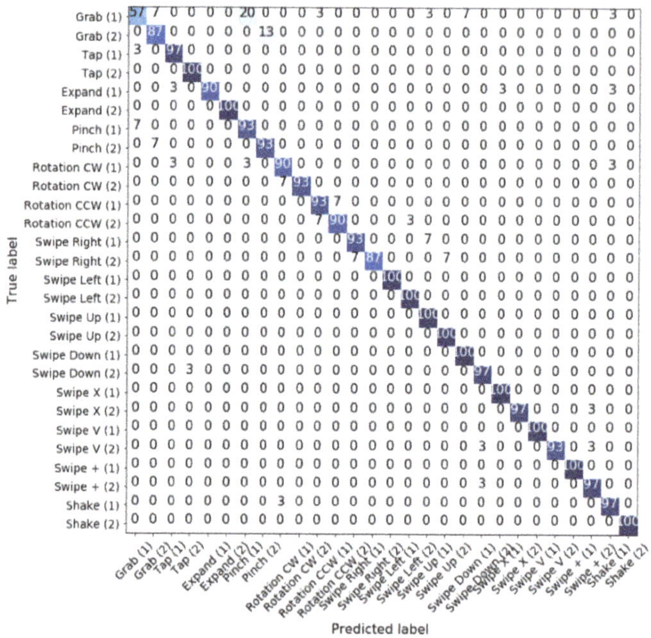

(**b**)

Figure 16. Confusion matrix of the best prediction results using depth + skeleton data: MLF LSTM (accuracy of 96.07%) (**a**) on DHG-14 and (accuracy 92.38%) (**b**) on DHG-28.

7. Conclusions

In this study, we build a novel method for benefiting from the MLF LSTM model from the 3D geometric transformation and displacement features in hand skeleton data, as well as the hand shape features in depth data from the hand component segmentation model. For the hand skeleton feature approach, we improve the handcrafted features in the motion features by adding the major axes and the skeleton shape through the displacement and rotation of the hand joints with respect to their neighbors. We propose using PointNet in the joint point-cloud model to exploit the 3D geometric transformation on the skeletal data. Our skeleton features improve the performance of our model over the state-of-the-art accuracy with 93.69% and 90.11% on DHG 14 and 28.

For the hand depth feature approach, we also propose using the hand component segmentation features from the depth shape model to recognize the hand shape. Our pre-trained depth shape model was based on U-Net with the Seresnext backbone. Our model using depth shape features gives improved performance with accuracies of 92.26% and 88.33% on DHG 14 and 28.

To learn from the two features, we propose MLF LSTM using Conv1D, the Conv2D pyramid block, and the LSTM block to exploit the hand features. Our model, using depth and skeleton data, gave the best performance with an accuracy of 96.07% and 94.4%. Upon comparison of our model with related works, our model achieves the best results.

In the future, we need to exploit the point-cloud features in the whole hand and enhance the LSTM model to natively integrate the diversity of features from the handcrafted features in the time-series the feature vector from the visual deep learning model and the point-cloud model.

Author Contributions: Conceptualization, N.-T.D. and S.-H.K.; Funding acquisition, S.-H.K., G.-S.L. and H.-J.Y.; Investigation, N.-T.D.; Methodology, N.-T.D.; Project administration, S.-H.K., G.-S.L. and H.-J.Y.; Supervision, S.-H.K.; Validation, N.-T.D.; Writing—original draft, N.-T.D.; Writing—review & editing, N.-T.D. and S.-H.K. All authors have read and agreed to the published version of the manuscript.

Funding: This work was supported by the Basic Science Research Program through the National Research Foundation of Korea (NRF) funded by the Ministry of Education under Grant NRF-2018R1D1A3A03000947 and Grant NRF-2020R1A4A1019191.

Conflicts of Interest: The authors declare no conflict of interest.

References

1. Huang, J.; Zhou, W.; Li, H.; Li, W. Sign Language Recognition using 3D convolutional neural networks. In Proceedings of the IEEE International Conference on Multimedia and Expo (ICME), Turin, Italy, 29 June–3 July 2015; pp. 1–6. [CrossRef]
2. Tan, T.D.; Guo, Z.M. Research of hand positioning and gesture recognition based on binocular vision. In Proceedings of the IEEE International Symposium on Virtual Reality Innovations (ISVRI), Singapore, 19–20 March 2011; pp. 311–315. [CrossRef]
3. Raheja, J.L.; Rajsekhar, G.A.; Chaudhary, A. Controlling a remotely located robot using hand gestures in real time: A DSP implementation. In Proceedings of the 2016 5th International Conference on Wireless Networks and Embedded Systems (WECON), Rajpura, India, 14–16 October 2016; Curran Associates, Inc.: New York, NY, USA, 2017; pp. 1–5. [CrossRef]
4. Lee, S.-H.; Sohn, M.-K.; Kim, D.-J.; Kim, B.; Kim, H. Smart TV interaction system using face and hand gesture recognition. In Proceedings of the IEEE International Conference on Consumer Electronics (ICCE), Las Vegas, NV, USA, 11–14 January 2013; pp. 173–174. [CrossRef]
5. Rautaray, S.S.; Agrawal, A. Interaction with virtual game through hand gesture recognition. In Proceedings of the IEEE International Conference on Multimedia, Signal Processing and Communication Technologies, Aligarh, India, 17–19 December 2011; pp. 244–247. [CrossRef]
6. Feix, T.; Pawlik, R.; Schmiedmayer, H.B.; Romero, J.; Kragi, D. A comprehensive grasp taxonomy. In Proceedings of Robotics, Science and Systems Conference: Workshop on Understanding the Human Hand for Advancing Robotic Manipulation, Seattle, WA, USA, 28 June–1 July 2009; pp. 2–3.
7. Wang, R.Y.; Popović, J. Real-time hand tracking with a color glove. *ACM Trans. Graph.* **2009**, *28*, 1–8. [CrossRef]
8. Schroder, M.; Elbrechter, C.; Maycock, J.; Haschke, R.; Botsch, M.; Ritter, H. Real-time hand tracking with a color glove for the actuation of anthropomorphic robot hands. In Proceedings of the 12th IEEE-RAS International Conference on Humanoid Robots (Humanoids), Osaka, Japan, 29 November–1 December 2012; pp. 262–269. [CrossRef]
9. Shotton, J.; Sharp, T.; Fitzgibbon, A.; Blake, A.; Cook, M.; Kipman, A.; Finocchio, M.; Moore, R. Real-Time human pose recognition in parts from single depth images. *Commun. ACM* **2013**, *56*, 116–124. [CrossRef]
10. Potter, L.E.; Araullo, J.; Carter, L. The leap motion controller: A view on sign language. In Proceedings of the 25th Australian Computer-Human Interaction Conference: Augmentation, Application, Innovation, Collaboration, Adelaide, Australia, 25–29 November 2013; pp. 175–178. [CrossRef]
11. Lu, W.; Tong, Z.; Chu, J. Dynamic Hand Gesture Recognition with Leap Motion Controller. *IEEE Signal Process. Lett.* **2016**, *23*, 1188–1192. [CrossRef]
12. De Smedt, Q.; Wannous, H.; Vandeborre, J.P. Skeleton-Based Dynamic Hand Gesture Recognition. In Proceedings of the IEEE Computer Society Conference on Computer Vision and Pattern Recognition Workshops, Las Vegas, NV, USA, 26 June–1 July 2016; pp. 1206–1214. [CrossRef]
13. Zhang, C.; Tian, Y. Histogram of 3D Facets: A depth descriptor for human action and hand gesture recognition. *Comput. Vis. Image Underst.* **2015**, *139*, 29–39. [CrossRef]
14. Ohn-Bar, E.; Trivedi, M.M. Joint angles similarities and HOG2 for action recognition. In Proceedings of the IEEE Computer Society Conference on Computer Vision and Pattern Recognition Workshops, Portland, OR, USA, 23–28 June 2013; pp. 465–470. [CrossRef]
15. Oyedotun, O.K.; Khashman, A. Deep learning in vision-based static hand gesture recognition. *Neural Comput. Appl.* **2017**, *28*, 3941–3951. [CrossRef]
16. Molchanov, P.; Gupta, S.; Kim, K.; Kautz, J. Hand gesture recognition with 3D convolutional neural networks. In Proceedings of the IEEE Conference on Computer Vision and Pattern Recognition Workshops, Boston, MA, USA, 7–12 June 2015; pp. 1–7. [CrossRef]

17. Wang, J.; Chen, Y.; Hao, S.; Peng, X.; Hu, L. Deep learning for sensor-based activity recognition: A survey. *Pattern Recognit. Lett.* **2019**, *119*, 3–11. [CrossRef]
18. Kuznetsova, A.; Leal-Taixé, L.; Rosenhahn, B. Real-time sign language recognition using a consumer depth camera. In Proceedings of the IEEE International Conference on Computer Vision, Sydney, Australia, 2–8 December 2013; pp. 83–90. [CrossRef]
19. Pugeault, N.; Bowden, R. Spelling It Out: Real – Time ASL Fingerspelling Recognition University of Surrey. In Proceedings of the 2011 IEEE International Conference on THE Hand: Computer Vision Workshops (ICCV Workshops), Barcelona, Spain, 6–13 November 2011; pp. 1114–1119.
20. Dong, C.; Leu, M.C.; Yin, Z. American Sign Language alphabet recognition using Microsoft Kinect. In Proceedings of the IEEE Computer Society Conference on Computer Vision and Pattern Recognition Workshops, Boston, MA, USA, 7–12 June 2015; pp. 44–52. [CrossRef]
21. Ren, Z.; Yuan, J.; Meng, J.; Zhang, Z. Robust part-based hand gesture recognition using kinect sensor. *IEEE Trans. Multimed.* **2013**, *15*, 1110–1120. [CrossRef]
22. Oreifej, O.; Liu, Z. HON4D: Histogram of oriented 4D normals for activity recognition from depth sequences. In Proceedings of the IEEE Computer Society Conference on Computer Vision and Pattern Recognition, Portland, OR, USA, 23–28 June 2013; pp. 716–723. [CrossRef]
23. Monnier, C.; German, S.; Ost, A. A multi-scale boosted detector for efficient and robust gesture recognition. In *Lecture Notes in Computer Science (Including Subseries Lecture Notes in Artificial Intelligence and Lecture Notes in Bioinformatics)*; Springer: Berlin, Germany, 2015; pp. 491–502. [CrossRef]
24. Krizhevsky, A.; Sutskever, I.; Hinton, G.E. ImageNet classification with deep convolutional neural networks. *Commun. ACM* **2017**, *60*, 84–90. [CrossRef]
25. Ren, S.; He, K.; Girshick, R.; Sun, J. Faster R-CNN: Towards Real-Time Object Detection with Region Proposal Networks. *IEEE Trans. Pattern Anal. Mach. Intell.* **2017**, *39*, 1137–1149. [CrossRef]
26. Badrinarayanan, V.; Kendall, A.; Cipolla, R. SegNet: A Deep Convolutional Encoder-Decoder Architecture for Image Segmentation. *IEEE Trans. Pattern Anal. Mach. Intell.* **2017**, *39*, 2481–2495. [CrossRef]
27. Asadi-Aghbolaghi, M.; Clapes, A.; Bellantonio, M.; Escalante, H.J.; Ponce-Lopez, V.; Baro, X.; Guyon, I.; Kasaei, S.; Escalera, S. A Survey on Deep Learning Based Approaches for Action and Gesture Recognition in Image Sequences. In Proceedings of the 2017 12th IEEE International Conference on Automatic Face & Gesture Recognition (FG 2017), Washington, DC, USA, 30 May–3 June 2017; pp. 476–483. [CrossRef]
28. Sun, L.; Jia, K.; Yeung, D.Y.; Shi, B.E. Human action recognition using factorized spatio-temporal convolutional networks. In Proceedings of the IEEE International Conference on Computer Vision, Santiago, Chile, 7–13 December 2015. [CrossRef]
29. Varol, G.; Laptev, I.; Schmid, C. Long-Term Temporal Convolutions for Action Recognition. *IEEE Trans. Pattern Anal. Mach. Intell.* **2018**, *40*, 1510–1517. [CrossRef]
30. Neverova, N.; Wolf, C.; Taylor, G.W.; Nebout, F. ModDrop: Adaptive multi-modal gesture recognition. *IEEE Trans. Pattern Anal. Mach. Intell.* **2014**, *38*, 1692–1706. [CrossRef]
31. Garcia-Hernando, G.; Yuan, S.; Baek, S.; Kim, T.K. First-Person Hand Action Benchmark with RGB-D Videos and 3D Hand Pose Annotations. In Proceedings of the IEEE Computer Society Conference on Computer Vision and Pattern Recognition, Honolulu, HI, USA, 21–26 July 2017; pp. 409–419. [CrossRef]
32. Chen, X.; Guo, H.; Wang, G.; Zhang, L. Motion feature augmented recurrent neural network for skeleton-based dynamic hand gesture recognition. In Proceedings of the 2017 IEEE International Conference on Image Processing (ICIP), Beijing, China, 17–20 September 2017; pp. 2881–2885. [CrossRef]
33. De Smedt, Q. Dynamic Hand Gesture Recognition—From Traditional Handcrafted to Recent Deep Learning Approaches. Ph.D. Thesis, Université de Lille 1, Sciences et Technologies, Lille, France, 2017.
34. De Smedt, Q.; Wannous, H.; Vandeborre, J.P. Heterogeneous hand gesture recognition using 3D dynamic skeletal data. *Comput. Vis. Image Underst.* **2019**, *181*, 60–72. [CrossRef]
35. Ge, L.; Cai, Y.; Weng, J.; Yuan, J. Hand PointNet: 3D Hand Pose Estimation Using Point Sets. In Proceedings of the 2018 IEEE/CVF Conference on Computer Vision and Pattern Recognition, Salt Lake City, UT, USA, 18–23 June2018; pp. 8417–8426. [CrossRef]
36. Moon, G.; Chang, J.Y.; Lee, K.M. V2V-PoseNet: Voxel-to-Voxel Prediction Network for Accurate 3D Hand and Human Pose Estimation from a Single Depth Map. In Proceedings of the 2018 IEEE/CVF Conference on Computer Vision and Pattern Recognition, Salt Lake City, UT, USA, 18–23 June 2018; Volume 2. [CrossRef]

37. Cherabier, I.; Hane, C.; Oswald, M.R.; Pollefeys, M. PointNet: Deep Learning on Point Sets for 3D Classification and Segmentation. In Proceedings of the 2016 4th International Conference on 3D Vision, Stanford, CA, USA, 25–28 October 2016; pp. 601–610. [CrossRef]
38. Arun, K.S.; Huang, T.S.; Blostein, S.D. Least-Squares Fitting of Two 3-D Point Sets. *IEEE Trans. Pattern Anal. Mach. Intell.* **1987**, 698–700. [CrossRef]
39. Ronneberger, O.; Fischer, P.; Brox, T. U-Net: Convolutional Networks for Biomedical Image Segmentation. In *Medical Image Computing and Computer-Assisted Intervention—MICCAI*; Springer: Cham, Switzerland, 2015; pp. 1–8. [CrossRef]
40. Simonyan, K.; Zisserman, A. Very deep convolutional networks for large-scale image recognition. In Proceedings of the 3rd International Conference on Learning Representations (ICLR 2015), San Diego, CA, USA, 7–9 May 2015; pp. 1–14.
41. He, K.; Zhang, X.; Ren, S.; Sun, J. Deep Residual Learning for Image Recognition. In Proceedings of the IEEE Conference on Computer Vision and Pattern Recognition (CVPR), Las Vegas, NV, USA, 27–30 June 2016; pp. 770–778. [CrossRef]
42. Sharp, T.; Keskin, C.; Robertson, D.; Taylor, J.; Shotton, J.; Kim, D.; Rhemann, C.; Leichter, I.; Vinnikov, A.; Wei, Y.; et al. Accurate, robust, and flexible realtime hand tracking. In Proceedings of the Conference on Human Factors in Computing Systems, Seoul, Korea, 18–23 April 2015; pp. 3633–3642. [CrossRef]
43. Sudre, C.H.; Li, W.; Vercauteren, T.; Ourselin, S.; Jorge Cardoso, M. Generalised Dice Overlap as a Deep Learning Loss Function for Highly Unbalanced Segmentations. In *Lecture Notes in Computer Science (Including Subseries Lecture Notes in Artificial Intelligence and Lecture Notes in Bioinformatics)*; Springer: Berlin, Germany, 2017; pp. 240–248. [CrossRef]
44. Hochreiter, S.; Schmidhuber, J. Long Short-Term Memory. *Neural Comput.* **1997**, *9*, 1735–1780. [CrossRef]
45. Kingma, D.P.; Ba, J. Adam: A method for stochastic optimization. In Proceedings of the International Conference on Learning Representations (ICLR), San Diego, CA, USA, 7–9 May 2015.
46. Schaul, T.; Zhang, S.; LeCun, Y. No more pesky learning rates. In Proceedings of the 30th International Conference on Machine Learning (ICML 2013), Atlanta, GA, USA, 16–21 June 2013.
47. Sandler, M.; Howard, A.; Zhu, M.; Zhmoginov, A.; Chen, L.C. MobileNetV2: Inverted Residuals and Linear Bottlenecks. In Proceedings of the 2018 IEEE/CVF Conference on Computer Vision and Pattern Recognition, Salt Lake City, UT, USA, 18–23 June 2018; pp. 4510–4520. [CrossRef]
48. Hu, J.; Shen, L.; Albanie, S.; Sun, G.; Wu, E. Squeeze-and-Excitation Networks. *IEEE Trans. Pattern Anal. Mach. Intell.* **2020**. [CrossRef]
49. Devanne, M.; Wannous, H.; Berretti, S.; Pala, P.; Daoudi, M.; Del Bimbo, A. 3-D Human Action Recognition by Shape Analysis of Motion Trajectories on Riemannian Manifold. *IEEE Trans. Cybern.* **2015**, *45*, 1340–1352. [CrossRef]
50. Evangelidis, G.; Singh, G.; Horaud, R. Skeletal Quads: Human Action Recognition Using Joint Quadruples. In Proceedings of the 2014 22nd International Conference on Pattern Recognition, Stockholm, Sweden, 24–28 August 2014; pp. 4513–4518. [CrossRef]
51. Xu, Y.; Wang, Q.; Bai, X.; Chen, Y.L.; Wu, X. A novel feature extracting method for dynamic gesture recognition based on support vector machine. In Proceedings of the 2014 IEEE International Conference on Information and Automation (ICIA), Hailar, China, 28–30 July 2014; pp. 437–441. [CrossRef]
52. De Smedt, Q.; Wannous, H.; Vandeborre, J.P.P.; Guerry, J.; Le Saux, B.; Filliat, D.; Saux, B.L.; Filliat, D. 3d hand gesture recognition using a depth and skeletal dataset: Shrec'17 track. In Proceedings of the Workshop on 3D Object Retrieval, Lyon, France, 23–24 April 2017; pp. 33–38. [CrossRef]
53. Núñez, J.C.; Cabido, R.; Pantrigo, J.J.; Montemayor, A.S.; Vélez, J.F. Convolutional Neural Networks and Long Short-Term Memory for skeleton-based human activity and hand gesture recognition. *Pattern Recognit.* **2018**, *76*, 80–94. [CrossRef]
54. Hou, J.; Wang, G.; Chen, X.; Xue, J.H.; Zhu, R.; Yang, H. Spatial-temporal attention res-TCN for skeleton-based dynamic hand gesture recognition. In *Lecture Notes in Computer Science (Including Subseries Lecture Notes in Artificial Intelligence and Lecture Notes in Bioinformatics)*; Springer: Berlin, Germany, 2019; pp. 273–286. [CrossRef]
55. Devineau, G.; Moutarde, F.; Xi, W.; Yang, J. Deep learning for hand gesture recognition on skeletal data. In Proceedings of the 13th IEEE International Conference on Automatic Face and Gesture Recognition, Xi'an, China, 15–19 May 2018; pp. 106–113. [CrossRef]

56. Ma, C.; Wang, A.; Chen, G.; Xu, C. Hand joints-based gesture recognition for noisy dataset using nested interval unscented Kalman filter with LSTM network. *Vis. Comput.* **2018**, *34*, 1053–1063. [CrossRef]
57. Li, Y.; He, Z.; Ye, X.; He, Z.; Han, K. Spatial temporal graph convolutional networks for skeleton-based dynamic hand gesture recognition. *EURASIP J. Image Video Process.* **2019**, 78. [CrossRef]
58. Chen, Y.; Zhao, L.; Peng, X.; Yuan, J.; Metaxas, D.N. Construct Dynamic Graphs for Hand Gesture Recognition via Spatial-Temporal Attention. In Proceedings of the 30th British Machine Vision Conference 2019, Cardiff, UK, 9–12 September 2019; pp. 1–13.
59. Ma, C.; Zhang, S.; Wang, A.; Qi, Y.; Chen, G. Skeleton-based dynamic hand gesture recognition using an enhanced network with one-shot learning. *Appl. Sci.* **2020**, *10*, 3680. [CrossRef]

applied sciences

MDPI

Article

Recognition of Hand Gesture Sequences by Accelerometers and Gyroscopes

Yen-Cheng Chu [1,†], Yun-Jie Jhang [2,†], Tsung-Ming Tai [3,†] and Wen-Jyi Hwang [1,*,†]

1 Department of Computer Science and Information Engineering, National Taiwan Normal University, Taipei 117, Taiwan; 60647009s@ntnu.edu.tw
2 Andro Video, Taipei 115, Taiwan; taco.chang@androvideo.com
3 NVIDIA AI Technology Center, NVIDIA Taiwan, Taipei 114, Taiwan; ntai@nvidia.com
* Correspondence: whwang@ntnu.edu.tw
† These authors contributed equally to this work.

Received: 25 August 2020; Accepted: 16 September 2020; Published: 18 September 2020

Abstract: The objective of this study is to present novel neural network (NN) algorithms and systems for sensor-based hand gesture recognition. The algorithms are able to classify accurately a sequence of hand gestures from the sensory data produced by accelerometers and gyroscopes. They are the extensions from the PairNet, which is a Convolutional Neural Network (CNN) capable of carrying out simple pairing operations with low computational complexities. Three different types of feedforward NNs, termed Residual PairNet, PairNet with Inception, and Residual PairNet with Inception are proposed for the extension. They are the PairNet operating in conjunction with short-cut connections and/or inception modules for achieving high classification accuracy and low computation complexity. A prototype system based on smart phones for remote control of home appliances has been implemented for the performance evaluation. Experimental results reveal that the PairNet has superior classification accuracy over its basic CNN and Recurrent NN (RNN) counterparts. Furthermore, the Residual PairNet, PairNet with Inception, and Residual PairNet with Inception are able to further improve classification hit rate and/or reduce recognition time for hand gesture recognition.

Keywords: hand gesture recognition; human–machine interface; artificial intelligence; feedforward neural networks

1. Introduction

Hand gesture recognition is a technique for the mathematical interpretation of hand movements by computers. It can be used to facilitate the interaction between human and computer. Gesture recognition has been found to be effective for applications such as devices control, entertainment, health care, and education. Hand gesture recognition approaches can be separated into two classes: Vision-Based Recognition (VBR) algorithms [1–4] and Sensor-Based Recognition (SBR) algorithms [5–16]. The VBR algorithms perform gesture recognition from images captured by a camera. Although accurate classification is possible, high computation efforts may be required to extract information from images for both training and inference operations.

The SBR algorithms are based on sensors other than camera. Commonly used sensors include accelerometers [5,6], gyroscopes [7,8], photoplethysmography (PPG) [9], flex sensors [10], electromyography (EMG) [11], Radio Frequency IDentification (RFID) [12,13], WiFi Channel State Information (CSI) [14–16], and the fusion of these sensors. In some SBR-based studies, high classification accuracy for gesture recognition has been observed. However, many of these techniques do not support the recognition of a sequence of gestures. Only isolated gestures can be recognized. Furthermore, some SBR techniques are based on Dynamic Time Warping (DTW) [7] technique for gesture recognition, which may incur high computation complexities. The Recurrent Neural Networks (RNNs) [17] and its variants such as Long Short Term Memory (LSTM) algorithm [18] have been found to be effective [8,19] for the recognition of gesture sequences with low computation costs. Nevertheless, because of the inherent gradient vanishing problem, the algorithms may not be able to effectively exploit the long term dependency of the sensory data. The dependency may be beneficial for accurate recognition when slow gesture movements are observed.

In addition to RNNs, feedforward networks such as Convolutional Neural Networks (CNNs) [17,20] can be used for hand gesture recognition. The One-Dimensional (1D) CNNs have been found to be effective for a number of applications such as the classifications of ECG signals [21], human activities [22], and internet traffic [23]. Gesture recognition based on basic 1D CNNs may achieve high classification accuracy when the kernel sizes and/or the depth of the network are large. However, in these cases, the computation complexities for inference may also be high. Although the computation load can be alleviated by lowering the kernel sizes and/or the depth of the network, the coverage area of receptive field will then become small. Consequently, the classification error may become large because the long term dependency may not be effectively exploited with a small receptive field. The WaveNet [24,25] can be used to solve the problem. It is based on dilated convolution operations for the growth of receptive field with low computational complexities. However, the WaveNet is used for signal generation with additional autoregression operations. Direct applications of WaveNet to hand gesture recognition may then be difficult.

The objective of this study is to present novel SBR algorithms and systems based on feedforward neural networks for effective recognition of a sequence of hand gestures. The proposed algorithms have the advantages of high classification accuracy and low computation complexities. They are based on basic accelerometers and gyroscopes commonly used in smart phones or devices. This may facilitate the deployment of the algorithms to large varieties of the applications in smart devices.

The proposed feedforward algorithms are based on the PairNet [26] featuring large receptive field and simple inference process. The PairNet is implemented by configuring the CNN with a kernel size of two and a stride size of two. The corresponding CNN operations can be regarded as the simple pairing operations, where the input sequences to each convolution layer can be separated into non-overlapping pairs. The operations for each pair are determined by the weights associated with the kernels. Because of the simplicity of the operations, a deep CNN based on PairNet algorithm can then be easily formed for high classification accuracy with low computation time as compared with the basic CNN.

Although the PairNet has a simple structure, it may have large depth for full coverage of a single gesture. However, the classification accuracy may be saturated as the depth increases. One approach to further improve the performance is the employment of short-cut connections [27–29] for some layers. This may be beneficial for providing a good reference for effective training and inference at these layers [27]. The resulting network, termed Residual PairNet, has superior classification accuracy over the PairNet and basic CNN. Similar to the PairNet, the Residual PairNet has low computational complexities as compared with basic CNN.

The performance of the feedforward algorithms can be further improved. This is based on the observation that sensory data produced by accelerator and gyroscope may possess both slow and rapid variations dependent on the hand movements. Consequently, the employment of inception modules [20,30–32] consisting of different sizes of kernels at the same layer may be beneficial for efficient capturing of gesture features. Furthermore, while having superior classification accuracy, the PairNet with inception is able to lower the computation costs of the PairNet. This is because dimensionality reduction can be carried out in the inception modules for lowering the number of weights [30]. Similarly, the residual PairNet can also operate in conjunction with inception. The employment of short-cut connections and/or inception modules to PairNet provides flexibilities to the feedforward networks for varying requirements on classification accuracy, computation costs, and design complexities.

To demonstrate the effectiveness of the proposed feedforward algorithms, a smart home system was built, where the home appliances can be controlled remotely by hand gestures. The sequences of hand gestures were acquired by smart phones, and were delivered to the embedded systems in the home appliances for gesture recognition. The trained neural network models were deployed in the embedded systems so that sequences of hand gestures can be recognized in real-time. Experimental results reveal that the proposed algorithms are able to carry out real-time gesture recognition on embedded devices with only limited computation capacity. Furthermore, they have superior classification accuracy over existing works. They are therefore well-suited for the implementation of light-weight smart human–computer interfaces where the only sensors used are for hand gesture recognition.

The remaining parts of this paper are organized as follows. Section 2 reviews some advanced works for gesture recognition. Section 3 presents the proposed gesture recognition systems. The PairNet algorithm is studied in detail. The augmentation of short-cut connections and inception modules to the PairNet are also discussed. Experimental results of the proposed feedforward networks are then presented in Section 4. Finally, Section 5 includes some concluding remarks of this work.

2. Related Works

VBR approaches can be separated into two classes: data-driven and model-based. The data-driven methods [1–4] are based on the appearance of gesture images. The gestures are classified by relating the appearance of any gesture to the appearance of pre-defined gestures. Basic model-based methods [33] can be adopted for 3D hand tracking/recognition and 3D gesture estimation. Hand model fitting operations are usually required for the tracking of hand motion. As cameras are required for VBR approaches, computational complexities for these approaches may be high.

Many existing SBR methods are data-driven. A common feature of some SBR-based techniques is that they are based on wearable sensors such as PPG sensors [9], flex sensors [10], and EMG sensors [11] for attaining high recognition accuracy. However, the systems with wearable sensors demand users to take extra devices for gesture recognition. These techniques may then be intrusive, because the devices sometimes are obstructive and inconvenient for gesture actions. This could undermine the quality of user experience.

An alternative to wearable sensors is based on RFIDs for the location-sensing of hand gestures. The techniques in [12] carry out the tracking of motion patterns of RFID tags for gesture recognition. Although high accuracy can be achieved, users are still required to wear tags. To alleviate user intrusiveness, the RFID tags and readers were deployed in a fixed location in the study [13]. The recognition is then based on the fact that when gestures are performed in front of tags, the movement information of the gestures can be extracted by the resulting phases of the RFID signals. Although there is no demand for wearing sensors, gesture recognition can only take place in locations where RFID tags and readers are properly deployed.

Another approach for gesture recognition with no user intrusiveness is based on the CSI information from commercial WiFi. It is known that fluctuation of WiFi signals can be observed by human actions. Variation statics such as variance and correlation coefficients [14] have been found to be beneficial for the detection of human activities. However, the detection performance may be limited by the random noises and/or WiFi channel variations. Furthermore, it may be difficult to carry out accurate gesture classifications from the simple statics. Approaches based on deep learning in [15,16] are proposed to solve the problem. By the incorporation of RNN and its variants such as LSTM, the features of gestures can be effectively extracted, and accurate recognition can be achieved at the presence of noises. However, similar to their RFID counterparts [12,13], the performance of WiFi CSI-based techniques may be dependent on the deployments of WiFi transmitters and receivers in an indoor environment.

As compared with existing SBR-based algorithms, the proposed algorithms have a number of advantages. The first is that they are not user intrusive. Sensors commonly available in smart phones such as accelerometers and gyroscopes are adopted for gesture recognition. Therefore, only a single smart phone is needed for gesture recognition without additional wearable sensors and/or devices. Furthermore, the proposed algorithms provide higher flexibilities for the deployment of gesture recognition systems as compared with the existing RFID and WiFi techniques. The acquisition of sensory data is performed in the smart phone locally from its accelerometers and gyroscopes. The acquired sensory data can be delivered to any external devices accessible by Internet for gesture inference. Therefore, in the proposed algorithms, gesture recognition can take place in either indoor or outdoor environments, wherever Internet access is available.

3. The Proposed Gesture Recognition Algorithms and Systems

In this section, we first give an overview of the proposed gesture algorithms and systems. The four feedforward neural networks (i.e., PairNet, Residual PairNet, PairNet with Inception, and Residual PairNet with Inception) are presented separately in the subsequent subsections. We then consider the issues for the training data collection for the algorithms, which is followed by the applications of the algorithms such as the remote control systems for home appliances. To facilitate the understanding of the discussions in this study, Table 1 includes a list of frequently used symbols.

Table 1. A list of frequently used symbols in this study.

Symbol	Meaning
A	The path of the sensory data. Each individual index along the path is the index of the gesture having highest probability at its corresponding time step.
a_t	The index of the gesture having the largest probability at time step t. It is the t-th component of $A, t = 1, ..., T$.
K	Number of different gestures in the input sensory data sequence X.
M	Number of elements in each input sample $x_t, t = 1, ..., T$.
N	Length of input window X_t.
Q	Number of gesture classes.
R	Classification results.
r_k	The k-th element of classification results $R, k = 1, ..., K$.
T	Length of the input sequence X and output sequence Y.
X	Input sensory data sequence.
X_t	Input window with central component x_t.
x_t	The sample at time step t of input sensory data sequence $X, t = 1, ..., T$.
Y	Gesture spotting results.
y_t	The sample at time step t of gesture spotting results $Y, t = 1, ..., T$.
$y_{t,j}$	The j-th element of $y_t, j = 1, ..., Q, t = 1, ..., T$.

3.1. Overview

As shown in Figure 1, the proposed gesture recognition algorithms can be separated into two parts: feedforward neural networks and Maximum A Posteriori (MAP) estimation. The feedforward networks take sensory data as input. Examples of sensory data include the data produced by accelerometers and/or gyroscopes. Based on the input data, the feedforward neural networks carry out the gesture spotting operations. Given the spotting results, the MAP estimation is then performed to obtain the final classification outcomes.

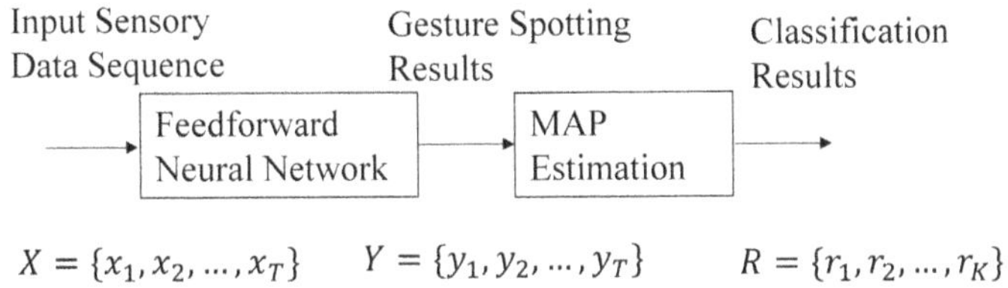

Figure 1. The overview of the proposed gesture recognition system, where X, Y, and R denote the input sensory data, gesture spotting results, and classification results, respectively.

Let $X = \{x_1, \ldots, x_T\}$ be an input sensory data sequence for the recognition of a hand gesture sequence containing K different gestures back-to-back. Each of the K gestures belongs to one of the pre-defined Q gesture classes. Each x_t, $t = 1, 2, \ldots, T$, is the sample of the sequence X acquired at time step t, and T is the length of of the sequence. There are M elements in each sample x_t of the data sequence X. The dimension M is dependent on the sensors for the gesture recognition. For example, when both 3-axis accelerometer and 3-axis gyroscope are used, each sample x_t contains six elements, and therefore $M = 6$. However, when only one of the sensors is used, $M = 3$. Let $Y = \{y_1, \ldots, y_T\}$ be the results produced by a feedforward neural network, where y_t is the output of the network at time step t. There are Q elements in each output y_t. Let $y_{t,j}$ be the j-th element of y_t, $t = 1, \ldots, Q$. The activation function associated with y_t is the softmax activation function. We then can view $y_{t,j}$, $j = 1, \ldots, Q$, as the probability of the occurrence of gesture class j at the time step t.

Figure 2 shows the operations of a feedforward network on X for producing Y. Starting from $t = 1$, the feedforward network produces y_t based on X_t for each t until $t = T$ is reached, where X_t is an input window with length N. The central component of X_t is x_t. This allows for the full exploitation of correlation among neighboring samples of x_t at expense of higher latency for acquiring the input window X_t. Although causal operations are possible, only the correlation in the past neighboring samples of x_t is exploited. Because the full utilization of correlation among neighboring samples is important for accurate gesture spotting, causal operations are not considered. Note that, when $t < N/2$ or $t > T - N/2$, parts of X_t is outside X. These parts are filled with zeros.

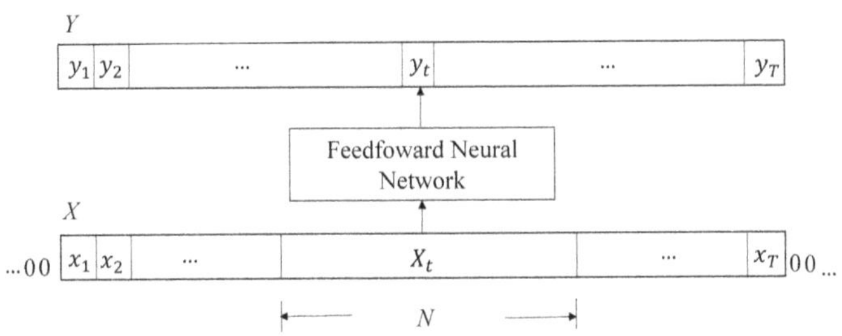

Figure 2. The feedforward operations on X for producing Y.

After $Y = \{y_1, \ldots, y_T\}$ is available, post-processing operations are required to produce the final classification results. With the a priori knowledge of the number of gestures K contained in the input sensory data $X = \{x_1, \ldots, x_T\}$, the goal of the post-processing operations is to find the set of indices $R = \{r_1, \ldots, r_K\}$ of the gestures, where r_k is the index of the k-th gesture appears in the sensory data $X = \{x_1, \ldots, x_T\}$. The selection of K is dependent on the number of different gesture sequences required for an application. Under the restriction of no occurrence of repetitive gestures, there are at most $Q \times (Q-1) \times \ldots \times (Q-(K-1))$ different gesture sequences.

Let a_t be the index of the gesture having the largest probability at time step t. That is,

$$a_t = \underset{1 \le j \le Q}{\operatorname{argmax}}\, y_{t,j}. \quad (1)$$

We call $A = \{a_1, ..., a_T\}$ the path of the sensory data, where each individual index along the path is the index of the gesture having highest probability at its corresponding time step. The computation of final classification outcome R is based on the probability model

$$Prob(C/A) = \prod_{k=1}^{K} Prob(c_k/A), \quad (2)$$

where $C = \{c_1, ..., c_K\}$ is a possible classification outcome, and c_k is the index of the k-th gesture of C. The probability $Prob(c_k/A)$ is estimated by

$$Prob(c_k/A) = \frac{|I_{c_k}|}{T}, \quad (3)$$

where

$$I_{c_k} = \{a_t : a_t = c_k\}, \quad (4)$$

and the size of I_{c_k} is denoted by $|I_{c_k}|$. The final classification result R is then the C maximizing $Prob(C/A)$. That is,

$$R = \underset{C \in S}{\operatorname{argmax}}\, P(C/A), \quad (5)$$

where S denotes the set of all possible classification outcomes.

From (2) and (3), the search process in (5) is equivalent to the identification of gestures having top-K occurrence. The classification results $R = \{r_1, ..., r_K\}$ are then obtained from these gestures according to their locations in the path A. Figure 3 shows an example of mapping from the path A to the classification results R for the recognition of three gestures (i.e., $K = 3$). In this example, based on the results of the feedforward neural network, the operations in (1) produces index values $1, 3, 4, 6$, or 7. The locations of the corresponding intervals I_1, I_3, I_4, I_6, and I_7 are revealed in Figure 3. The top-three largest intervals are I_1, I_3, and I_6. The set of gestures having the top-three occurrences then contains Gesture 1, Gesture 3, and Gesture 6. Their order of occurrence along the path A is Gesture 3, Gesture 6, and Gesture 1. Consequently, $r_1 = 3$, $r_2 = 6$, and $r_3 = 1$.

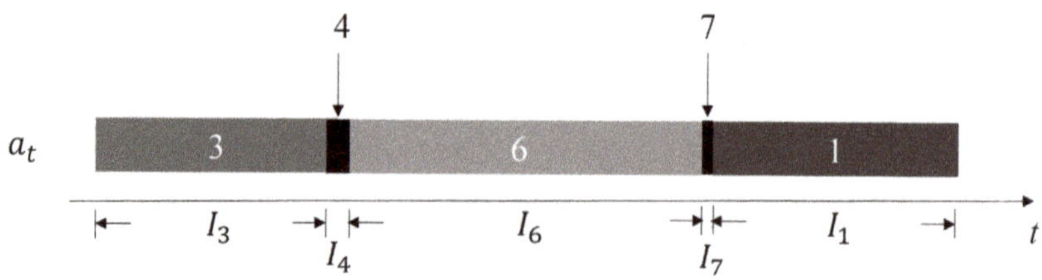

Figure 3. An example of mapping from path $A = \{a_1, ..., a_T\}$ to final classification outcome $R = \{r_1, r_2, r_3\}$ for the recognition of three gestures (i.e., $K = 3$). In this example, $r_1 = 3$, $r_2 = 6$, and $r_3 = 1$.

The feedforward neural network in the proposed system could be a basic CNN, or the PairNet, Residual PairNet, PairNet with Inception, or Residual PairNet with Inception. They are presented separately in the following subsections.

3.2. *PairNet*

As shown in Figure 4, the PairNet is a 7-layer CNN, where layers 1–5 are convolutional layers. Layer 1 is the convolutional layer with stride size 1 and kernel size 1×3. We can observe from Figure 4 that the input data to layer 1 is X_t with length $N = 50$. Recall that X_t is a window from the sensory data X, where each sample contains $M = 6$ elements. We therefore view X_t as a set of six 1D sequences, where the i-th 1D sequence is formed by the i-th element of samples of the sensory data in $X_t, i = 1, \ldots, 6$. In this study, each 1D sequence is termed a channel. Therefore, the number of input channels is six. Layer 1 produces a set of 128 1D sequences. The number of output channels for layer 1 is then 128. From Figure 4, we see that the length of each 1D sequence (i.e., channel) is 48.

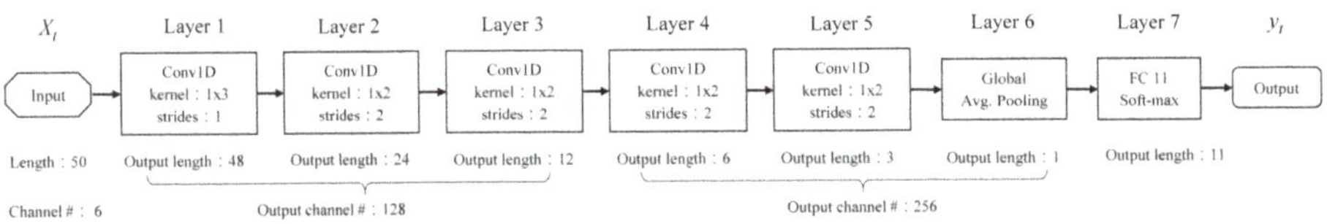

Figure 4. The parameters of the PairNet. The dimension of each sample of X_t is 6 ($M = 6$). Therefore, there are six channels. Each channel has a length of 50. Layer 1 is a convolution layer with kernel size 1×3 and stride size 1. The output length therefore is 48. Layers 2–5 are also convolution layers with kernel size 1×2 and stride size 2 for pairing operations. The output lengths of Layers 2–5 are then 24, 12, 6, and 3, respectively. Layer 6 is an average pooling layer averaging three elements to a single one. The number of gestures to be classified is $Q = 11$. Layer 7 is a fully connected layer producing 11 outputs.

The output sequences produced by layer 1 then serve as the input sequences to layer 2, which in turns propagates its computation results to subsequent layers. For layers 2–5, the stride size is 2, and kernel size is 1×2. The corresponding convolution operations can be viewed as a pairing operations, where each of the input 1D sequences are separated into disjoint pairs (due to a stride size of 2), and each pair operates independently in accordance with the kernel weights (due to kernel size 1×2). The pairing operations then produces output channels half the length of the input channels to that layer. Although the length of channels decrease as the data propagate through the network, we retain or increase the number of channels so that sufficient features can be extracted for final classification. In fact, there are 128 output channels at layers 2 and 3, and there are 256 output channels at layers 4 and 5.

When we only consider the length of channels at each layer, the structure of PairNet therefore is pyramid-like, as shown in Figure 5. The output channels produced by the layer 5 are located on the top of the pyramid. At the layer 6, the average pooling operations are then carried out over these output channels. The resulting data are subsequently flattened, and served as inputs to the fully connected layer with softmax activation function for the final classification results at layer 7.

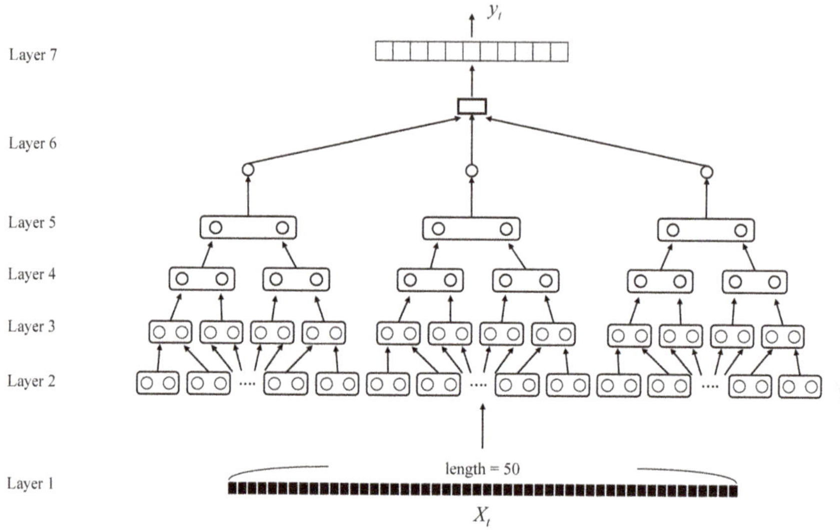

Figure 5. The structure of PairNet. In this example, the length of X_t is $N = 50$. There are seven layers in the network. Layer 1 is a basic convolution layer. Layers 2–5 are convolution layers supporting pairing operations. Layers 6 and 7 are pooling layer and fully-connected layers, respectively.

3.3. Residual PairNet

The performance of the PairNet can be improved by the employment of short-cut connections in the network. The resulting network is termed residual PairNet in this study. Figure 6 shows an example of the Residual PairNet accommodating short-cut connections. By comparing Figure 4 with Figure 6, we can see that the Residual PairNet has two short-cut connections. The first short-cut connection is from the output of Layer 1 to the output of Layer 3. The second one is from the input of Layer 4 to the output of Layer 5.

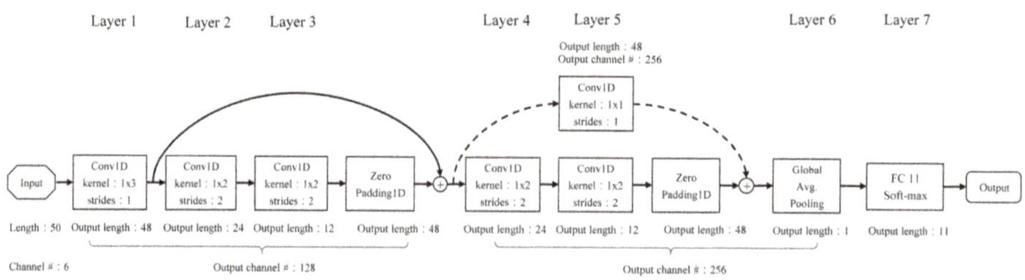

Figure 6. The parameters of the Residual PairNet. The input X_t has the same dimension and number of channels as those of the input of PairNet. The network contains seven layers and two short-cut connections. Two zero-padding modules are also included for the combination of the short-cut connections with the convolutional layers. Layers 1–5 are convolution layers. Layer 6 is the average pooling layer. The number of gestures to be classified is $Q = 11$. Layer 7 is a fully connected layer producing 11 outputs.

It can be observed from Figure 6 that the short-cut connections should be combined with the convolution layers for enhancing the performance of the network. To implement the combination, the length of the short-cut connections and the length of output of convolution layers should be identical. Because the pairing operations in the convolution layers reduce the length of their outputs, zero padding operations are included for the compensation of the length reduction in our design.

For example, the length of the first short-cut connection is 48. The length of the output of the Layer 3 is only 12. Therefore, a module is included to increase the length of the output of Layer 3 from 12 to 48 by padding zeros to the end of the output of Layer 3. In this way, the path can be combined with the first short-cut connection, as shown in Figure 6. Similarly, another zero padding module is employed at the output of Layer 5 to facilitate the combination of this path with the second short-cut connection.

In addition to the path length of the short-cut connection, we may need to take the number of channels into consideration for the combination of short-cut connection and convolution layers. This is particularly true for the second short-cut connection. It originates at the path with only 128 channels. Furthermore, it needs to be combined with the path containing 256 channels. This discrepancy can be solved by the employment of the convolution module with kernel size 1×1. There are 128 input channels and 256 output channels for the module. We can then see from Figure 6 that the number of channels at the output of the module is the same as that of the target path for combination. The integration of the short-cut connections to the PairNet can then be carried out.

3.4. PairNet with Inception

Another technique for the improvement of the PairNet technique is the incorporation of an inception module. As shown in Figure 7, the inception module is located at layer 5 of the network. The architecture of the inception module is revealed in Figure 8, which contains four paths. Each path is associated with convolution modules, average pooling module, and/or a zero-padding module. The convolution modules at different paths have different kernel sizes. In this way, features of hand gestures with slow or fast movements can be effectively captured by the modules.

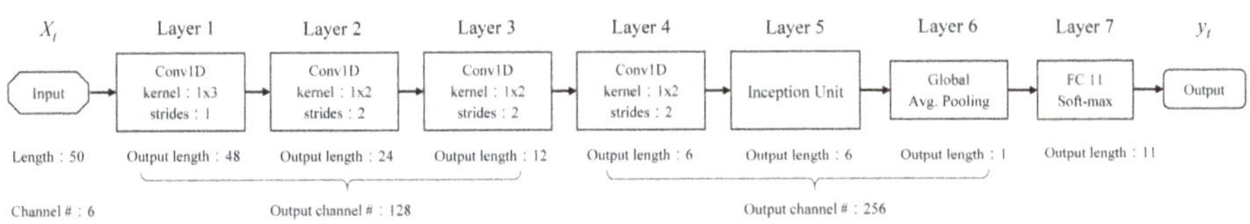

Figure 7. The parameters of the PairNet with Inception. The input X_t has the same dimension and number of channels as those of the input of PairNet. The network contains seven layers, where layers 1–4 are convolution layers. The inception module is located at layer 5. The architecture of inception module is revealed in Figure 8. Layer 6 is the average pooling layer. The number of gestures to be classified is $Q = 11$. Layer 7 is a fully connected layer producing 11 outputs.

An additional advantage of the inception module is that it has lower size of weights. Consequently, its computation complexity is lower than that of the other feedforward counterparts. The superior computation efficiency can be observed from Figure 8 that each convolution module in the inception module carries out the dimension reduction operations. That is, each convolution modules has lower number of output channels as compared with its number of input channels. In this way, the number of weights of the network may be effectively lowered. This may be beneficial for reducing the number of addition and multiplications.

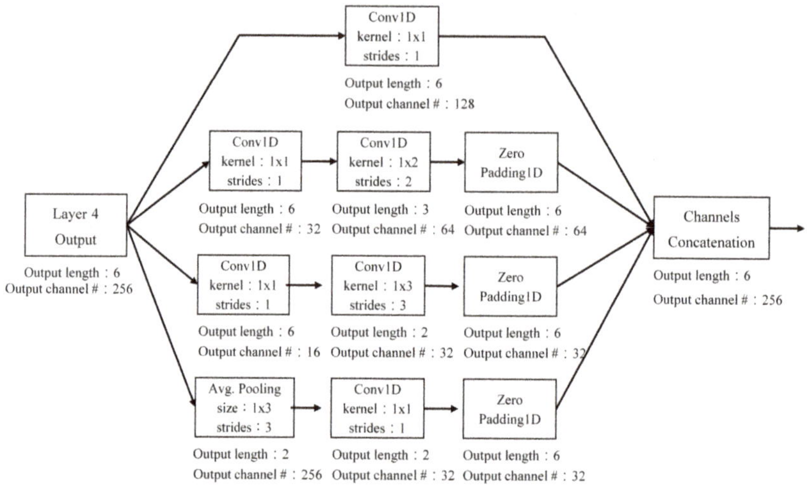

Figure 8. The parameters of the Inception module, which consists of four paths. Each path is associated with convolution modules, average pooling module, and/or a zero-padding module.

3.5. Residual PairNet with Inception

The employment of short-cut connections and inception module may have the advantages of both high classification accuracy and low computational complexities. An example of the design is shown in Figure 9, where both the inception module and short-cut connection are located at layer 5 of the network. Figure 10 shows the architecture of the inception module with short-cut. There are three paths in the module, where one of the paths serves as the short-cut. This allows layer 4 to be directly connected to layer 6. The remaining two paths contain convolution modules and a zero-padding module.

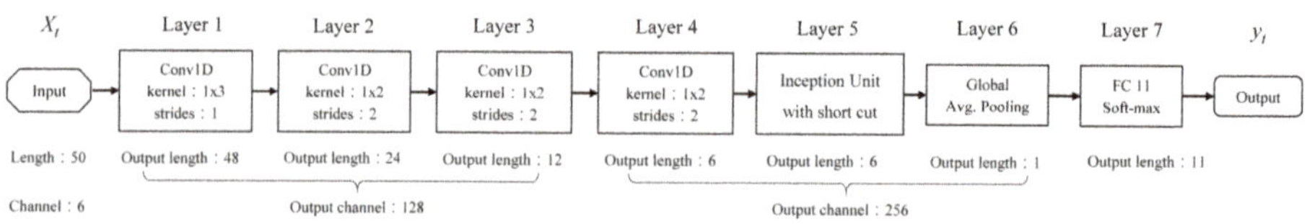

Figure 9. The parameters of Residual PairNet with inception. The input X_t has the same dimension and number of channels as those of the input of PairNet. The network contains seven layers, where the inception module with short-cut connection is located at layer 5. The architecture of inception module is revealed in Figure 10. The number of gestures to be classified is $Q = 11$. Layer 7 is a fully connected layer producing 11 outputs.

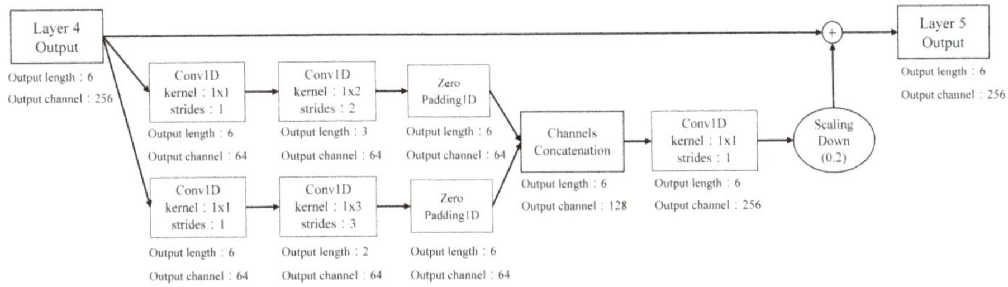

Figure 10. The parameters of the inception module with short-cut. It contains three paths. The top path is the short-cut for connecting the output of layer 4 to the input of layer 6. The remaining two paths are convolution modules and zero padding operations.

By comparing Figures 5, 7 and 9, it can be observed that the major difference among PairNet, Pairnet with Inception, and Residual PairNet with Inception is at layer 5. For the PairNet, only a single convolution module is employed at that layer. By contrast, inception module is adopted at layer 5 for Pairnet with Inception and Residual PairNet with Inception. Therefore, the PairNet may have inferior classification accuracy with larger size of weights as compared with the other two. Furthermore, because the Residual PairNet with inception contains short-cut connections, it may has highest classification accuracy among the three algorithms.

3.6. Gesture Data Collection for Training

Because this study aims to recognize a sequence of K gestures, a basic approach for the collection of training sequences for the proposed algorithms is to require each sensory data sequence for training should contain K different gestures. A drawback of this approach is that the proposed algorithms should be re-trained when the applications for other lengths of gestures K are desired. In addition, the collection of the training sets should also be carried out again when different number of gesture classes Q are adopted for the new applications. The reusability of the neural network models and/or the training data are low. Furthermore, large efforts are required for accurate labeling of K gestures for each sensory data sequence.

To improve the reusability of neural network models and training data for gesture sequences with different lengths of gestures K and/or different number of gesture classes Q, each sensory data sequence for training contains only a single gesture. Since there are Q gesture classes, the training set can be separated into Q subsets. The sensory data sequences belonging to the same subset represent the same gesture. All the training sequences in the Q subsets should be involved in the training of the proposed algorithms. The resulting neural network models can be applied to the recognition of gesture sequences with different K values, where K is known a priori. The same neural network models can then be re-used even though K varies. When the incorporation of new gesture classes or removal of existing gesture classes are desired for the new applications, the training subsets common to the new and original applications can be re-used. Consequently, the reusability for the neural network models and training sets can be improved. Moreover, because each training sensory sequence represents only a single hand gesture, the efforts for labeling can be minimized.

3.7. Gesture-Based Remote Control System for Home Appliances

The proposed algorithms can be effectively used for SBR applications. Figure 11 shows an example of the applications, where the home appliances such as a TV, air conditioner, and music player in a smart home can be remotely controlled by the sensory data of hand gestures acquired by mobile phones. It is assumed that the mobile phones are equipped with accelerators and gyroscope. We developed a mobile application (APP) for the smart phones for capturing the sensory data, and deliver the data via Wifi to external devices.

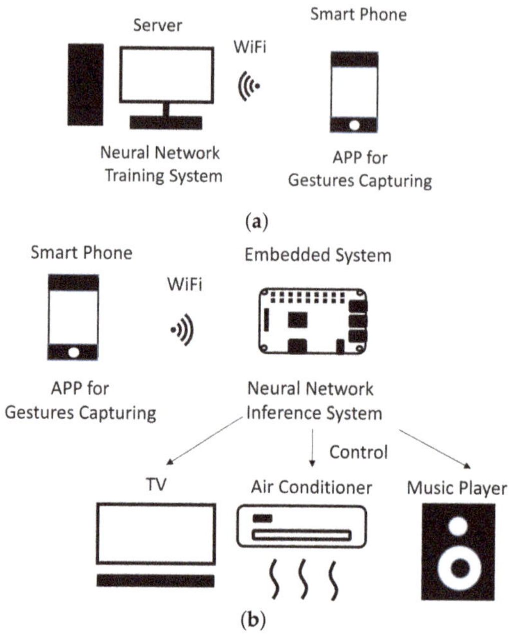

Figure 11. The proposed gesture-based remote control system for home appliances: (**a**) the training system for dynamic gesture recognition; (**b**) real-time gesture recognition system for the remote control of home appliances.

During the training phase, the sensory data acquired by smart phones serve as the training data. The data are delivered to a dedicated server for subsequent annotation and algorithm training. After the training operations are completed, the resulting neural network models are stored in an embedded system for online gesture recognition. It is preferable that the embedded system has a small size and low power consumption so that it can be easily attached to home appliances for the action control. A typical example would be a Raspberry Pi computer. During the inference phase, the sensory data captured by smart phones are sent to the embedded systems, which carry out the gesture recognition based on the trained neural models. The recognition results are then translated into action commands for home appliances.

Because the APP responsible for sampling sensory data is deployed in the smart phone, our system is able to carry out the recording of sensory data in parallel with human gesture actions. In fact, the APP is able to directly acquire the samples produced by accelerometers and gyroscopes of the smart phone, and deliver the data immediately over WiFi to the embedded system responsible for inference. After the human actions are completed, the embedded system has all the sensory data for classification. Inference time is then simply the computation time of neural networks producing the classification results. No additional waiting time for acquiring sensory data at sampling rate is required.

To implement the gesture-based remote control system in a smart home, a set of Q gesture classes needs to be pre-defined. Figure 12 shows an example of 11 gesture classes (i.e., $Q = 11$) for the implementation of the system. The gestures in each class actually involve entire arm motions, not only upper limb ones. Based on the set, Table 2 shows examples of the gesture sequences and their actions for various home appliances. For each gesture sequence, its order revealed in Table 2 should be followed. As shown in these examples, each sequence for actions contains two gestures (i.e., $K = 2$). In addition, the sequences used as Personal Identification Numbers (PINs) for home appliances authentication contain three or four gestures (i.e., $K = 3$ or $K = 4$), dependent on the home appliances.

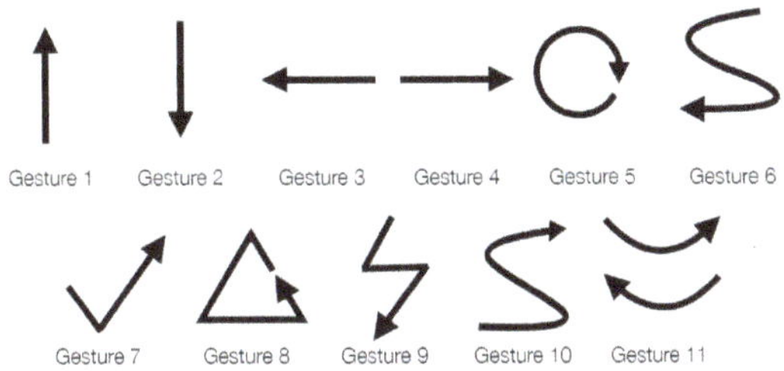

Figure 12. The eleven gesture classes considered in the smart home system in this study.

Table 2. Examples of the gesture sequences and their actions for various home appliances. Each sequence of actions contains two gestures. Each sequence has a personal identification number (PIN) for home appliances' authentication contains three or four gestures. The definition of gestures are shown in Figure 12.

	Gesture Sequences	Actions	Gesture Sequences	Actions
TV	Gest. 5+1	Volume Up	Gest. 5+2	Volume Down
	Gest. 5+3	Prev. Chan.	Gest. 5+4	Next Chan.
	Gest. 5+6	Power On/Off	Gest. 5+7	Record On/Off
PIN for TV	Gest. 11+ $i+j, i \neq j \neq 11$	Authentication		
Air	Gest. 8+1	Temp. Up	Gest. 8+2	Temp Down
Cond.	Gest. 8+3	Air Vol. Up	Gest. 8+4	Air Vol. Down
(AC)	Gest. 8+6	Power On/Off	Gest. 8+7	Func. Sel.
PIN for AC	Gest. 3+ $i+j+k, i \neq j \neq k \neq 3$	Authentication		
Music	Gest. 9+1	Volume Up	Gest. 9+2	Volume Down
Player	Gest. 9+3	Prev. Song	Gest. 9+4	Next Song
(MP)	Gest. 9+6	Power On/Off	Gest. 9+7	Source Sel.
PIN for MP	Gest. 11+ $i+j, i \neq j \neq 11$	Authentication		

4. Experimental Results

This section presents some experimental results for the proposed gesture recognition algorithms and systems. We implemented a gesture-based remote control system for home appliances, shown in Figure 11. The evaluation is then based on the system. There are eleven gesture classes (i.e., $Q = 11$), where each gesture class is defined in Figure 12. The gesture sequences to be classified are listed in Table 2. They have lengths of two, three, or four (i.e., $K = 2, 3$, or 4) dependent on the types of home appliances, actions, or PIN. To facilitate the evaluation, a JAVA-based APP was built on the smartphones for gesture capturing and delivery. We adopted a Samsung Galaxy S8 and HTC ONE M9 for the experiments.

The training of feedforward neural networks was carried out offline by Keras [34] with backend TensorFlow. The server for the training was a personal computer with an Intel I7 CPU and NVIDIA GTX 1070 GPU. Raspberry Pi 3 computers are adopted as the embedded systems attached to the home appliances. A python-based inference system is deployed in each embedded system. It is built from the neural network model acquired from Keras after training operations are completed. The inference system is capable of receiving the sensory data from smart phones for real-time dynamic gesture recognition.

All the gesture sequences for training and testing were captured by accelerometers and gyroscopes associated with the smart phones. The sensors were capable of measuring acceleration and angular velocity in three orthogonal axes, respectively. Therefore, the dimension of each sample x_t was $M = 6$. Figures 13 and 14 show the samples of waveforms produced by gyroscopes and accelerometers of the smart phones for each gesture class in Figure 12, respectively. The sampling rate was 50 samples/s. Although a large number of classes was considered in our experiments, it can be observed from Figures 13 and 14 that different gesture classes have different waveforms produced from gyroscopes and/or accelerometers. The employment of the sensors would then be beneficial for the accurate classification of the gestures. Furthermore, we can also see from Figures 13 and 14 that some of the waveforms exhibit small and fast vibrations. These may be due to unstable and/or shaking hands. It would then be essential for the proposed algorithm to accurately classify the gestures with the presence of the vibrations.

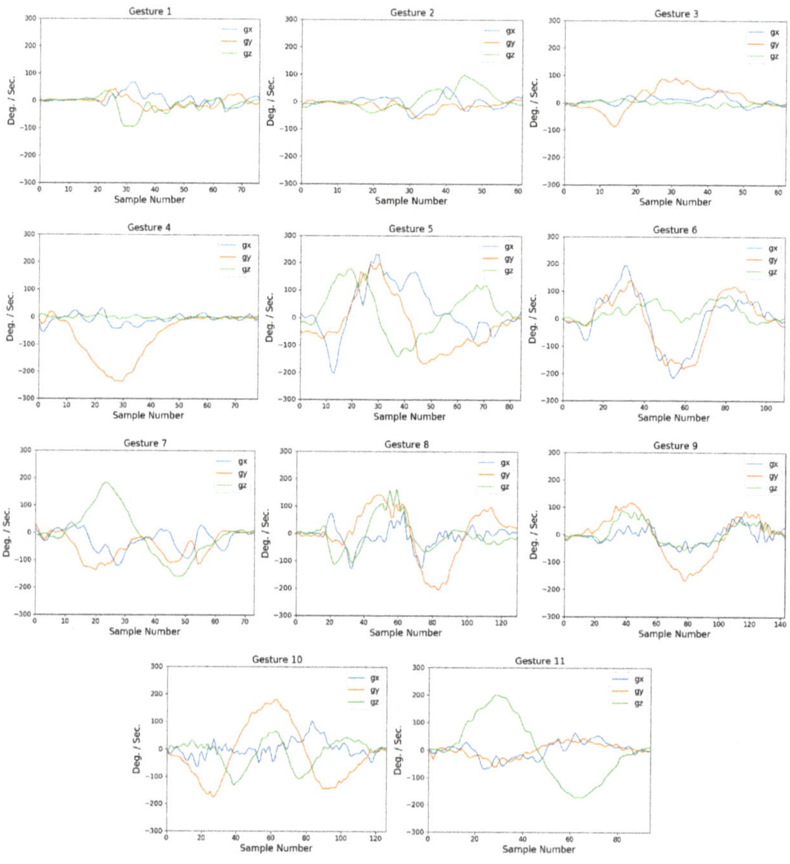

Figure 13. Samples of waveforms produced by gyroscopes in three axes (gx, gy, and gz) for each gesture.

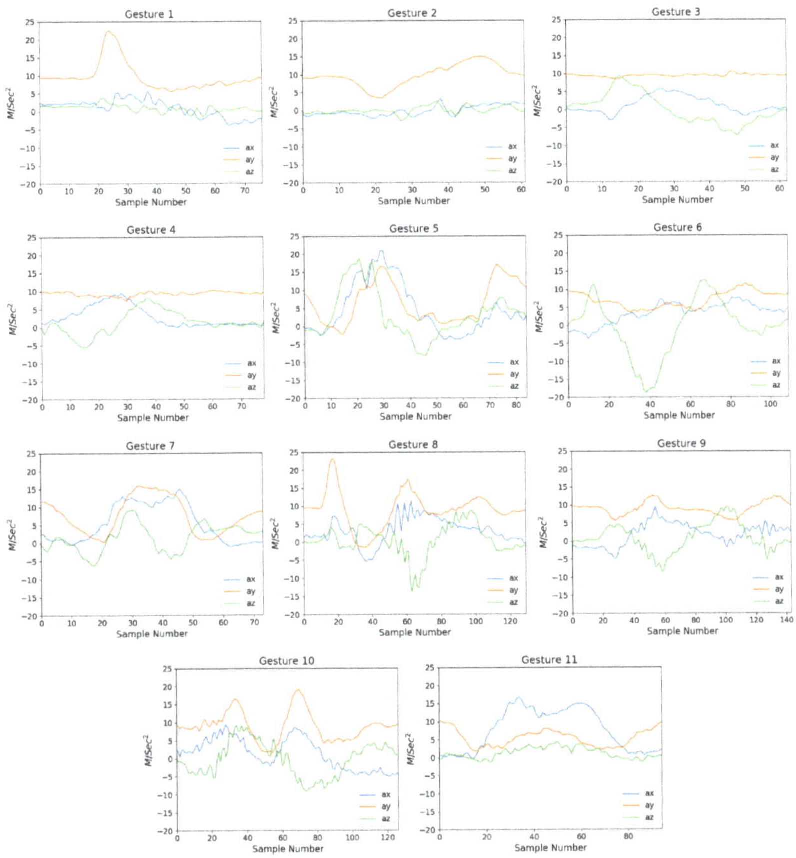

Figure 14. Samples of waveforms produced by accelerometers in three axes (gx, gy, and gz) for each gesture.

In the experiments, each training gesture sequence represents only a single gesture (i.e., $K = 1$). There were 100 training sequences for each gesture class. Because the number of gesture classes is 11 (i.e., $Q = 11$), the training set of the experiments consisted of 1100 training gesture sequences. They were acquired from two participants. The sequences were collected on different days for capturing different variations in gesturing, such as variations in lengths of gestures and/or amplitudes of samples. The testing set was different from the training set. It contained 3404 gestures from six participants. For some test sequences, vibrations due to shaking or unstable hands were included for testing the robustness of the proposed algorithm. The initial orientation of smart phones for data acquisition of both training and testing sequences was in the portrait orientation. Each test sequence may contain two, three, or four hand gestures (i.e., $K = 2$, 3, or 4). Therefore, although the proposed neural networks were trained by sequences with $K = 1$, they can be applied to sequences with larger K values. The high reusability of the proposed models for different K values is an advantage of the proposed algorithms.

Examples of a testing sequences captured by gyroscopes consisting of three (Gesture 11, Gesture 8, and Gesture 10) and four hand gestures (Gesture 3, Gesture 7, Gesture 5, and Gesture 11) are revealed in Figures 15 and 16, respectively. The results of gesture spotting by PairNet are annotated in the bottom of the figures. For the sake of brevity, the waveforms produced by the accelerometers are not included. Small and fast vibrations on some of the waveforms can be found in the figures because of unstable and/or shaking hands. Therefore, gesture spotting may be difficult even by direct visual inspection. Nevertheless, it can be observed from the bottom of Figure 15 that the size of sets I_{11}, I_8, and I_{10} are the largest as compared with other sets. The results are consistent with the ground truth. Furthermore, the sets I_3, I_7, I_5, and I_{11} have the largest size in Figure 16. Accurate classification is then possible based on the gesture spotting results provided by the PairNet.

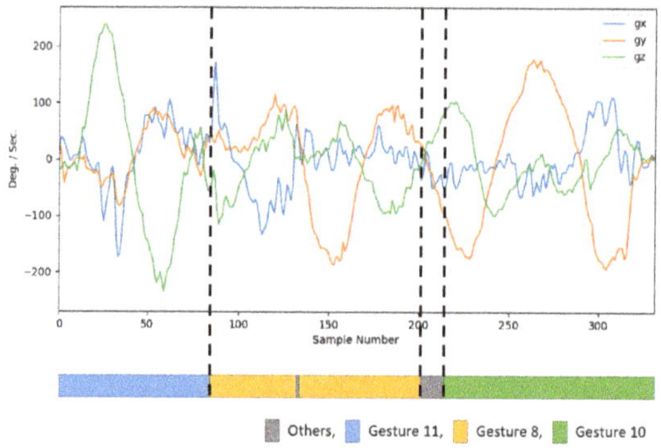

Figure 15. An example of test sequence produced by a gyroscope containing three gestures (Gesture 11, Gesture 8, and Gesture 10). The bottom of the figure reveals the gesture-spotting results by PairNet.

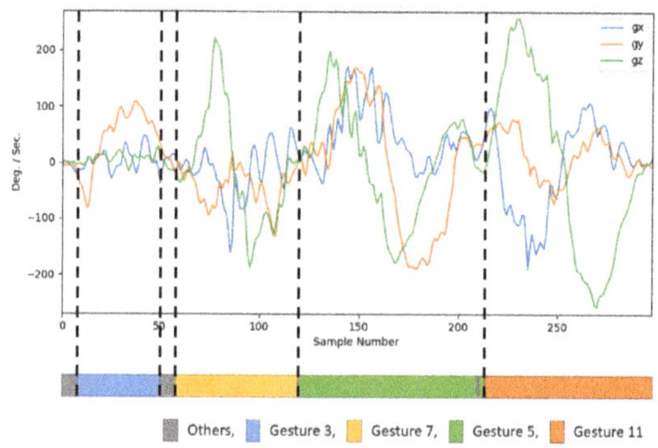

Figure 16. An example of test sequence produced by a gyroscope containing four gestures (Gesture 3, Gesture 7, Gesture 5, and Gesture 11). The bottom of the figure reveals the gesture-spotting results by PairNet.

To evaluate the performance of the proposed feedforward algorithms, we first consider the classification accuracy of each gesture class for the algorithms. Let H_i of an algorithm be the hit rate of gesture class i for the algorithm in the testing set. In the experiments, the hit rate H_i of an algorithm was equal to the number of gestures in class i, which are correctly classified by the algorithm divided by the total number of gestures in class i in the testing set. Tables 3 and 4 contain the model summary of the proposed algorithms and their classification accuracy, respectively. For comparison purposes, the 1D CNN [22], LSTM [8], and Bidirectional LSTM (Bi-LSTM) [19] algorithms are also considered. For each algorithm, models from 20 independent training operations were acquired. The hit rates of the 20 models of each given algorithm over the testing set were measured. The hit rates of the best model of each algorithm are reported in Table 4.

Table 3. The model summary of various algorithms.

Model	Summary
PairNet	5 Conv. layers, 1 Avg. polling layer, 1 FC layer Stride Size 1 and kernel size 1 × 3 for Conv. layer 1 Stride Size 2 and kernel size 1 × 2 for Conv. layers 2–5 Softmax output
Residual PairNet	5 Conv. layers, 1 Avg. polling layer, 1 FC layer Stride Size 1 and kernel size 1 × 3 for Conv. layer 1 Stride Size 2 and kernel size 1 × 2 for Conv. layers 2–5 2 Short-cut connections, Softmax output
PairNet with Inception	4 Conv. layers, 1 Inception layer 1 Avg. polling layer, 1 FC layer Stride Size 1 and kernel size 1 × 3 for Conv. layer 1 Stride Size 2 and kernel size 1 × 2 for Conv. layers 2–4 Softmax output
Residual PairNet with Inception	4 Conv. layers, 1 Inception layer 1 Avg. polling layer, 1 FC layer Stride Size 1 and kernel size 1 × 3 for Conv. layer 1 Stride Size 2 and kernel size 1 × 2 for Conv. layers 2–4 1 Short-cut connection, Softmax output
CNN	5 Conv. Layers, 2 Max. polling layer, 1 FC layer Stride Size 1 and kernel size 1 × 3 for Conv. layers 1–5 Softmax output
LSTM	Hidden states and mem. cells for single direction Hidden states and mem. cells dimension = 32 1 FC layer with Softmax output
Bi-LSTM	Hidden states and mem. cells for dual directions Hidden states and mem. cells dimension = 32 1 FC layer with Softmax output

We can see from Table 3 that all five feedforward algorithms (i.e., CNN, PairNet, Residual Pairnet, PairNet with inception, and Residual PairNet with inception) have five convolution/inception layers. However, it can be observed from Table 4 that the CNN has a lower classification accuracy when compared to the other four models. In particular, the average hit rates of CNN and PairNet are 90.42% and 92.36%, respectively. The PairNet outperforms the CNN in average hit rate by 1.94%. The CNN is inferior because it is based on convolution operations with a stride size of 1 for all the convolution layers. As a result, it has smaller receptive field for classification. By contrast, the convolution layers of the proposed algorithms have a stride size of 2. They may then have a larger receptive field for attaining better classification accuracy.

Table 4. The comparisons on hit rates (in percentage) of various algorithms. Both gyroscope and accelerometer are used for the corresponding implementations.

	H_1	H_2	H_3	H_4	H_5	H_6	H_7	H_8	H_9	H_{10}	H_{11}	**Ave.**
PairNet	70.35	90.19	72.14	86.43	99.12	97.93	98.23	98.16	94.05	100.00	99.54	92.36
Res. PairNet	81.86	96.98	82.66	97.29	99.12	97.93	98.99	96.63	92.57	100.00	99.54	95.30
PairNet w. Incep.	75.66	91.70	75.85	91.09	99.12	99.11	97.73	99.69	92.57	100.00	99.54	94.00
Res. PairNet w. Incep.	77.88	95.47	75.23	92.25	98.95	98.82	98.99	97.55	94.05	100.00	99.54	94.10
CNN [22]	69.03	74.72	70.90	85.66	99.12	97.34	96.97	96.93	92.94	99.07	99.54	90.42
LSTM [8]	70.35	72.83	65.63	78.29	98.60	97.04	89.14	93.25	79.55	99.54	99.07	86.87
Bi-LSTM [19]	70.80	88.68	73.07	81.01	97.90	97.34	87.12	90.18	84.01	97.22	99.54	88.66

In addition to larger stride sizes, it is revealed from Table 4 that the incorporation of short-cut connections and/or inception modules are also beneficial for the improvement of classification accuracy. That is, Residual Pairnet, PairNet with inception, and Residual PairNet with inception have superior hit rates over those of PairNet. In fact, the experimental results show that the average hit rates of PairNet, Residual Pairnet, PairNet with inception, and Residual PairNet with inception are 92.36%, 95.30%, 94.00%, and 94.10%, respectively. The improvement in average hit rate over the PairNet by Residual Pairnet, PairNet with inception, and Residual PairNet with inception are then 2.94%, 1.64%, and 1.74%, respectively. The improvements are due to the facts that the employment of short-cut connections may be able to provide more reliable references for training. Furthermore, the inception modules are able to effectively capture both slow and fast hand movements in a single layer.

Note that the hit rates in Table 4 are the best hit rates for the corresponding algorithms. To further assess Residual Pairnet, PairNet with inception, and Residual PairNet with inception algorithms, the statistical evaluations over the hit rates for each algorithm are included in Table 5. The evaluation includes the measurements of highest, mean, and lowest average hit rates over the 20 models associated with each algorithm. It can be observed from Table 5 that these statistical measurements for PairNet are inferior to those for Residual Pairnet, PairNet with inception, and Residual PairNet with inception algorithms. These results confirm the superiority of the proposed algorithms over their baseline PairNet counterpart.

Table 5. Statistical evaluation on the hit rates (in percentage) of various algorithms.

Algorithm	**Highest Hit Rate**	**Mean Hit Rate**	**Lowest Hit Rate**	**Standard Deviation**
PairNet	92.36	90.19	87.39	1.48×10^{-2}
Res. PairNet	95.30	93.60	91.36	1.22×10^{-2}
PairNet w. Incep.	94.00	91.37	89.62	1.54×10^{-2}
Res. PairNet w. Incp.	94.10	91.93	89.86	1.34×10^{-2}

An additional advantage of the proposed algorithms is that they outperform recurrent algorithms such as LSTM and Bi-LSTM, as shown in Table 4. The model summary of the LSTM and Bi-LSTM in our experiments can also be found in Table 3. The LSTM has inferior hit rates because of the gradient vanishing problem. The performance can be improved by bidirectional training and inference.

However, it can be observed in Table 4 that the proposed algorithms still has superior hit rates over the Bi-LSTM.

To further elaborate the effectiveness of the proposed algorithms, the confusion matrix of the Residual PairNet is revealed in Table 6. The confusion matrix reveals information about actual and predicted gesture classifications by the proposed algorithm. In the confusion matrix, the cell located at row i and column j represents the percentage in which the gesture in the row i is classified as the gesture in the column j. The hit rate H_i is then the value of the cell at row i and column i of the confusion matrix. It can be shown in Table 6 that simple gestures such as Gesture 1 and Gesture 3 have slightly lower hit rate because they may be misclassified as a part of other more complicated gestures. Nevertheless, the hit rates of these classes are still above 80%. For complicated gestures such as Gesture 5, Gesture 7, Gesture 10, and Gesture 11, the corresponding hit rates are above 99%.

Table 6. The confusion matrix of the Residual PairNet over the testing dataset. The cell located at row i and column j of the matrix represents the percentage in which the gesture in the row i is classified as the gesture in the column j.

	Gest. 1	**Gest. 2**	**Gest. 3**	**Gest. 4**	**Gest. 5**	**Gest. 6**	**Gest. 7**	**Gest. 8**	**Gest. 9**	**Gest. 10**	**Gest. 11**
Gest. 1	81.86	0.00	0.40	0.00	0.00	0.40	4.40	5.83	1.81	0.00	5.30
Gest. 2	0.00	96.98	0.00	0.38	0.00	0.77	0.79	0.00	0.00	0.00	1.08
Gest. 3	0.00	0.62	82.66	0.91	0.00	5.31	0.00	2.19	4.60	3.71	0.00
Gest. 4	0.00	0.39	0.35	97.29	0.00	0.75	0.00	0.00	1.21	0.00	0.00
Gest. 5	0.00	0.00	0.18	0.00	99.12	0.00	0.00	0.35	0.00	0.35	0.00
Gest. 6	0.00	0.00	0.29	0.00	0.00	97.93	0.00	0.00	0.89	0.89	0.00
Gest. 7	0.04	0.23	0.03	0.00	0.00	0.24	98.99	0.01	0.00	0.22	0.23
Gest. 8	0.00	0.00	0.00	0.00	0.00	2.11	0.62	96.63	0.00	0.00	0.64
Gest. 9	0.00	0.35	0.00	0.00	0.00	5.15	0.00	1.93	92.57	0.00	0.00
Gest. 10	0.00	0.00	0.00	0.00	0.00	0.00	0.00	0.00	0.00	100.0	0.00
Gest. 11	0.00	0.00	0.00	0.00	0.00	0.00	0.00	0.00	0.46	0.00	99.54

As compared with the basic CNN, the proposed algorithms may have both the advantages of higher classification accuracy, lower storage overhead, and lower computation complexity. In this study, we define the storage overhead and computation complexity of a neural network as the parameter size and the number of multiplications for the inference operations of that network, respectively. Table 7 shows the average hit rate, parameter size, and the number of multiplications of the algorithms considered in Table 3. It can be observed from the figure that all the proposed feedforward models have higher average hit rate, lower weight size, and lower number of multiplications as compared with basic CNN. This is because all the proposed models have a common feature that the intermediate convolution layers are with kernel size 1×2 and a stride size of 2. By contrast, all the convolution layers of the CNN have kernel size 1×3 and a stride size of 1. Smaller kernel sizes adopted by the proposed networks are beneficial for lowering the parameter sizes and computation complexities. Furthermore, larger stride sizes could enhance classification accuracy.

Table 7. The comparisons on average hit rates, number of weights, and computation complexity of various neural networks. The computation complexity of a neural network is the total number of multiplications required for the inference operations of the neural network.

	PairNet	CNN [22]	LSTM [8]	Bi-LSTM [19]	Res. PairNet	PairNet w. Incep.	Res. PairNet w. Incep.
Average Hit Rate	92.36%	90.42%	86.87%	88.66%	95.30%	93.63%	94.10%
Weights Size	271,116	426,764	5,388	10,764	304,396	200,844	244,876
Number of Multiplications	2,079,488	10,064,640	256,363	506,304	6,011,648	1,988,352	2,130,176

The improvement in classification accuracy of the PairNet and its variants over basic CNN is due to the employment of both a stride size of 2 and a kernel size of 1×2. The improvement would be marginal when only a stride size of 2 is adopted, while the kernel size retains the same as 1×3. To elaborate this fact, a new CNN with a stride size of 2 is considered, where its kernel sizes are 1×3 for all the convolution layers. To achieve meaningful comparisons, the new CNN and PairNet have the same number of convolution layers, and the same number of output channels associated with each convolution layer. Furthermore, 20 models of the new CNN algorithms were trained, and the best model only achieved the average hit rate of 90.78%. By contrast, the best average hit rate of PaiNet is 92.36% . The new CNN does not perform well because the large kernel size of 1×3 introduces a large number of parameters (i.e., 400,369) so that overfitting may be likely. To provide better generalization for the training, smaller kernel size with short-cut connections and/or inception modules would be more effective.

Among the proposed feedforward algorithms, it can be observed from Table 7 that the PairNet with Inception has lowest parameter size and computation complexity. This is because the dimension reduction is carried out by front kernel in each path of the inception module shown in Figure 8. The total number of kernels can then be effectively reduced. This in turn may lower the parameter size and computation complexity. Because inception modules may be beneficial for reducing the model complexity, the Residual PairNet with inception also has lower complexity as compared with its Residual PairNet counterpart without inception, as shown in Table 7.

It can also be observed from Table 7 that the LSTM algorithm has the lowest computational complexity. Although there are four pairs of matrices used in the cell and hidden state calculations in the LSTM, the dimension of cells and hidden states is only 32. This is beneficial for maintaining small number of multiplications for the algorithm. However, the inference operations of the LSTM should be carried out in a recurrent fashion. Therefore, the reduction in computation time may not be significant as compared with the feedforward algorithms. To elaborate this fact, we have measured the computation time of LSTM, PairNet and CNN on the Raspberry Pi 3 platform, where the computation time of a network is the average CPU time for the inference of a single gesture from the testing set of that network. Based on the experiments, the computation time of LSTM, PairNet, and CNN are 102, 162, and 410 ms, respectively. The computation time of PairNet is only slightly longer than that of LSTM. This is because the inference operations of the feedforward networks can be computed in parallel. Therefore, the real-time inference may still be possible even the system is deployed in the embedded systems with Raspberry Pi 3 platform.

Finally, we evaluate the performance of the proposed algorithms when only one of the sensors is adopted. Table 8 shows the results of the experiments for the PairNet. We can see from Table 8 that the hit rates of most gesture classes are degraded without the employment of both sensors. These results show that the employment of both gyroscope and accelerometer is beneficial for hand-gesture recognition.

Table 8. The hit rates (in percentage) of the proposed PairNet algorithm with the employment of only accelerometer, gyroscope, or both.

	H_1	H_2	H_3	H_4	H_5	H_6	H_7	H_8	H_9	H_{10}	H_{11}	**Ave.**
Both	70.35	90.19	72.14	86.43	99.12	97.93	98.23	98.16	94.05	100.00	99.54	92.36
Gyroscope	69.03	74.72	70.90	85.66	99.12	97.34	96.97	96.93	92.94	99.07	99.54	90.42
Accelero.	70.35	72.83	65.63	78.29	98.60	97.04	89.14	93.25	79.55	99.54	99.07	86.87

5. Conclusions

The proposed feedforward algorithms have been found to be effective for dynamic hand gesture recognition. In our experiments, smart phones equipped with accelerometers and gyroscopes were adopted for the data acquisition of hand gestures. The resulting implementations can be deployed for the remote control of devices such as home appliances. It is observed from the experiments that the PairNet algorithm attains average hit rate of 92.36% for 11 gesture classes over 3404 test gestures. The hit rate of the PairNet is 1.94%, 5.49%, and 3.70% higher than those of the basic CNN, LSTM, and Bi-LSTM, respectively. Furthermore, the proposed Residual PairNet, PairNet with inception, and Residual PairNet with inception have superior performance over PairNet. In fact, the improvement in average hit rate over the PairNet by Residual Pairnet, PairNet with inception, and Residual PairNet with inception are then 2.94%, 1.64%, and 1.74%, respectively. The proposed feedforward algorithms have superior hit rate because they have a large receptive field for accurate gesture spotting. Furthermore, the algorithms also have lower weight sizes as compared with its basic CNN counterpart. In particular, the average computation time on the Raspberry Pi 3 platform for the inference of a single gesture is only 162 ms for PairNet, while the basic CNN needs 410 ms. The proposed work is therefore beneficial for human–machine interface applications where reliable continuous hand gesture recognition with real-time computation on the embedded platform is desired.

Author Contributions: Conceptualization, Y.-C.C., Y.-J.J., and T.-M.T.; methodology, Y.-C.C. and Y.-J.J.; software, Y.-C.C. and Y.-J.J.; visualization, Y.-C.C., Y.-J.J., and T.-M.T.; validation, Y.-C.C., Y.-J.J., and T.-M.T.; supervision, T.-M.T. and W.-J.H.; project administration, W.-J.H.; writing—original draft, W.-J.H. All authors have read and agreed to the published version of the manuscript.

Funding: This research was funded by the Ministry of Science and Technology, Taiwan, under Grant MOST 107-2221-E-003-001-MY2.

Conflicts of Interest: The authors declare no conflict of interest. The funders had no role in the design of the study; in the collection, analyses, or interpretation of data; in the writing of the manuscript, or in the decision to publish the results.

Abbreviations

The following abbreviations are used in this manuscript:

Bi-LSTM	Bidirectional Long Short Term Memory
CNN	Convolution Neural Network
CSI	Channel State Information
LSTM	Long Short Term Memory
MAP	Maximum A Posteriori
PIN	Personal Identification Number
RNN	Recurrent Neural Network
SBR	Sensor-Based Recognition
VBR	Vision-Based Recognition

References

1. Wang, C.; Liu, Z.; Chan, S.-C. Superpixel-based hand gesture recognition with kinect depth camera. *IEEE Trans. Multimed.* **2015**, *17*, 29–39. [CrossRef]
2. Bobic, V.; Tadic, P.; Kvascev, G. Hand gesture recognition using neural network based techniques. In Proceedings of the 2016 13th Symposium on Neural Networks and Applications, Belgrade, Serbia, 22–24 November 2016.
3. Zhu, G.; Zhang, L.; Shen, P.; Song, J. Multimodal gesture recognition using 3-D convolution and convolutional LSTM. *IEEE Access* **2017**, *5*, 4517–4524. [CrossRef]
4. Oyedotun, O.K.; Khashman, A. Deep Learning in Vision-Based Static Hand Gesture Recognition. *Neural Comput. Appl.* **2017**, *28*, 3941–3951. [CrossRef]
5. Xu, R.; Zhou, S.; Li, W.J. MEMS Accelerometer Based Nonspecific-User Hand Gesture Recognition. *IEEE Sens. J.* **2012**, *12*, 1166–1173. [CrossRef]
6. Xie, R.; Sun, X.; Xia, X.; Cao, J. Similarity Matching-Based Extensible Hand Gesture Recognition. *IEEE Sens. J.* **2015**, *15*, 3474–3483. [CrossRef]
7. Gupta, H.P.; Chudgar, H.S.; Mukherjee, S.; Dutta, T.; Sharma, K. A Continuous Hand Gestures Recognition Technique for Human-Machine Interaction Using Accelerometer and Gyroscope Sensors. *IEEE Sens. J.* **2016**, *16*, 6425–6432. [CrossRef]
8. Tai, T.M.; Jhang, Y.J.; Liao, Z.W.; Teng, K.C.; Hwang, W.J. Sensor-Based Continuous Hand Gesture Recognition by Long Short-Term Memory. *IEEE Sens. Lett.* **2018**, *2*, 6000704. [CrossRef]
9. Zhao, T.; Liu, J.; Wang, Y.; Liu, H.; Chen, Y. PPG-Based Finger-Level Gesture Recognition Leveraging Wearables. In Proceedings of the IEEE Conference on Computer Communications, Honolulu, HI, USA, 15–19 April 2018; pp. 1457–1465.
10. Pathak, V.; Mongia, S.; Chitranshi, G. A Framework for Hand Gesture Recognition Based on Fusion of Flex, Contact and Accelerometer Sensor. In Proceedings of the Third International Conference on Image Information Processing, Waknaghat, India, 21–24 December 2015; pp. 312–319.
11. Zhang, X.; Chen, X.; Li, Y.; Lantz, V.; Wang, K.; Yang, J. A Framework for Hand Gesture Recognition Based on Accelerometer and EMG Sensors. *IEEE Trans. Syst. Man Cybern. Part A Syst. Humans* **2011**, *41*, 1064–1076. [CrossRef]
12. Asadzadeh, P.; Kulik, L.; Tanin, E. Gesture Recognition Using RFID Technology. *Pers. Ubiquit. Comput.* **2012**, *16*, 225–234. [CrossRef]
13. Zhou, Y.; Xiao, J.; Han, J.; Wu, K.; Li, Y.; Ni, L.M. GRfid: A Device-Free RFID-Based Gesture Recognition System. *IEEE Trans. Mob. Comput.* **2017**, *16*, 381–393. [CrossRef]
14. Lv, J.; Yang, W.; Gong, L.; Man, D.; Du, X. Robust WLAN-based indoor fine-grained intrusion detection. In Proceedings of the 2016 IEEE Global Communications Conference, Washington, DC, USA, 4–8 December 2016; pp. 1–6.
15. Chen, Z.; Zhang, L.; Jiang, C.; Cao, Z.; Cui, W. WiFi CSI Based Passive Human Activity Recognition Using Attention Based BLSTM. *IEEE Trans. Mob. Comput.* **2019**, *18*, 2714–2724. [CrossRef]
16. Ding, J.; Wang, Y. WiFi CSI-Based Human Activity Recognition Using Deep Recurrent Neural Network. *IEEE Access* **2019**, *7*, 174257–174269. [CrossRef]
17. Goodfellow, I.; Bengio, Y.; Courville, A. *Deep Learning*; MIT Press: Cambridge, MA, USA, 2016.
18. Hochreiter, S.; Schmidhuber, J. Long Short-Term Memory. *Neural Comput.* **1997**, *9*, 1735–1780. [CrossRef] [PubMed]
19. Lefebvre, G.; Berlemont, S.; Mamalet, F.; Garcia, C. Inertial Gesture Recognition with BLSTM-RNN. In *Artificial Neural Networks, Springer Series in Bio-/Neuroinformatics*; Springer: Berlin/Heidelberg, Germany, 2015; Volume 4, pp. 393–410.
20. Gu, J.; Wang, Z.; Kuen, J.; Ma, L.; Shahroudy, A.; Shuai, B.; Liu, T.; Wang, X.; Wang, G.; Cai, J.; et al. Recent advances in convolutional neural networks. *Pattern Recognit.* **2018**, *77*, 354–377. [CrossRef]
21. Li, D.; Zhang, J.; Zhang, Q.; Wei, X. Classification of ECG Signals Based on 1D Convolution Neural Network. In Proceedings of the 2017 IEEE 19th International Conference on e-Health Networking, Applications and Services (Healthcom), Dalian, China, 12–15 October 2017.

22. Lee, S.M.; Yoon, S.M.; Cho, H. Human Activity Recognition From Accelerometer Data Using Convolutional Neural Network. In Proceedings of the 2017 IEEE International Conference on Big Data and Smart Computing (BigComp), Jeju, Korea, 13–16 February 2017; pp. 131–134.
23. Wang, W.; Zhu, M.; Wang, J.; Zeng, X.; Yang, Z. End-to-end Encrypted Traffic Classification with One-dimensional Convolution Neural Networks. In Proceedings of the 2017 IEEE International Conference on Intelligence and Security Informatics (ISI), Beijing, China, 22–24 July 2017; pp. 43–48.
24. Van den Oord, A.; Dieleman, S.; Zen, H.; Simonyan, K.; Vinyals, O.; Graves, A.; Kalchbrenner, N.; Senior, A.; Kavukcuoglu, K. Wavenet: A Generative Model for Raw Audio. *arXiv* **2016**, arXiv:1609.03499.
25. Engel, J.; Resnick, C.; Roberts, A.; Dieleman, S.; Norouzi, M.; Eck, D.; Simonyan, K. Neural audio synthesis of musical notes with WaveNet autoencoders. In Proceedings of the 34th International Conference on Machine Learning, Sydney, Australia, 6–11 August 2017.
26. Jhang, Y.J.; Chu, Y.C.; Tai, T.M.; Hwang, W.J.; Cheng, P.W.; Lee, C.K. Sensor-Based Dynamic Hand Gesture Recognition by PairNet. In Proceedings of the 2019 International Conference on Internet of Things (iThings) and IEEE Green Computing and Communications (GreenCom) and IEEE Cyber, Physical and Social Computing (CPSCom) and IEEE Smart Data (SmartData), Atlanta, GA, USA, 14–17 July 2019; pp. 994–1001.
27. He, K.; Zhang, X.; Ren, S.; Sun, J. Deep Residual Learning for Image Recognition. *arXiv* **2015**, arXiv:1512.03385.
28. He, K.; Zhang, X.; Ren, S.; Sun, J. Identity Mappings in Deep Residual Networks. *arXiv* **2016**, arXiv:1603.05027.
29. Wu, Z.; Shen, C.; van den Hengel, A. Wider or Deeper: Revisiting the ResNet Model for Visual Recognition. *Pattern Recognit.* **2019**, *90*, 119–133. [CrossRef]
30. Szegedy, C.; Liu, W.; Jia, Y.; Sermanet, P.; Reed, S.; Anguelov, D.; Erhan, D.; Vanhoucke, V.; Rabinovich, A. Going Deeper with Convolutions. *arXiv* **2014**, arXiv:1409.4842.
31. Szegedy, C.; Ioffe, S.; Vanhoucke, V. Inception-v4, Inception-Resnet and the impact of residual connections on learning. *arXiv* **2016**, arXiv:1602.07261.
32. Sun, Y.; Liang, D.; Wang, X.; Tang, X. DeepID3: Face Recognition with Very Deep Neural Networks. *arXiv* **2015**, arXiv:1502.00873.
33. Chua, C.-S.; Guan, H.; Ho, Y.-K. Model-Based 3D Hand Posture Estimation form a Single 2D Image. *Image Vis. Comput.* **2002**, *20*, 191–202. [CrossRef]
34. Chollet, F. Keras. Available online: http://github.com/fchollet/keras (accessed on 3 August 2020).

applied sciences

MDPI

Article

Translating Videos into Synthetic Training Data for Wearable Sensor-Based Activity Recognition Systems Using Residual Deep Convolutional Networks

Vitor Fortes Rey [1,*], Kamalveer Kaur Garewal [2,*] and Paul Lukowicz [1,3,*]

1 Embedded Intelligence, German Research Center for Artificial Intelligence (DFKI), 67663 Kaiserslautern, Germany
2 Augmented Vision, DFKI, 67663 Kaiserslautern, Germany
3 Informatics Department, University of Kaiserslautern, 67663 Kaiserslautern, Germany
* Correspondence: vitor.fortes@dfki.de (V.F.R.); kamalveerkaur.garewal@dfki.de (K.K.G.); paul.lukowicz@dfki.de (P.L.)

Citation: Fortes Rey, V.; Garewal, K.K.; Lukowicz, P. Translating Videos into Synthetic Training Data for Wearable Sensor-Based Activity Recognition Systems Using Residual Deep Convolutional Networks. *Appl. Sci.* **2021**, *11*, 3094. https://doi.org/10.3390/app11073094

Academic Editor: Hyo Jong Lee

Received: 28 February 2021
Accepted: 19 March 2021
Published: 31 March 2021

Abstract: Human activity recognition (HAR) using wearable sensors has benefited much less from recent advances in Deep Learning than fields such as computer vision and natural language processing. This is, to a large extent, due to the lack of large scale (as compared to computer vision) repositories of labeled training data for sensor-based HAR tasks. Thus, for example, ImageNet has images for around 100,000 categories (based on WordNet) with on average 1000 images per category (therefore up to 100,000,000 samples). The Kinetics-700 video activity data set has 650,000 video clips covering 700 different human activities (in total over 1800 h). By contrast, the total length of all sensor-based HAR data sets in the popular UCI machine learning repository is less than 63 h, with around 38 of those consisting of simple mode of locomotion activities like walking, standing or cycling. In our research we aim to facilitate the use of online videos, which exist in ample quantities for most activities and are much easier to label than sensor data, to simulate labeled wearable motion sensor data. In previous work we already demonstrated some preliminary results in this direction, focusing on very simple, activity specific simulation models and a single sensor modality (acceleration norm). In this paper, we show how we can train a regression model on generic motions for both accelerometer and gyro signals and then apply it to videos of the target activities to generate synthetic Inertial Measurement Units (IMU) data (acceleration and gyro norms) that can be used to train and/or improve HAR models. We demonstrate that systems trained on simulated data generated by our regression model can come to within around 10% of the mean F1 score of a system trained on real sensor data. Furthermore, we show that by either including a small amount of real sensor data for model calibration or simply leveraging the fact that (in general) we can easily generate much more simulated data from video than we can collect its real version, the advantage of the latter can eventually be equalized.

Keywords: activity recognition; data augmentation; pose estimation; deep learning

1. Introduction and Related Work

Human activity recognition (HAR) using wearable sensors has been a successful research field [1,2] for nearly two decades. There has been the expectation that as large amounts of sensor data become available, HAR will be able to benefit from the advances in Deep Learning techniques in the same way as fields, such as computer vision and speech recognition [3,4]. However, while the application of deep learning methods to HAR has produced undeniable advances [2], the progress has so far been far less significant than computer vision or NLP. Thus for example in recent HAR competitions, most deep learning entries are out-performed by classical machine learning methods, namely random forests [5], while in computer vision, deep learning techniques have dominated since

2012 [6] and beyond [7]. To a large degree this can be attributed to the difficulty of acquiring **labeled** sensor data from complex realistic scenarios. Consider for example the UCI machine learning repository, a popular machine learning repository. While it contains many datasets for HAR using Inertial Measurement Units (IMUs), their combined volume is less than 63 h, with around 38 of those consisting of simple mode of locomotion activities like walking, standing or cycling. Probably the most popular benchmark datasets in the wearable sensing HAR community are PAMAP2 [8,9] and Opportunity [10,11] with, respectively, less than 8 h of labeled data and less than 14 even if we count all activity granularities. By contrast a well known computer vision repository, ImageNet [12] has images for around 100'000 categories (based on WordNet) with on average 1000 images per category (making it up to 100'000'000 samples). A big HAR related video data set, the Kinetics-700 [13] has 650,000 video clips covering 700 different human activities. With each clip lasting 10 s this is over 1800 h!

There are two main reasons why there are much more image/video data than sensor-based HAR data. First, while more and more people are wearing sensor enabled devices (e.g., fitness trackers) and often uploading to respective platforms, this pales in comparison to those just snapping a picture or making a quick video clip and uploading it to an online repository. Thus, today there are billions of images and videos freely available online, but much less sensor data, even less publicly available. Second, videos can be quickly and easily labeled using crowdsourcing services such as Amazon Mechanical Turk (this is how most ImageNet images were labeled), by leveraging captions, or known topics of image collections. By contrast, labeling sensor data is much more difficult, as interpreting raw sensor data can be challenging even for experts. As a consequence, using crowdsourcing for labeling large amounts of sensor data is not a viable option.

The top level goal of our work is to develop a methodology to allow converting videos of human activities into synthetic sensor inertial measurement unit (IMU) data and thus potentially make all the video based HAR datasets usable for training sensor-based HAR systems.

1.1. Paper Contributions

Recently the idea of using labeled videos to generate "synthetic" sensor data has been proposed as a solution for the HAR training data problem. As one of the first published approaches in this direction we have previously shown [14] initial results on how regression models can be trained to simulate motion sensor data from videos of the respective activities. In that preliminary work we relied on recordings that contained sensor data and videos *of the same activities* that we later wanted to have the system generate the sensor data for. In this paper, we describe the full development and evaluation of that initial idea making the following contributions:

1. We have adapted the method to generalise outside the target exercises. Our model can now be trained on other subjects performing a set of "generic" motions selected to be representative of a broad domain. Using this generic regression model, we can generate synthetic training data for a variety of different activities.
2. We have performed and described an in depth analysis of the physical background of deriving different components of the IMU signal and have shown how to simulate different sensor signals beyond acceleration norm.
3. We have performed an analysis of the influence of various signal processing techniques on the quality of the simulated sensor signal.
4. We have proposed and implemented a new deep neural network based regression model for the generation of simulated sensor data from videos. Compared to our previous work, we now have one regression network per sensor position, which helped reduce overfitting the training motions.
5. We have identified a set of generic motions suitable for training the regression model for the broad domain of aerobics like physical exercises, recorded a dataset containing appropriate video footage and sensor data and used it to train the above model.

6. We have performed an evaluation on an activity recognition task from the above broad domain of aerobic like exercises for which we have also recorded our own dataset. We have both collected and used as source of simulated training data appropriate online videos. Our analysis includes testing the influence of simulating different sensor modalities and adding small amounts of real sensor data to the simulated one to further improve accuracy. Compared to our previous preliminary work [14], we achieve significantly better results, although the problem that we tackle (using **generic motions** for training the regression model) is a harder one. Overall we show that
   (a) Systems trained on simulated data generated by our regression model and tested on real sensor data can achieve a recognition performance that is within 10% of the recognition performance achieved by a model trained on the corresponding real sensor data.
   (b) By increasing the amount of simulated data, we can match the performance of a real signal based model trained on less data. Results so far indicate about a factor of 2 to be needed. This is in line with the motivation behind this work, which is the fact that a very large amount of videos are available for many relevant activities online, which in turn means that we can get much more training data if we can generate it from such videos (which we aim to enable).
   (c) Adding even small amounts of real sensor data to fine tune the model generated on the basis of simulated data can improve the performance significantly. In our experiments we found that real data from only 1 or 2 real users can already make the performance of a system trained on simulated data comparable to one fully trained on real data.

Our method for using videos for training sensor-based activity recognition systems follows the steps of Figure 1. First we record a dataset that contains a broad set of generic motions typical for a given domain (e.g., physical exercise). This dataset contains both an appropriate video recording and the corresponding data from on-body sensors. We then train a regression model to map the videos onto the corresponding sensor data. The development of this model is the main contribution of our work. Note that a regression model needs to be generated only one time for a given broad domain. Once it is trained, it can be repeatedly used to generate simulated sensor data for various arbitrary activity recognition tasks from that respective domain. For each new recognition task a set of labeled videos of the respective activities is first acquired (e.g., from YouTube). The videos are then fed to the regression model (step 2) which converts it to simulated sensor data. Obviously the labels for the sensor data are identical to the video labels. Thus we have labeled training sensor data. We next use it to train a recognition model for the required activities (step 3). Finally the trained model is used to recognize those activities from real sensor data in the end application (step 4).

1.2. Related Work

Today in the wearable sensing community there is a broad consensus that obtaining sufficient labeled training data is a key challenge facing the discipline. The use of online information to generate such training data has recently been proposed as a promising approach to address this challenge. While itself not solved and as yet not widely studied, the problem of generation of sensor data from online information sources is closely related to a number of established research fields. These include:

1. Simulation of IMU data directly from virtual environments (see Section 1.2.1).
2. Generation of 3D animations from 2D monocular videos (see Section 1.2.2).
3. Machine learning (ML) methods for generating signal variations such as Generative Adversarial Networks (GANs) (see Section 1.2.3).
4. ML methods for predicting signal values in physical systems (see Section 1.2.4).

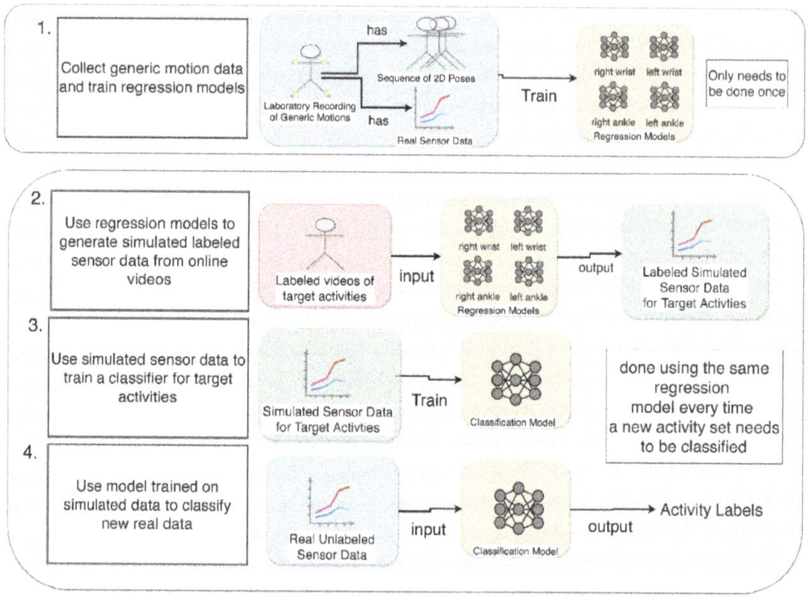

Figure 1. The steps involved in applying our overall method. In step 1 we record video and sensor data for a broad set of generic motions typical for a given domain. Using this data we train models to map the videos onto the corresponding sensor data. Step 2 consists of acquiring for each new recognition task a set of labeled videos of the respective activities from online repositories. The videos are then fed to the regression models which generate their simulated sensor data. Since the labels for the sensor data are identical to the video labels, we now have labeled training sensor data, which we use to train a recognition model for the required activities (step 3). The final step is 4, where we use the trained model to recognize those activities from real sensor data in the end application.

1.2.1. Simulating IMU Data Directly from 3D Trajectories in Virtual Environments

Many methods exist for simulating IMU data. For example, ref. [15–17] provide a simulation environment where one can model human subjects with virtual sensor models to allow experimentation and exploration in virtual space of scenarios in inertial sensing. Depending on the application, specialised tools can be used. For example, there is ample work in simulating foot IMU data for pedestrian dead reckoning [18]. In general many tools exist in this field as simulation is useful for many projects [19,20], but one may also use any existing physical simulator.

While many simulators exist, in general they fail to capture the diversity of human motion and all the different ways different people can perform a given activity. Such diversity can on the other hand be captured by extracting sensor data from videos of real people executing the respective activities as proposed by our work.

1.2.2. Obtaining 3D Poses from Videos

Recently, pose estimation has become a very interesting and successful area of research, with many applications. Obtaining 3D poses is possible with a calibrated system of more than one camera [21]. Commercial systems that can extract poses using 2 cameras (among other things) includes also kinect, which was used in other work [22] to obtain the 3D poses and with it simulate IMU signals. Using a single camera is obviously more challenging. Works in this area include [23–28]. Some works [24,27] predict the 2D joints and then fit the 3D skeleton using different strategies, others [23,25,26] are trained to predict 3D joints directly from the image. Even more interestingly, some methods predict both 2D and 3D keypoints [28].

While methods have achieved impressive results, the field is still evolving and the task itself is far from easy due to occlusion, clothing, lighting, and the inherent ambiguity in inferring 3D from 2D [29]. Thus, for the generation of IMU sensor data going directly from 2D poses to IMU, values as proposed by this work is a promising alternative that needs to be investigated.

1.2.3. Deep Learning Based Generation of Signal Variations

While more known in the computer vision community, GANs have shown great promise also in mobile sensors where they have been used to augment datasets and even improve models using semi-supervised learning [30,31]. Those works can generate new sensor data for a specific dataset and label, relying on the data itself and not on the motions of the subject. In other words, in a dataset where we are simulating a person walking, they can generate more data for the sensors available in that dataset, but cannot simulate what the data would be for sensors that were not deployed. This is very relevant as there are many possible sensor placements and possible target activities, and thus being able to simulate, based on the person's poses, many different placements are necessary, especially if one wants to use online video repositories to obtain sensor data for the target activities. Another case where GANs provide sensor data is through domain adaptation [32], by simulating the readings of one sensor, based on the reading of another one present. While this can alleviate in part the cited restrictions, it requires the source sensor to contain enough information to simulate the target one and is limited to a specific scenario. In the context of poses, GANs have been used to augmented vision datasets by generating subjects in novel poses [33] for better re-identification.

Despite the undisputed potential of the technology, so far there is no work on using GANs to generate sensor data based on video input which is what this work does.

1.2.4. Deep Learning Methods for the Prediction of Physical Parameters

There has recently been increased interest in exploiting the neural networks capability as universal function approximators for the prediction of the dynamics of physical systems, in particular the solution of differential equations [34] replacing traditional methods such as FEMs. Initially, the amount of training data required to accurately model complex physical systems was a problem. As a solution, so called Physically Informed Neural Networks [35] were proposed, which incorporate physical knowledge in the network either as regularisation terms or by leveraging the neural networks automatic differentiation capabilities to capture the information contained in the differential equations.

Obviously, the mapping of poses and pose changes to acceleration and angular velocity values can be seen as a type of physical parameters prediction task where the concept of PINNs can be beneficial. However, so far such an application of the PINN concept approach has not been studied. It is something that we will investigate in future work.

1.2.5. Overall Positioning with Respect to State of the Art

With respect to simulating HAR sensor data, we have [15] doing it directly in simulated environments. Other approaches rely on on-body markers [36,37] or systems with multiple cameras such as kinect [22]. Our approach does not use any special markers, as we would not have access to such markers in video sources such as YouTube. Instead, it relies directly on the position of the body's joints. Moreover, we do not assume we have more than a single monocular camera, as that is the case for most datasets and video sources. Compared to works using forward kinematics, our approach does not directly perform physical simulation, as it is only feasible if the 3D poses of the subject are estimated. Obtaining those 3D poses is possible with a calibrated system of more than one camera [21] or even using a single one [23–28]. Thus, one could extract those poses and then use the cited methods to obtain the sensor values, but their quality depends on the quality of poses that can be obtained from a monocular camera. Recent work that follows this approach is [38], which uses off the shelf methods to predict 3D human poses for monocular images and then forward kinematics to simulate sensor data. They have demonstrated the feasibility of the approach in generating acceleration data without a static camera, but their pipeline works best when predicting modes of locomotion and not complex activities that do not happen in a 2D plane [38]. This is likely due to accumulated distortions in the predicted 3D poses. Extracting those poses from a monocular camera is no easy task due to occlusion, clothing, lighting, and the inherent ambiguity in inferring 3D from 2D [29]. Our approach is different as we predict sensor data directly from the 2D poses using a deep neural network without forward kinematics or a complex pipeline that includes several deep neural networks. We situate our work in the area of applying advances in vision to the HAR field. More recently vision has been shown to be successful for both labeling and knowledge transfer in mobile sensor-based HAR [39], indicating that there are many benefits in using video information too for HAR and many cases where it can aid in data acquisition.

2. Problem Description and Design Considerations

A video (at least when taken from the right perspective) clearly contains elaborate motion information. On the other hand, signals produced by the sensors that we are considering (IMUs and their components: accelerometers and gyroscope) are in effect a representation of motion. The generation of IMU signals from known 6DOF (x,y,z coordinates plus the yaw, roll pitch angles) trajectories is a well understood and solved task. Thus in simplified terms we consider the mapping from one representation of relevant motion (video) to another (IMU sensors). The difficulty that we need to address stems from three considerations:

1. **Fundamental incompleteness of 2D video information.** Monocular video (which is what we need to work with to be able to harvest online video sources) does not contain complete 6-DOF information about objects (in our case human body parts) that it shows. Instead for each object a single video frame provides 2D coordinates in its own frame of reference which are essentially the x- and y-coordinates in pixels. When looking at such a frame humans can translate such 2D image coordinates into information about physical coordinates (which in some cases may include all 6 DOFs) through semantic analysis of the picture and putting the results of such semantic analysis in the context of their understanding of the world. Thus when seeing a man waving his arms, we can estimate the physical coordinates of various body parts with respect to each other just from our knowledge of human physiology (degrees of freedom of joints, typical motions typical size and proportions of human body etc.). Clearly not knowing the exact dimensions of the specific person, this can only be an approximation.
2. **Inherent inaccuracies in 2D video information.** Even when physical coordinates of relevant body parts can in principle be inferred from the semantic information, the achievable accuracy can vary greatly depending on the camera angle and position. Thus, for example, given frontal view of the user (as for example in Figure 2 the x-y position of the wrist, which is point 4 in Figure 3) in respect to the hip (point 8) can be estimated with reasonable accuracy. On the other hand, the angle of rotation of the wrist around the lower arm axis is much more difficult to estimate accurately. This is because in a typical video, the wrist itself can be just a few pixels wide.
3. **Sensor specific physical effects.** Sensor signals are influenced by factors which do not directly manifest themselves in an image. Best example of this is gravity. While the direction of "down" can mostly be inferred from semantic analysis of an image, it is not always enough. Consider the example of the wrist rotation above in the context of a wrist worn acceleration sensor. If the lower arm is parallel to the ground then the rotation determines how the gravity vector is projected onto the sensor axis perpendicular to the lower arm. Thus, the rotation has a very significant influence on the value of the acceleration signals on those two axes. On the other hand, as described above, wrist rotation tends to produce a "weak signal" in an image so that it can only be roughly estimated. Another example is high frequency "ringing" of the

acceleration signals, when for example the user stamps his feet on the ground. This results from high frequency vibration of the device caused by the strong impact of the foot on a hard surface and is a very characteristic feature in the acceleration signals, but mostly invisible on the video signal as the amplitude of the vibration is too small and affects the device only (not body parts as a whole) which may be invisible (e.g., hidden under clothing).

4. **Frame of reference and scaling.** IMUs typically use the "world" coordinate system which relies on magnetic sensors to determine geographic north and derives the direction of the gravity vector as down. The signals are given in standard units (e.g., $\frac{m}{s^2}$ for acceleration). In principle gravity can be derived through semantic analysis of an image, which is however not necessarily trivial and not always reliable. Obviously in most images "north" is not given. All motions are given in pixels per frame with the relationship between pixels and physical units depending on the camera position and settings. Overall the translation between the image frame of reference and pixel units and the sensor frame of reference and physical units is a non trivial challenge, even more in videos obtained from online repositories where cameras are not calibrated.

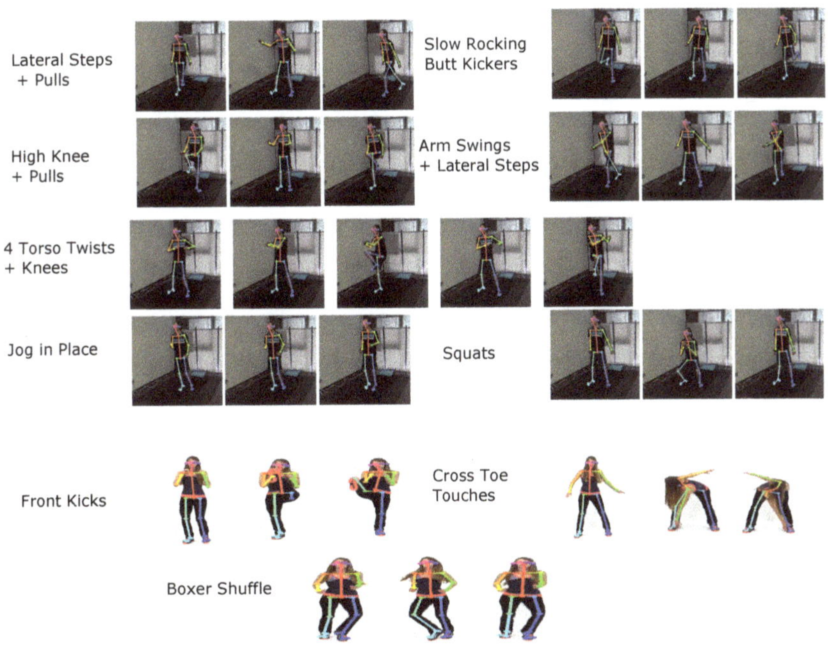

Figure 2. Short depiction of the motions present in the 10 exercises of the Drill dataset, in which participants performed the exercises of [40]. The first examples are of one of our subjects, while the last are from the original YouTube video each subject tries to replicate. In this dataset, each exercise is repeated once for 30 s.

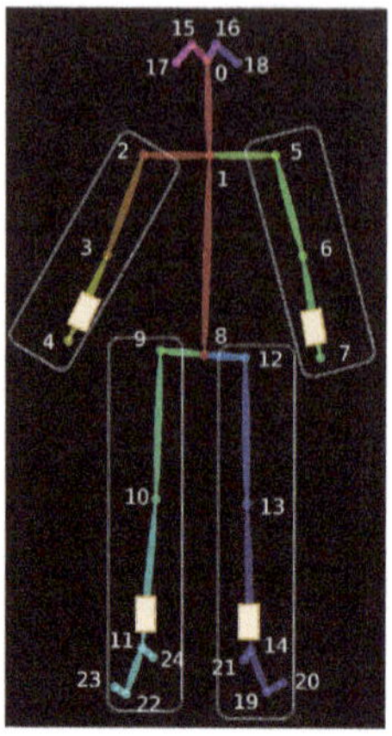

Figure 3. Joints used for each of the 4 regression models. Those models also receive the scaled hip speed and the change in scale. Orange boxes represent where sensors were placed during the exercises, while their surrounding white boxes describe the joints relevant for each of their regression models.

In summary the generation of IMU sensor data from videos cannot be accomplished with an exact physical model. This is a fundamental difference to the well understood and largely solved problem of generating IMU values from 6DOF trajectories. Instead a heuristic approach is needed that minimizes the unavoidable errors with respect to the needs of a given domain. Here a key question is whether we narrowly optimize the system to a specific domain or follow a more general domain agnostic approach. Other core design concerns are how much, at which stage in the process and in what form can physical models be included and what they can accomplish and which part of the task will be delegated to what type of data driven ML approaches. The most obvious choice is between (1) using existing ML methods for estimating 3D poses from 2D videos [23–27] and then using bio mechanical models [15] to generate the sensor data and (2) training an end-to-end ML model that goes directly from 2D video to sensor data. For the ML approach (this work), questions include how to accomplish the conversion between image frame of reference and pixel coordinates, how to handle the sensor frame of reference and its physical units and whether to explicitly filter, scale and normalize the signal.

3. Method

In this section we will explain in detail our method starting with how it is trained, which is shown in Figure 4. The first training step consists of extracting the poses from video and translating them into appropriate parameters:

1. The application of standard tools (specifically Open Pose [21,41]) to identify humans in videos and extraction of their **2D** poses. The poses are defined in terms of joint coordinates in pixels.
2. Optional: low pass filtering.
3. A sliding window based translation of the raw 2D pixel coordinates into appropriately filtered and scaled values in the body centered (hip as origin) frame of reference. This window considers 1.5 s in the past and 1.5 in the future, with a total length of 3 s.
4. Generating sliding windows of the pre-processed pose data with window size W and step S.

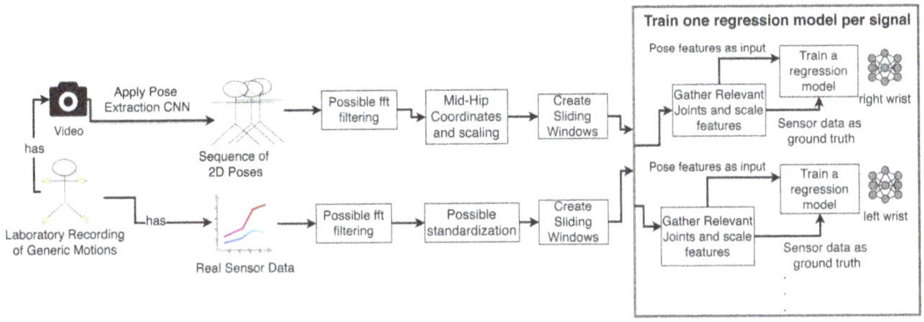

Figure 4. Procedure for training our regression models, which translates sequences of poses to the sensor values they generate. Using a dataset that includes both video and sensor data, we train for each position and sensor channel a regression model that uses the relevant joints and other features to predict the sensor values for that channel and place.

The system is trained on a dedicated dataset that contains synchronized sensor and video data for a set of motions typical for a broad domain. The sensor data collected is used as ground truth for training the regression model. Thus, any pre-processing of the sensor data performed before it is used to compute the loss will be reflected in the synthetic sensor data that the model will learn to generate. This means that the aim of the pre-processing must be to remove from the sensor signal any information that will hamper learning by "confusing" the model (noise, frequencies too high to be contained in the video data), while retaining as much useful information as possible. Examples of pre-processing strategies that we investigated were low pass filtering with different cut off frequencies and scaling. After pre-processing we aggregate the ground truth sensor data into sliding windows with window size W and step S to correspond to the partitioning of the pose parameters. The sliding windows are the input to the regression model. After some experimentation we have settled on a model based on residual deep convolution neural networks that uses dilated convolutions, similar to a Temporal Convolutional Network (TCN) [42]. We train a separate model for each sensor position and channel. Once trained, the system is presented with a video of the activity and produces a corresponding sequence of simulated sensor signals.

In the remainder of this Section we will explain in detail the training steps of our method. In Section 3.1 we go into depth about how we obtain poses for each video as well as explain the procedures and rationales behind our sensor and pose pre-processing. The motivation and configuration of our deep learning model is explained in Section 3.2, while the dataset that was used to train it, is described in Section 3.3.

3.1. Obtaining 2D Poses

We used the OpenPose library [21,41] to extract 2D poses of subjects in each frame. The poses are given in terms of rigid body segments connected by respective joints as shown in Figure 3. We accept all joints detected by OpenPose with at least 0.0002 confidence. We track each subject in a video using their Mid Hip coordinates. For each subject, missing joints are filled using linear interpolation. This is done for two reasons: First, not all joints are detected in all frames. Second, videos vary in fps, so we reach 50 poses per second after linear interpolation. This is used for all video sources as the target YouTube videos we used vary in fps from 24 to 60, but the training ones we recorded had 50 fps.

3.1.1. Pose Translation and Processing

For every video frame the pose extraction system produces the pixel coordinates of the individual joints (and/or the angles between the respective rigid body segments). The values depend not only on the motion of the user, but also on the camera perspective. The same posture at the top left corner of an image will produce different coordinate values than at the bottom right. For a given motion the position difference between two frames expressed in pixels will be different when executed close to the camera than when executed far from the camera (and will also vary depending on the angle and of course the resolution of the video). On the other hand the sensor values that we want to generate are expressed in absolute physical units and are related only to the user action (and of course sensor placement on the body). In summary we need to address (1) the translation of the poses into coordinates that are independent of the position of the user in the image and (2) a scaling making the magnitude of pose changes between frames invariant with respect to the size of the user's body in the image. Given the power of modern deep learning based computer vision systems one could in principle attempt to have the system learn the corresponding transformation from data. However, this would require a very large amount of data on top of the data needed to learn the dependence of the motions observed in the video and the sensor signals as such. As a consequence our method contains an explicit pre-processing step before feeding the data into the regression model.

As a scaling factor to achieve motion amplitudes independent of the distance from the camera we use body dimensions in the picture. In other words we express the magnitude of motions in "units" of body size as seen in the image. To this end we selected the euclidean distance between Neck and MidHip (joints 8 and 1 in Figure 3) as a scaling factor for the motions expressed in pixels. Since the distance between the camera and the user can change we recompute the scaling factor for every frame. Let's call the euclidean distance at time t the $dist_t$. In order to avoid scaling outliers, we are going to use a sliding window of size 3 s, 1.5 back and 1.5 forward, and compute the median. This value we call the $scale_t$ which is

$$scale_t = median(dist_{t-75}, \ldots, dist_{t+75}) \tag{1}$$

and the function to scale any value v_t is

$$scale(v_t, scale_t) = -1.0 + \frac{v_t}{scale_t} * 2.0 \tag{2}$$

We represent the joint position in a frame of reference centered in the MidHip joint, that is, joint 8 in Figure 3. This means that for each joint other than the MidHip we compute its new x coordinate at time t ($NewJoint_x^t$) and y coordinate at time t ($NewJoint_y^t$) using

$$\begin{aligned} NewJoint_x^t &= scale(Joint_x^t - MidHip_x^t, scale_t) \\ NewJoint_y^t &= scale(Joint_y^t - MidHip_y^t, scale_t) \end{aligned} \tag{3}$$

This provides us with a coordinate system that is independent of the location of the person in the image. Note, that a single scale is used for both x and y, as the aim is to match real space and not to do standard machine learning scaling. In other words, the scale translates pixel distances to joint relative ones, so it should be the same for both axes in order to not deform the poses.

For the MidHip joint we do not perform this procedure, as all joints are now relative to it. In order to capture the overall movements of the human in the picture plane (which can have influence on the acceleration values), we also include the scaled x and y speed of the MidHip, which means adding to our input

$$\begin{aligned} HipSpeed_x^t &= scale(MidHip_x^{t+1} - MidHip_x^t, scale_t) \\ HipSpeed_y^t &= scale(MidHip_y^{t+1} - MidHip_y^t, scale_t) \end{aligned} \tag{4}$$

Using joint positions relative to MidHip also obscures motion of the body towards and away from the camera and its influence on the sensor signals. Fortunately, we can retrieve this information by giving the model access to two extra values related to the $scale_t$ representing the relative speed of scale change

$$speed_t = \frac{scale_{t+1} - scale_t}{scale_t} \tag{5}$$

and the first derivative of this scale speed. To distinguish between the camera moving and the motion of the user in the environment, we can consider the optical flow within the picture as a whole (which we do not do yet as in our sample data the camera was static). As we explained, changes in scale can be a proxy for absolute movement as the subject is coming closer to/moving away from the camera regardless of specific initial coordinates.

3.1.2. Additional Signal Processing and Selection

Signal processing beyond the translation and scaling of the features described above can be considered in two categories. The first is further improvement of the quality of the pose information extracted from video. Here, smoothing of the scaling using the median to remove small shifts in joint position and removing outliers in cases where the network predicts a distorted pose in some of the frames have proven to be most effective. Note that these steps are applied to the video signal only. With respect to the sensor signal, the only prepossessing we have found to be consistently effective was linear interpolation of the sensor data to 50 Hz to match the video frame rate.

The second is the removal from the signal of components that are unlikely to be reliably translated from video to sensor signals (see also Section 2). In essence the idea is to entirely remove certain types of information from our classification process, opting to have less, but more reliable information. Examples are: computing the norm of 3D sensor signal. gravity removal ("linear acceleration") and low pass filtering. The latter was motivated by the insight (see Section 2 point 3) that phenomena such as high frequency "ringing" of the signal caused by sudden, strong impacts that are fundamentally invisible in the video data. This means that the simulated sensor data generated from videos will not contain such "ringing".

These steps need to be applied to both the sensor signals and the video signals and done during: the training of the regression model, the generation of the synthetic training data, and the classification of real sensor data with the model generated using the simulated data. Obviously, in cases where the information is not present in the video signal at all, it is sufficient to remove it from the sensor.

3.2. *Regression Models*

As regression model we selected a residual deep convolutional neural network that uses dilated convolutions, similar to *TCN* [42]. It has been shown to be successful in many sequence modeling tasks, outperforming even long short-term memory models [42]. Our architecture (Figure 5) relies on dilated convolutions and residual connections present in the TCN blocks. Residual connections help train deeper networks by adding the output of convolutions to an identity function [43], while dilated convolutions increase the receptive field by orders of magnitude without increasing the computational cost. By employing the same padding in all convolutions, we maintain the temporal size of the inputs, allowing us to go deeper even when using small temporal windows and map the sequence of poses to the sequence of sensors signals. In order to prevent overfitting, we apply dropout in all TCN blocks except the first and reduce the number of parameters by decreasing the number of filters in the 4th TCN block by applying convolutions with filter size 1. We train one regression model per sensor position and feature. Each model was trained using the mean squared error loss and the Adam optimizer [44] with 0.001 learning rate and 0.9 and 0.999 for β_1 and β_2, respectively. The model was trained for 500 epochs with early stopping using a patience of 25 to avoid overfitting.

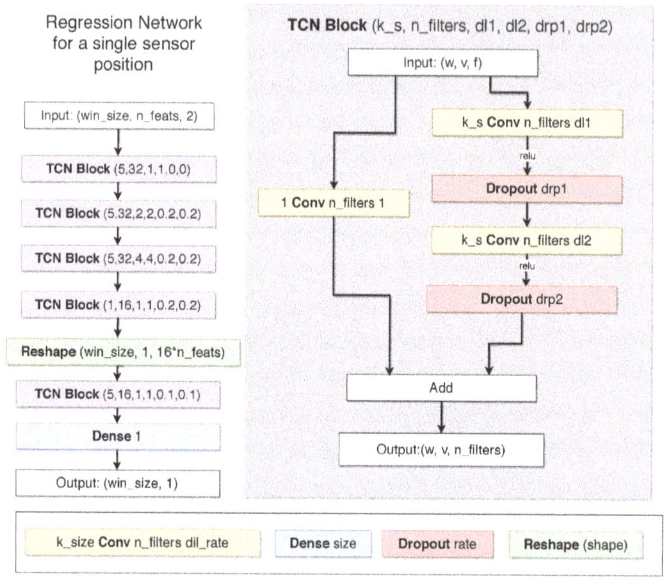

Figure 5. Architecture for the neural network used for regression of a single sensor.

It is important to discuss the motivation for our training regimen. A key question for the training is which information from the video to include in training which sensor.

1. Providing information about the motion of body parts that we know, from physical consideration, to be irrelevant for the signal of a given sensor at a given location will "confuse" the model. Eventually, given enough data, good models will likely learn to distinguish relevant from irrelevant information. However, getting rid of confusing information is known to significantly reduce the amount of required training data and speed up training.
2. Providing information about body parts that, from a bio-mechanical point of view do not influence each other, on the other hand can be counterproductive and lead to overfitting the specific training motion sequences and combinations. As an example, consider the motion of the left and right wrist. With the exception of some very extreme motions (e.g., extremely strong shaking of one wrist making the whole body shake), the motion of one wrist has no influence on the other. However, in many motion sequences there are sometimes strong correlations between the motion of both wrists. The best example is walking when people often swing their arms in sync. When training the system to recognize walking, we want the system to learn such correlations. However, when training a regression that should map motions in an image onto sensor data for arbitrary activities based on physical constraints only this would be undesirable overfitting of artefacts of the specific training sequence.
3. Finally there is the question if a joint model should be trained for all sensor signals, if separate models should be trained for all signals from one on-body location or if we should train a separate model for each signal at each location. Training one model for all locations and signals is not advisable due to the need of avoiding overfitting a specific training set because of spurious correlations between signals from different locations. In principle, training a model to generate all the signals from a given body location should allow it to capture the physical dependencies between those signals and should thus improve the quality of the results. However, in our experiments we have found individual models trained for each sensor modality at each location to perform best. This may be due to insufficient training data. It may also be necessary to introduce onto the model mechanisms for explicitly encoding

physical boundary conditions as proposed, e.g., by the concept of Physically Informed Neural Networks [45].

For each target sensor signal we trained one regression model per possible sensor placement using only the joints relevant for it. We placed sensors on the wrists and calves, as shown in Figure 3, which also shows which joints are used for each position. For each one of those inputs we built a regression model as depicted in Figure 5. Regression is done using a window of size 16 frames (around a third of a second) and step 1. This increases training data and was also selected to avoid overfitting specific movements. The regression output is 16 numbers representing the IMU values for one channel at those times. The input for each regression model using n_j joints is a tensor $(16, n_j + 2, 2)$, that is, the 2D coordinates of the joints involved and also the 2 extra values related to the scale changes and hip speeds mentioned earlier. For the joints involved for each position, see Figure 3.

3.3. Datasets for Training the Regression

Our aim is to train a model that can generate simulated sensor data not just for a single specific set of activities but for a broader domain. Thus having our regression model, application developers should be able to use our model to generate the required training data from online videos for any activity set they are interested in (as long as it is within the broad general domain). The idea is thus to train the regression model on a set of motions that are representative "basic components" of the activities of the respective broader domain. The core question is how to define the "broader domain" and how to identify the corresponding "basic components". For this work we have selected aerobic-like physical training. It is a very broad domain with a great variety of different exercises for which online videos are extremely popular. In terms of applications the ability to monitor the exercises is something that current fitness trackers lack, despite the popularity of such exercises and the associated potentially large user base. In technical terms the domain has the advantage that the corresponding videos tend to provide a stable, easy to process frontal perspective with the relevant parts being clearly visible most of the time (as the whole purpose of the videos is to show the viewers how to move). The activities tend to be characterized by distinct motion of the limbs and posture changes which means that recognition systems should be reasonably robust with respect to sensor data noise and inaccuracy.

With respect to the "basic components" of the motions we first started by showing the participants example videos of the domain and asking them to perform random motions related to those videos. This approach has not produced good results, however, as people tended to repeat the same, not necessarily representative motions. We then turned to high school video material explaining the degrees of freedom of human joints and the types of motions that different muscles can actuate. We combined parts of 7 such videos into an 11 minute compilation [46] and had 8 subjects re-enact the motions from the videos while wearing sensors on the locations for which we wanted to train our regression model (wrists and lower legs). We filmed each user on a different day and our camera placement and angle changed slightly across recordings. Examples of the recordings are shown in Figure 6. For sensor synchronization we begin the recording by holding all IMUs stacked on top of each other and moving them together up and down in front of the camera. After the synchronization gesture is performed at least 3 times, we start recording the generic motions. During a user session, subjects were instructed to try to follow in real time the motions present in our compilation video, which is playing in a monitor nearby. In order to increase the variability of motions, we told them that it was more important to move than to perform the motions correctly. For example, when quick motions happen in the videos, it is better to move quickly, even if one cannot follow the specific quick motion. Only one of the subjects performed those seed motions more than once. His second session was selected for validation, while all others were selected for training the regression.

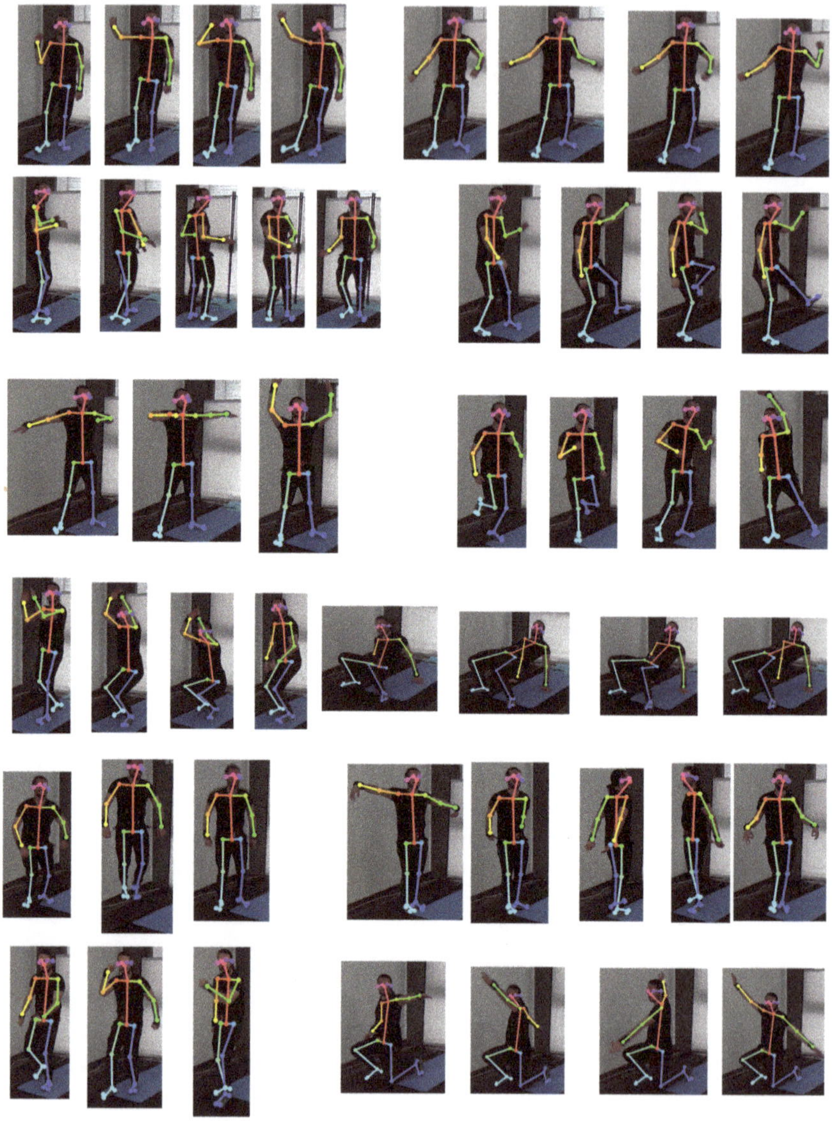

Figure 6. Example of a subject performing some of the movements present in our generic motions dataset used for training the regression model. The video followed can be found in [46].

4. Evaluation

The evaluation procedure was designed based on the understanding that our aim is not to generate signals that are as close as possible to the signals that real sensors would generate when performing the respective motions. Instead we want to facilitate the generation of training data that will allow activity recognition systems to achieve performance that is as close as possible to the performance achievable with training data recorded with real sensors. These goals are related, but not identical. Generating sensor data that is very similar or even identical is certainly a sufficient condition for getting recognition results that are close to what is achievable with real sensor-based training data. However, it is not a necessary condition. Thus, if the system can replicate enough distinct features of the signal to separate the classes in a particular application then the fact that it may miss or fail to reproduce other signal features is of no significance. This is illustrated in Figure 7. The signals shown in the figure are the actual signals we used.

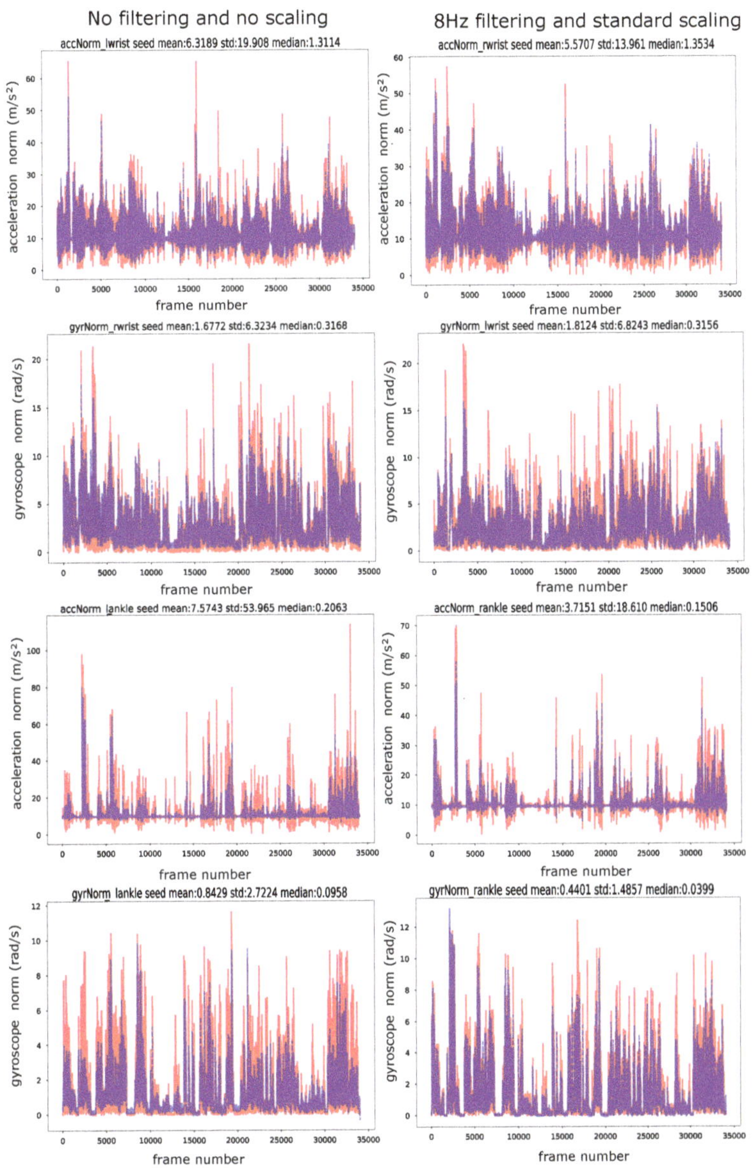

Figure 7. Real signals (red) versus simulated (blue) in the regression training set (generic motions) for the norm of signals.

As a consequence we focus on the evaluation of the performance of activity recognition trained using simulated sensor data generated by our system from respective videos and provide limited analysis of the quality of the generated sensor signal as such (see Section 5.1). The remainder of this section describes the dataset and approach used for the evaluation. The results are presented and discussed in the next section.

4.1. Data Collection for Activity Recognition Based Evaluation

We have selected the broad domain of aerobic like physical exercises for our evaluation since it is an interesting and relevant application area while at the same time being well suited for our approach (see Section 2). Specifically, we selected a set of 10 activities we refer to as the "Drill dataset" and that we have already used in our preliminary work described in [14]. Those are part of a light cardio workout taken from a popular YouTube fitness channel [40]. The recorded exercises can be seen in Figure 2. Altogether we have collected 12 YouTube videos of people doing those exercises. The full list of videos used can be seen in Table 1. In the videos individual exercises are performed sequentially and overall they contain about 24 min of usable footage of the 10 activities. In addition to the YouTube material we have recorded data with volunteers at our lab. To this end we have equipped them with IMUs (XSENSE) fixed to their lower legs and wrists (see Figure 3 for the sensor placement) and asked them to imitate the exercises in the video. We have recorded the subjects on video with a camera perspective similar to the online videos to ensure that we have not only real sensor data but can also generate simulated data for the subjects using our regression model. Overall we had 28 subjects performing the 10 exercises. Of those we have 17 users who performed a single session only, which are considered the training set. The test set consists of the 11 users who performed more than one session (5 performing 2 sessions, 4 doing 3 and 2 performing 4 for a total of 30 sessions). This gives us a diverse training set, a large number of testing sessions and the ability to test on the most difficult scenario: the training and testing sets being different users. For an overview of the datasets used and the length and number of users in each one, see Table 2.

Table 1. YouTube videos used to generate artificial sensor data. Each video has a single subject performing one or more activities and thus we have 14.28 min of the target activities in total. All accessed at 6 July 2020.

Video URL	Activity (ies)	Seconds Used	fps
https://www.youtube.com/watch?v=7X2Yx29DdBY	Cross Toe Touches	113	30
https://www.youtube.com/watch?v=8gLdmb9Ivkw&LateralSteps+Pulls	32	30	
https://www.youtube.com/watch?v=9-jBOcGeQcg	4 Torso Twists + Knees	16	30
https://www.youtube.com/watch?v=afghBre8NlI	Squats	83	30
https://www.youtube.com/watch?v=enz5TSRMmyM	High Knee + Pulls	13	25
https://www.youtube.com/watch?v=g-S1c-Scu3E	Lateral Steps + Pulls	33	30
https://www.youtube.com/watch?v=Kn621fAVEEI	Boxer Shuffle	9	30
https://www.youtube.com/watch?v=MG8DJpN-35g	4 Torso Twists + Knees	48	30
https://www.youtube.com/watch?v=mGvzVjuY8SY	Squats	100	30
https://www.youtube.com/watch?v=oMW59TKZvaI	Slow Rocking Butt Kickers	10	24
https://www.youtube.com/watch?v=ZiJdpPJbqYg	Front Kicks	4	30
https://www.youtube.com/watch?v=R0mMyV5OtcM	All of the Drill Activities	303.81	60

Table 2. Details about the datasets used. Every drill session lasts 5 min, while each of the generic motions last 11. For our generic motions, we use one session of each subject for training the regression and the second session of one for validation. For the Drill dataset, we use users with a single session as the training classification set and those with more than one as the test set. No user is shared between any of the datasets.

Dataset	Video	Sensor Data	Subjects	fps	Total min
Collected generic motions	yes	yes	8	50	99
YouTube videos	yes	no	12	24 to 60	14
Drill Dataset Training	yes	yes	17	50	85
Drill Dataset Test	no	yes	11	-	140

4.2. Evaluation Procedure

The evaluation procedure compares various ways of training the system using simulated sensor data generated by our method with the baseline of a system trained using real sensor data. For the baseline we use the sensor data from the volunteers in our Drill data with the 17 users who have recorded one session only being used for training and the remaining 11 users with their total of 30 sessions as testing set. For the generation of training data from videos using our regression model we have the 12 YouTube Videos and the videos of the volunteers of our Drill experiments. This way we test our regression model on activities which it has not seen before. In summary we consider three different sources of training data:

1. The real IMU data from our training users in the Drill dataset. This is the "gold standard". Systems trained with this data should perform best.
2. Simulated IMU data generated from the videos of our training users in the Drill dataset. Here the system trained on data generated from video is trained on exactly the same users and activity instances as the baseline system ensuring the "fairest" and most consistent comparison.
3. IMU data generated from the 12 YouTube videos.

When investigating how well the generated sensor data works for the training of new activity recognition applications we consider four more specific questions:

1. What is the effect of different pre-processing techniques described in Section 3.1 on the performance difference between a system trained on the real sensor data and one trained on data generated from video using our approach? In this context it is important to also consider the effect of the respective techniques on the absolute performance of the baseline system. Thus, we do not want to consider techniques that may reduce the difference between simulated and real sensor data based systems at the price of making both systems significantly worse.
2. How well does our approach perform for different types of sensor signals (see Section 2) ? Again we need to consider not only how the different sensor choices affect the difference between simulated and real sensor data but also how they impact the absolute performance.
3. Can we compensate for potentially inferior quality of the training data generated from videos by providing a larger amount of training data? Given that our approach gives the user access to huge amounts of data contained in online videos (much more than can realistically be collected as labeled data with on-body sensors), it is not necessary to match the performance of real sensor data in tests on the same sized training sets (as done with respect to points 1 and 2 above). Instead we need to investigate what happens when we provide the system, using simulated sensor data, with more training examples than the real sensor data based system.
4. Can the quality of training based on simulated sensor data generated from videos be improved by combining it with a small amount of real sensor data? The question is if small amounts of real sensor data, that can be easily collected, can "fine-tune" HAR systems trained on large amounts of simulated data to improve their performance.

4.3. Recognition Approach

Since the aim of this work is not to optimize the recognition of a specific set of activities but to evaluate the usefulness of simulated sensor data generated from video for wearable HAR systems in general, we use a standard architecture that has proven to be well suited for a variety of activity recognition tasks. The full network architecture can be seen in Figure 8, and, just like the regression model, is based on [42]. The input to the recognition system is 2.56 s (128 frames) long windows. This window size is sufficient to capture the motions of each exercise, which is repeated many times in a span of 30 s. Since all convolutions in this architecture use *same* padding, the output class predictions are on frame, not on window level. This allows the network to predict with finer granularity and better deal with cases where two activities are happening inside the same window (transition between activities). For the final prediction we apply majority voting to each window. We trained the network using the categorical cross-entropy loss function and the Adam optimizer [44] with 0.001 learning rate and 0.9 and 0.999 for β_1 and β_2, respectively. Each model is trained for 500 epochs with early stopping using a patience of 25 to avoid overfitting. If any real (not simulated) sensor data is included in a test, a stratified 10% random selection of it is used as the validation set. If no real data is included, the validation set consists of the same selection strategy but applied to all the simulated data. This guarantees the quality of the validation set, while avoiding unfair advantage to configurations that combine simulated and real data. If we simply selected a stratified 10% of all data for validation, tests that combine data (simulated and real) could benefit from having more real points in the training set.

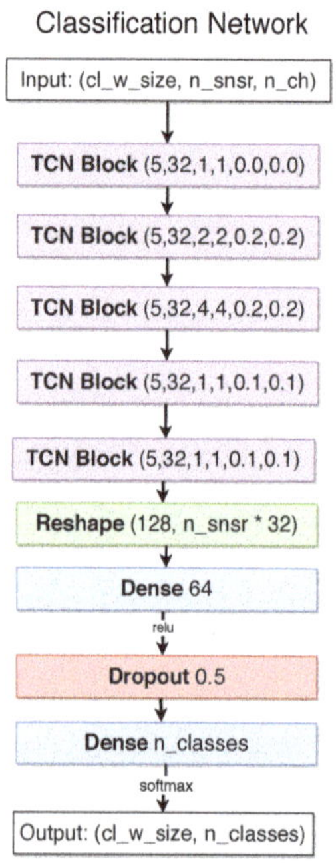

Figure 8. Architecture for the neural network used for classification. For a definition of the blocks see Figure 5. The size of the window for classification was 128, representing 2.56 s and the number of classes in our target dataset is 10.

5. Results

5.1. Signal Level Evaluation

Since the aim of our work is not to generate signals that mimic real sensor ones as exactly as possible, but to facilitate training of activity recognition systems that will later be applied to real signals, the key performance indicator is not some sort of numerical dissimilarity measure but the ability to replicate relevant, characteristic signal features that can be used for class discrimination. As the latter is difficult to quantify on signal level, we focus on the quantitative evaluation on the classification tasks with the results being presented in Sections 5.3 and 5.2.2. Nonetheless it is informative to have a qualitative look at the signals that our system generates for various sensors and how they compare to the real sensor signals generated in the same situation. To this end we consider situations where we have video and sensor data for the same activity. We then plot over each other the actual sensor signal (red in all figures) and the corresponding simulated sensor signal generated by our system (blue in all figures). We begin by looking at signals generated for the dataset that we used for training the regression. This is in a way the easiest task as we do not have to deal with the question of how representative the training dataset that we assembled for our regression model is. In Figure 7 we first consider examples of signals for the accelerometer and gyroscope norm $\sqrt{x^2 + y^2 + z^2}$. This avoids problems associated with the difficulty of estimating small rotations around the limb axis which have influence in particular on the projection of the gravity vector onto the individual sensor axis. The signals are shown for a period of time of around 10min. For each signal we show the unprocessed signal and a version filtered with 8 Hz (the impact of the filtering will be discussed later). General observations from Figure 7 are:

1. Overall the simulated signal has the same trends and large scale features as the original signal.
2. Many of the smaller features such as distinct peaks are also matched by the simulated signal. However, some are missed or have a much smaller amplitude in the simulated signal.
3. The main difference between the simulated and the real signal is in the amplitude. In most cases the simulated amplitude is smaller, but not by a constant factor that could be overcome with simple scaling. The underestimation is particularly pronounced for high frequency peaks. This can be explained by a number of factors. First of all high frequency components ("details") tend to be in general more difficult to reproduce and given the limited size of our training set, some limitations in this area are not surprising. Second, in particular for the acceleration sensors, some of the peaks are due to physical effects which are not present or very difficult to capture in the video (e.g., high frequency "ringing" after sudden impact, see Section 2).
4. Especially in the acceleration signals there is a "baseline shift"-like effect (especially visible in the third row of Figure 7). At times the system seems to completely ignore parts that are below a baseline situated just below 10 (corresponding to around 1 g). This can be attributed to downward motions where the earth gravity is subtracted from the acceleration. Thus a free falling object (accelerating downwards with 1 g) experiences no acceleration force (is weightless). This means that the acceleration norm is smaller than 1 g (actually 0 in free fall), at least as experienced by the real sensor. For all other motions on the other hand, the value of the acceleration is always equal to or above, at least in the norm, 1 g. Given the limited size of the training set for the regression model and the fact that we did not include any semantic analysis to detect the "down" direction it is not surprising that our model struggles to capture this phenomenon. One way of dealing with the problem is to use linear acceleration which can be derived by IMUs, which is discussed below.

To get a better understanding how the system handles detailed signal features Figures 9 and 10 show zoomed views of 2 signal segments that were selected to cover the "good" as well as the "bad" cases. Figure 9 shows for each signal a comparison of the acceleration and gyro norms. To explore in more detail the issue raised in point 4 above Figure 10 shows not just the acceleration norm, but also the norm of the linear acceleration which excludes the gravity contribution. The main observations are:

1. The left part of Figure 9 shows an example of a "good" simulation. It can be seen that the simulated gyro signal nearly perfectly tracks the real gyro, even through fairly subtle features. Except for the amplitude issues the same holds for the accelerometer signal. Note that while the signal segment contains subtle and fast components, it does not have very high frequency "singular" peaks, which is a key reason why the system works so well here.
2. On the right side of Figure 9 a case of much poorer performance is shown. This is largely due to the presence of many singular high frequency, high amplitude peaks which especially the acceleration signal fails to track correctly.
3. With respect to the linear acceleration Figure 9 shows two things. First we see on the right side that using the norm linear acceleration instead of the norm raw acceleration signal indeed does solve the problem of the system refusing to model signal values below the baseline of 1 g. On the other hand, as shown in the left part of the figure, using linear acceleration may lead to signal characteristics changing and acquiring additional high frequency peaks which can be difficult to model.

To probe further we have plotted signals from individual acceleration axes as both raw acceleration and linear acceleration in Figure 11 including a zoomed in version in Figure 12. The single axis acceleration signal is richer in features than both the linear acceleration on the same axis and the norm signals discussed above. This in itself is well known and not surprising. What is surprising is how well the simulated signals replicate the real ones given the issues described in Section 2 (e.g., the difficulty of estimating the projection of gravity into individual axis). This applies to both the overall, broad structure of the signal (Figure 11) and, in many cases to the detailed structure. An example is illustrated in Figure 12 on the left where some very subtle features have been matched nearly perfectly. Clearly, other examples exist, as shown on the right of the Figure where there is much more difference between the features of the simulated and the real signal.

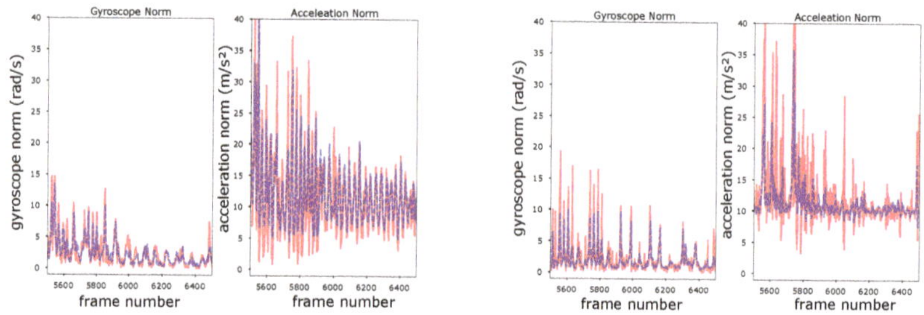

Figure 9. Real signals (red) versus simulated (blue) in the regression training set (generic motions) for acceleration and gyro norm. All cases trained without neither filtering nor standard scaling.

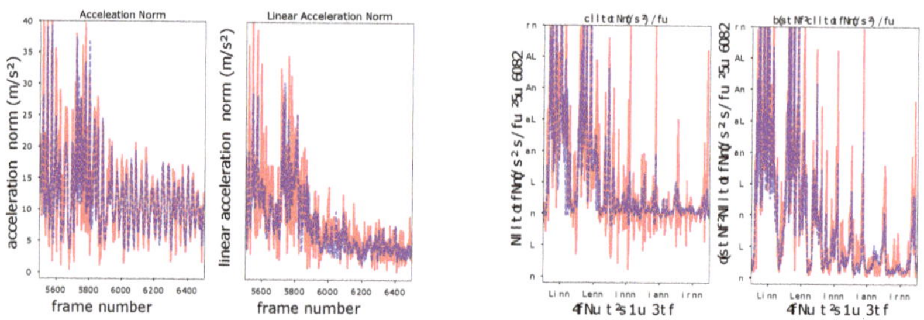

Figure 10. Real signals (red) versus simulated (blue) in the regression training set (generic motions) for acceleration norm, linear and not. All cases trained without neither filtering nor standard scaling.

5.2. Effects of Pre-Processing

From the above discussion it is apparent that the two main concerns are modelling high frequency components such as acceleration "ringing" caused by sudden forceful impacts and exact matching of the signal magnitude. This suggests the use of low pass filtering and scaling as obvious pre-processing approaches. Both are parts of our pipeline (Section 3.1). Note that we are talking about standard scaling when **learning the regression**; that is, with the mean and standard deviation computed on our dataset of generic motions being performed by other users. As the mean may be different in this dataset, we simply undo the standard scaling, which is fully reversible, when obtaining simulated signals.

Other pre-processing parameters within our pipeline are the sliding window sizes for training the regression. We experimented with window sizes of 50 and 16 values, representing 1 s and 0.32 s, respectively. In our early experiments training the regression models, it became clear that the smaller value was better, providing a smaller regression loss and better overall classification. Thus, for brevity, we report all results with the smaller window size. Regarding window step, we keep it at 1 to increase the amount of training data.

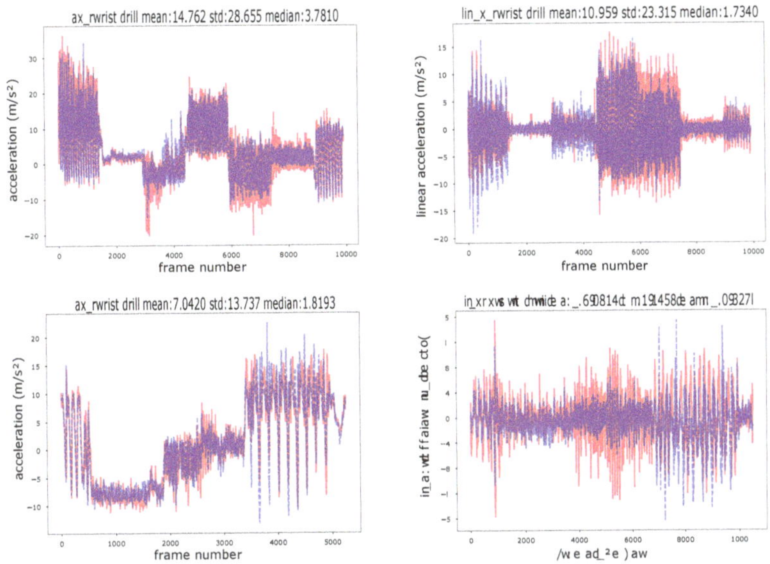

Figure 11. Examples of the signals generated by our system in the target dataset (fitness exercises). Specifically we show the acceleration parallel to the lower arm (left) and the linear acceleration along the same axis (acceleration without the gravity component that Inertial Measurement Units (IMUs) can derive with the help of gyro and magnetic field sensors).

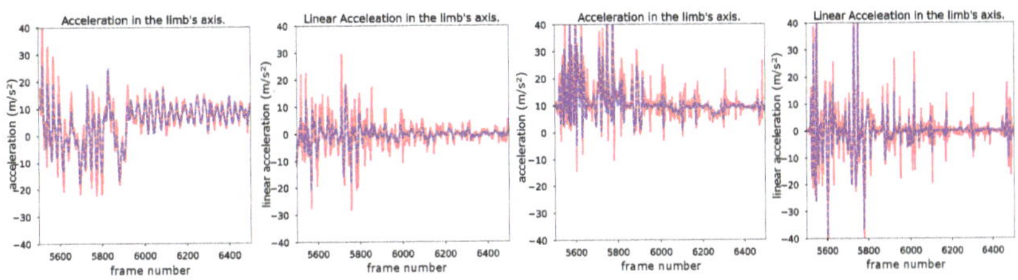

Figure 12. Real signals (red) versus simulated (blue) in the regression training set (generic motions) for acceleration and linear acceleration both in the limb's axis. All cases trained without neither filtering nor standard scaling.

5.2.1. Signal Level Effects of Filtering and Scaling

In Figure 13 we illustrate the effect of frequency filtering and scaling on the performance of our regression model. We consider low pass filtering with 12 Hz and 8 Hz (given the original rate of 50 Hz from the video signals). The low pass filtering itself was done using a butterworth low pass filter of the 6th order and standard scaling was done by removing the mean of the sensor channel in the training set and dividing by its standard deviation. Example comparisons of a signal with no filtering and no scaling on one hand and with 8 Hz filtering and scaling on the other are also shown in Figure 7 (left and right column respectively).

Key observations are:

1. Overall there is no dramatic effect. In all cases the simulated signal follows the structure of the real signal with reasonable accuracy while displaying the same types of problems.
2. Without scaling (left column on Figure 13) there are some pronounced amplitude outliers in the high frequency peaks amplitudes (between sample 1000 and 2000) that

are not influenced by the frequency filtering. These disappear in the scaled versions (right column); however, at the cost of significant underestimation of the amplitude in the same area and overestimation around 3000.

3. In Figure 13 frequency filtering has little effect, except for reducing the underestimation of the components below 1 g around sample 3000 in the no scaling case (which overall seems to be the best).

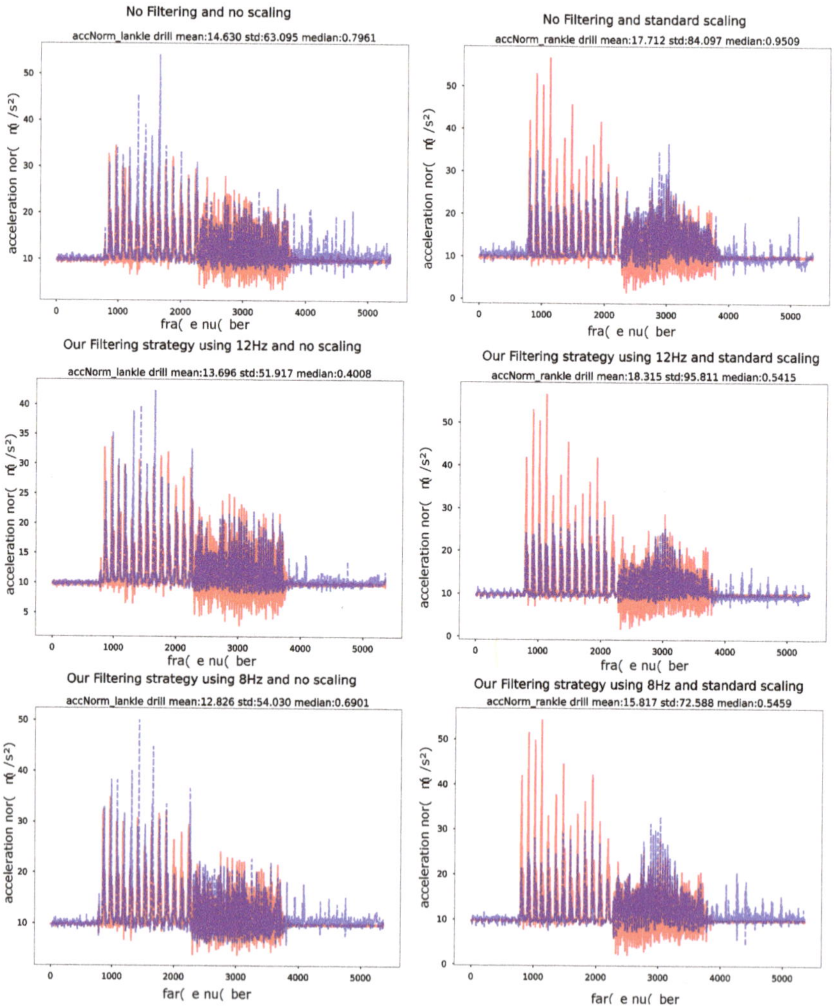

Figure 13. Examples signals generated by our method (blue) and their real counterparts (red) in the target dataset (fitness exercises) under different filter and scaling strategies.

In summary while both filtering and scaling do have benefits in some situations, qualitative signal examination does not indicate it to be decisive or obvious. To get a more quantitative idea, Figure 14 shows the Mean Squared Error for computed over all our samples for acceleration and gyroscope for different filter values. For acceleration the error is indeed smaller for 8 Hz (but not by much). For the gyroscope signal there is no significant difference.

5.2.2. Classification Level Effects of Filtering and Scaling

As a final evaluation step we did an analysis regarding the impact of filtering and scaling on the classification performance. To this end we need to consider not only how the pre-processing impacts the comparison between the classification within simulated sensor data, but also how it impacts the performance of the absolute recognition rates of the real sensor data based models. It makes little sense to apply pre-processing methods that make the results of the simulated signals based model equal to that of a real signals based one but at the cost of making both much worse. In Figure 15 we thus consider the effect of 8 Hz filtering and scaling in different combinations on the recognition for different placements (wrist and ankle) of the sensor for the real and the simulated data. The training was done on all the data designated for training and the testing on all testing data as described in Section 4.2. We can see that the recognition rates on the real data show little sensitivity to the filtering and scaling. The performance on the simulated data is also fairly invariant for the wrist sensor placement while for the ankle placement frequency filtering does indeed make a significant difference. However, it must also be noted that for the ankle placement the recognition rate with the simulated data is very poor to start with (around 40%).

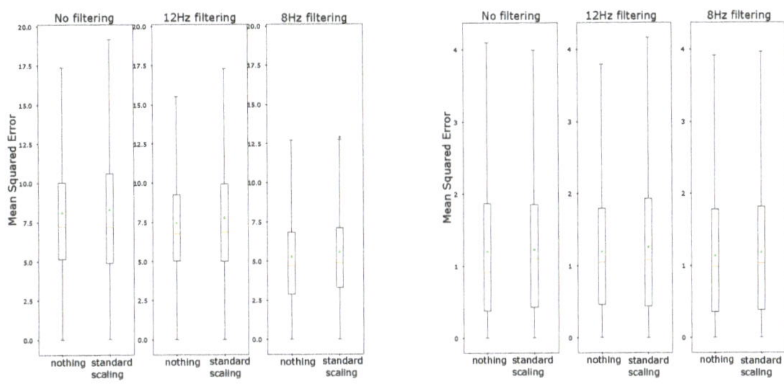

Figure 14. Loss for all windows using different training strategies for acceleration (**left**) and gyroscope norm (**right**).

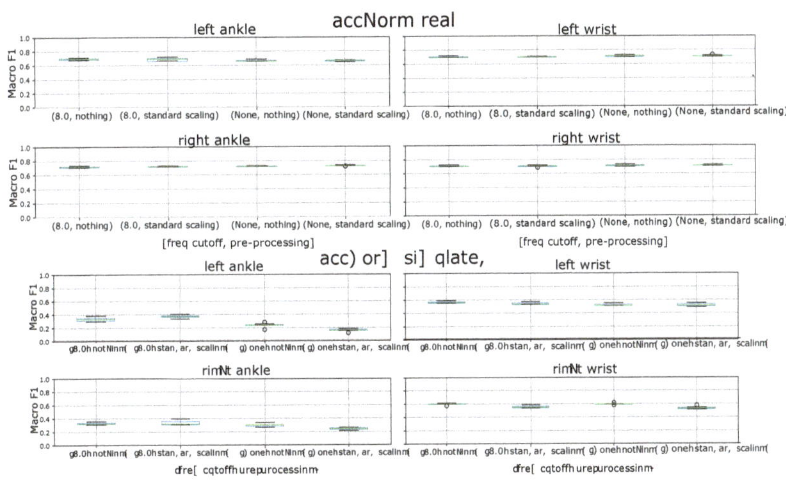

Figure 15. Comparison of classification performance for models trained with real (**upper**) or simulated data (**down**) under different preprocessing techniques using the acceleration norm. Results are shown for each different sensor placement separately.

5.3. Evaluation of Classification Performance Using the Simulated Signals

The discussion in the previous section has shown that, mostly on a qualitative level, our method generates signals that replicate a significant portion of the respective real sensor output within our application domain. We now proceed to quantitatively evaluate how well such simulated sensor signals are suited for training activity recognition that will later be applied to real sensor signals. The evaluation procedure has already been described in Section 4.2. Essentially we have recorded a dataset where for each activity and user we have both a video signal and sensor data. In addition, we have collected some YouTube videos for the same/similar activities. We then train classifiers using the real sensor data, simulated sensor data from the videos that we have recorded, simulated sensor data from the YouTube videos and various combinations thereof. The classifier trained on real sensor data is the one to which we compare the performance of the different combinations to evaluate the usefulness of our approach. Given the discussion in the previous section, we focus on using raw data without filtering and scaling with some examples of the impact of 8 Hz filtering. The results are shown in Figures 16–18. In each figure, we plot the recognition rate of each classifier vs. the number of users in the training set. The only exception is the case of sensor data generated from YouTube videos where we had different users perform different exercises from the set which means that the notion of "number of users" for the whole dataset makes no sense (see Table 1). Instead, we have generated training data from the entire 14 min of the YouTube data that we harvested and plotted it for comparison as a constant in the graphs.

1. **Baseline.** The recognition rates on the real data reach up to 90% which, given the fact that our aim was not to optimize a recognition system for a specific application, is an acceptable baseline to evaluate the performance of the simulated sensor data.
2. **Overall performance on simulated sensor data.** Systems trained on the simulated data reach up to 80% recognition rate, which is 10% below the results (see e.g., Figure 16 left where the acceleration norm based recognition is among the best performing variants overall). In general as the number of users is small the results for real and simulated data are very similar to the performance of the real data. The performance on the real sensor data improves faster with the amount of training data. This is to be expected as with a very small number of users (equal a small amount of training data) the recognition rate tends to be poor and limiting factor for the performance is the lack of diversity in the data and not the data quality. As the amount of data increases, the quality becomes the limiting factor, which is better for the real sensor data.
3. **Performance on YouTube data.** The performance of the system based on sensor data simulated from YouTube videos is in most cases slightly below the performance of the baseline on a single or two users. Given that the total amount of YouTube data (around 14 min) is in the order of magnitude of the length of the data from a single user this is not surprising.
4. **Compensating deficiencies of simulated data through training set size.** There is strong indication that deficiencies of the simulated sensor data can be compensated by the amount of such data. Thus in most cases (see discussion of different sensor combinations later on) systems based on simulated sensor data can match the performance of the real sensor data based system trained on about half as much data. Again looking at the acceleration norm from all sensors case in Figure 16 left top we see that the Mean F1 score for simulated data with 12 users is about the same as the score for 6 users with real data. Furthermore, while we see a slower growth of the F1 score with the number of users in the training set for the case of simulated data than for real data, in most cases growth exists.
5. **Effects of model fine tuning using real sensor data.** Another way to improve the performance of the simulated sensor data based systems is to use small amounts of real sensor data to fine-tune it. The idea is that collecting a small amount of real data is nearly always possible, it is the bulk of a large training data set that is difficult

to collect and that we want to get from existing videos. The results of this strategy are illustrated in Figure 17. We compare the baseline to the performance of a system trained on simulated data to which 1 (blue points) or 2 (yellow points) users with real data have been added. We can see that the simulated data is now much closer to the real one. In the top left graph showing the acceleration norm results it is virtually identical up to 10 users when the real data starts to overtake the simulated one. The effect is even more significant for the gyro norm based recognition (see Figure 17 right top) where the purely simulated signals based model trailed the real signals based ones by 30% and more (see Figure 16 left bottom) and are now not worse than about 10% worse (and up to 6 users nearly identical).

6. **Effects of adding simulated data from YouTube to small amounts of real data.** Another situation where a combination of real and simulated signals may be useful is when we have only a small amount of sensor-based training data with no easy possibility of collecting more. We then try to "top up" our training set with some simulated data from video. The effect of such a strategy is illustrated by the green points (real IMU data with YouTube) in Figure 17 where we combine the real sensor data with the YouTube data increasing the number of real data over the x axis. We can see that for small user numbers the strategy indeed helps.
7. **Performance of different sensors and sensor combinations.** Most of the discussion so far has been done with respect to the acceleration norm (Figures 16 and 17) applied to both wrists and both ankles together which has been the most effective modality with respect to the recognition performance both in the real sensor signal and with respect to how well the simulated signal can approximate it. Given the type of activities that we use as our test set this is not surprising. The activities are characterized by periodic, mostly strong motions of various limbs. In most cases the relative temporal pattern and relative intensity of the motions is a characteristic feature. At the same time, the acceleration norm (in particular when looking at the relative intensity and temporal patterns) is fairly robust against variations and noise. This also means that imperfections caused by the simulation of the signal from the video data using our regression model can be well tolerated.

   Further sensor modalities and their combinations that we investigated are:

   (a) *Acceleration norm vs. Gyro norm vs. combination.* In Figure 16 we have in addition to the acceleration norm the linear acceleration norm, the gyroscope norm, and the combination of gyroscope and acceleration norms (in all cases for all the 4 sensor placements). The first thing we see is that the linear acceleration leads to a poorer overall performance while having little impact on the relative performance of the simulated data. The former is not surprising since the gravity component (meaning vertical orientation) is a very important piece of information. The latter is contrary to what we might have expected since our model was not tuned to capture gravity effects (as discussed in Section 2). Apparently it was still able to capture a sufficient amount of orientation related information. Second, we see that the gyro signal has significantly worse relative performance for the simulated signal. Given the limitations on recognizing rotations around certain axes (in particular the limb axis) from the video signal discussed in Section 2 this was to be expected. The very poor performance of the simulated gyro signal also drags down the combined acceleration/gyro norm performance shown in the bottom right part of Figure 16.

   (b) *Acceleration and linear acceleration on the limb's axis.* Further looking at the impact of linear acceleration Figure 18 shows the results for raw acceleration and linear acceleration along the limb axis. The idea is that the acceleration along the axis is independent of the rotation around the axis, which, as already explained is hard to exactly capture from video. It can be seen that both have slightly lower performance than the acceleration norm (which is not surprising given that the acceleration norm is a good discriminator for our set of activities as described above). Within expected statistical bounds the difference between the simulated and the real sensor signal is about the same for both the raw and the linear acceleration.

(c) *Different sensor placements.* In Section 5.2.2 we have already considered the use of sensors from only a wrist or only the ankle (with respect to acceleration norm) as shown in Figure 15. The performance for the real sensor data on both the wrist and the ankle and the simulated sensor data on the wrist was with around 10–20% less than for the combination which is not a surprising result (both legs and arms are relevant for our test activities). What may be surprising at first is the fact that the performance for the ankle using simulated data is half that of the real data (30–40%). The explanation is that foot motions involve a lot of hard ground impacts that lead to "ringing" that has already often been mentioned as something that the simulated data cannot replicate (which is why 8 Hz filtering helps a lot here). In addition vertical orientation plays a bigger role than for the wrists and there are many motions "to the front" (towards the camera) which are more difficult to resolve due to the viewing angle.

8. **Effects of 8 Hz Filtering.** The effect of various pre-processing strategies has already been discussed in Section 5.2 establishing that while filtering and to some degree scaling did seem to be beneficial in some cases, overall the effect was not overwhelming given that these effects may be strongly application specific. This is why we have decided to do most of the analysis in this section on the unfiltered, not scaled signal. However, for comparison Figure 19 includes the same evaluation done in Figure 16 with 8 Hz filtering. We can see that, while for the gyro norm (which was the poorest one to start with) the filtering does indeed improve the relative performance of the simulated signal based model, it has little effect otherwise.

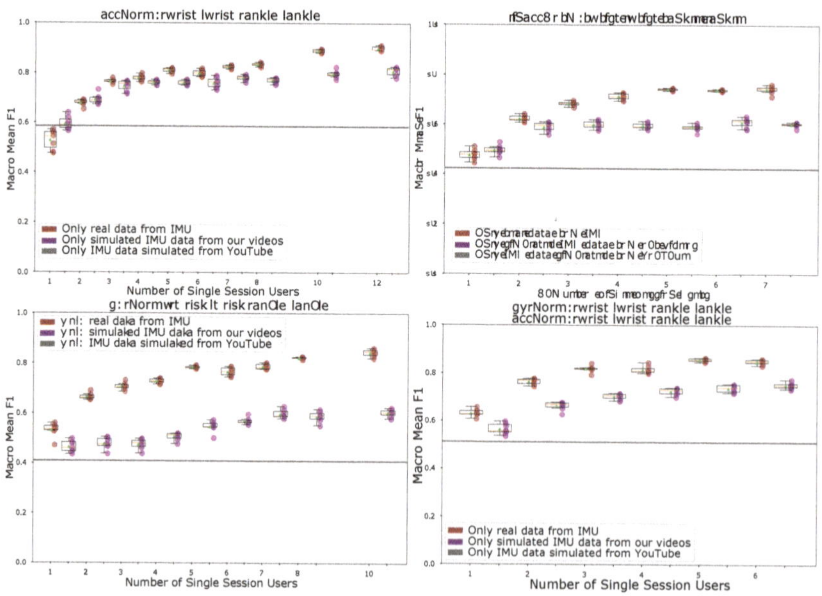

Figure 16. Results when using the accelerometer and gyro norm and the combination thereof.

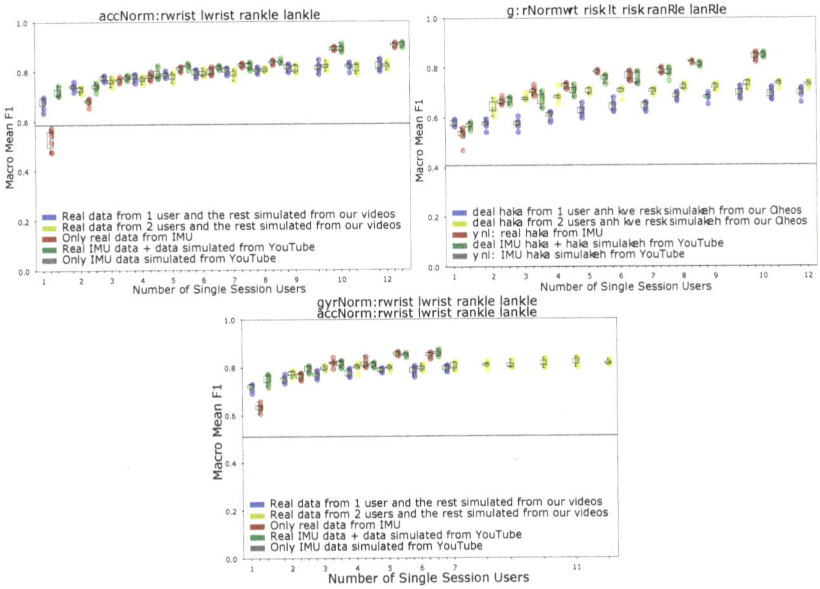

Figure 17. Results when using the gyroscope and accelerometer norms and adding simulated data to the real one.

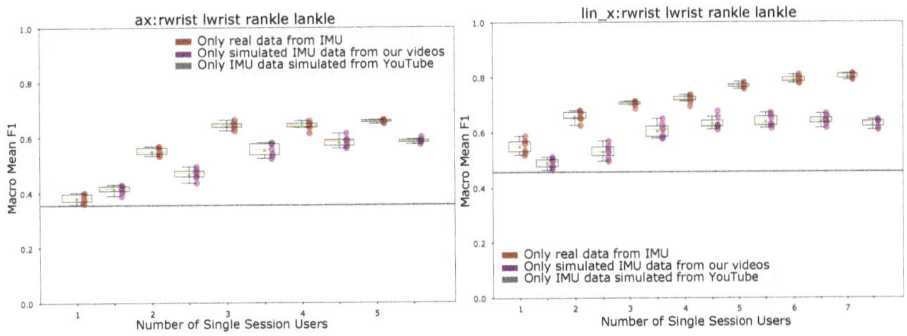

Figure 18. Results when using the linear acceleration in the limb axis.

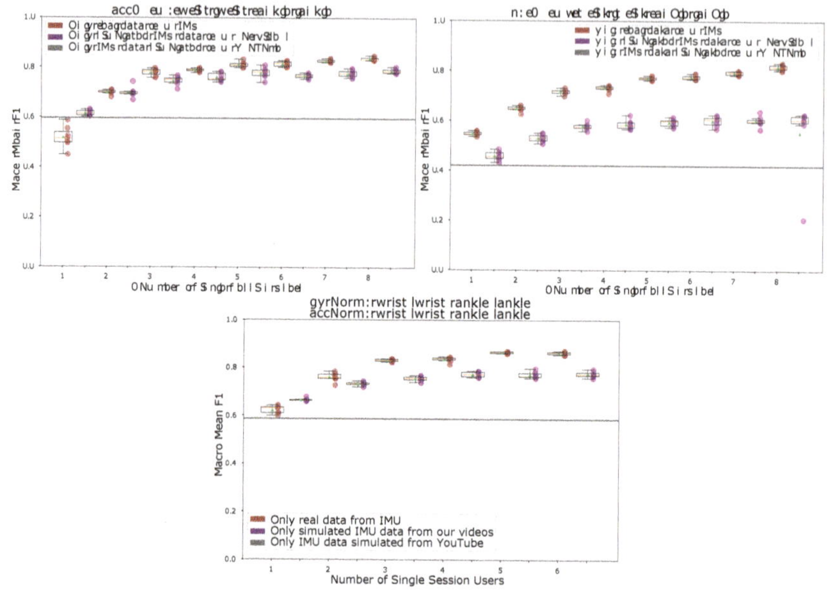

Figure 19. Results when using the gyroscope and accelerometer norms and also 8 Hz filtering.

6. Conclusions and Future Work

The main result of our work is the demonstration of the basic feasibility of using deep network regression models to generate simulated IMU signals directly from video extracted postures. This includes the detailed analysis of the influence of different design decisions on the performance and also includes discussion of the problems that need to be addressed to improve the results. In the long term, this work should contribute to the creation of training datasets that are comparable in size to what exists in computer vision today and thus facilitate a more profound impact of deep learning techniques on wearable sensors based HAR.

Clearly the work described here is just a first step towards this vision. Based on the results described and discussed in the previous section we consider the following to be the most promising next steps that we will investigate:

1. Extending the training data set for our regression model, in terms of the number of users, the variety of motions and the number of sensor placements. This includes also handling different camera positions and more complex poses. For example, lying on a mat or swimming.
2. Collecting a large set of online videos for a more diverse set of activities, and further testing and fine-tuning our model on them.
3. Extending the regression model to handle all types of sensor signals (e.g., acceleration and gyro on each axis) better. This includes considering relations between sensor signals e.g., the orientation of the wrist IMU in the frame of reference of the hip or back mounted IMU which is often used as relevant information in wearable activity recognition systems.
4. Implicit modeling of gravity by exploring image segmentation based detection of "down" direction and including the angle towards the down vector in the features that we feed our regression model. This will be particularly important as we start looking at more and more complex sensor signals as mentioned above.
5. Exploring training the regression using the concept of Physically Informed Networks [45] that incorporate physical constraints as regularization terms. The idea is to build a regression model that for example simultaneously trains the ax, ay and az component of the acceleration and the corresponding norm $|a|$ and embedding the

$|a| = \sqrt{ax^2 + ay^2 + az^2}$ in the regularization term. Relationships between the accelerometer and the gyroscope sensor signal at the same location and temporal conditions could also be considered. These are all especially relevant for generating more complex sensor signals as described in the previous points.

6. We could also explore applying our method to (more) easily identifiable locations where sensors should be placed on the body for a specific set of activities. Before recording sensor data, videos of those activities can be collected and sensor values simulated for different on-body locations, which then can help select the best locations to place them when recording training data.

Author Contributions: Conceptualization, V.F.R. and P.L.; Data curation, V.F.R. and K.K.G.; Formal analysis, V.F.R. and P.L.; Funding acquisition, P.L.; Investigation, V.F.R. and K.K.G.; Methodology, V.F.R. and P.L.; Project administration, V.F.R. and P.L.; Resources, P.L.; Software, V.F.R.; Supervision, V.F.R. and P.L.; Validation, V.F.R.; Visualization, V.F.R. and P.L.; Writing—original draft, V.F.R. and P.L.; Writing—review and editing, V.F.R. and P.L. All authors have read and agreed to the published version of the manuscript.

Funding: The simulations were executed on the high performance cluster "Elwetritsch" at the TU Kaiserslautern which is part of the "Alliance of High Performance Computing Rheinland-Pfalz" (AHRP). We kindly acknowledge the support under the AI Enhanced Cognition and Learning project, Social Wear and Humane AI Net in general.

Conflicts of Interest: The authors declare no conflict of interest.

References

1. Lara, O.D.; Labrador, M.A. A survey on human activity recognition using wearable sensors. *IEEE Commun. Surv. Tutor.* **2012**, *15*, 1192–1209. [CrossRef]
2. Wang, J.; Chen, Y.; Hao, S.; Peng, X.; Hu, L. Deep learning for sensor-based activity recognition: A survey. *Pattern Recognit. Lett.* **2019**, *119*, 3–11. [CrossRef]
3. Sun, C.; Shrivastava, A.; Singh, S.; Gupta, A. Revisiting unreasonable effectiveness of data in deep learning era. In Proceedings of the IEEE International Conference on Computer Vision, Venice, Italy, 22–29 October 2017; pp. 843–852.
4. Brain, D.; Webb, G. On the effect of data set size on bias and variance in classification learning. In *Proceedings of the Fourth Australian Knowledge Acquisition Workshop*; University of New South Wales: Sydney, Australia, 1999; pp. 117–128.
5. Wang, L.; Gjoreski, H.; Ciliberto, M.; Lago, P.; Murao, K.; Okita, T.; Roggen, D. Summary of the Sussex-Huawei Locomotion-Transportation Recognition Challenge 2020. In *Adjunct Proceedings of the 2020 ACM International Joint Conference on Pervasive and Ubiquitous Computing and Proceedings of the 2020 ACM International Symposium on Wearable Computers*; Association for Computing Machinery: New York, NY, USA, 2020; pp. 351–358. [CrossRef]
6. Russakovsky, O.; Deng, J.; Su, H.; Krause, J.; Satheesh, S.; Ma, S.; Huang, Z.; Karpathy, A.; Khosla, A.; Bernstein, M.; et al. ImageNet Large Scale Visual Recognition Challenge. *Int. J. Comput. Vis. (IJCV)* **2015**, *115*, 211–252. [CrossRef]
7. Knoll, F.; Murrell, T.; Sriram, A.; Yakubova, N.; Zbontar, J.; Rabbat, M.; Defazio, A.; Muckley, M.J.; Sodickson, D.K.; Zitnick, C.L.; et al. Advancing machine learning for MR image reconstruction with an open competition: Overview of the 2019 fastMRI challenge. *Magn. Reson. Med.* **2020**, *84*, 3054–3070. [CrossRef] [PubMed]
8. Reiss, A.; Stricker, D. Creating and benchmarking a new dataset for physical activity monitoring. In Proceedings of the 5th International Conference on PErvasive Technologies Related to Assistive Environments, Crete, Greece, 6–8 June 2012; pp. 1–8.
9. Reiss, A.; Stricker, D. Introducing a new benchmarked dataset for activity monitoring. In Proceedings of the 2012 16th International Symposium on Wearable Computers, Newcastle, UK, 18–22 June 2012; pp. 108–109.
10. Chavarriaga, R.; Sagha, H.; Calatroni, A.; Digumarti, S.T.; Tröster, G.; Millán, J.d.R.; Roggen, D. The Opportunity challenge: A benchmark database for on-body sensor-based activity recognition. *Pattern Recognit. Lett.* **2013**, *34*, 2033–2042. [CrossRef]
11. Roggen, D.; Calatroni, A.; Rossi, M.; Holleczek, T.; Förster, K.; Tröster, G.; Lukowicz, P.; Bannach, D.; Pirkl, G.; Ferscha, A.; et al. Collecting complex activity datasets in highly rich networked sensor environments. In Proceedings of the 2010 Seventh International Conference on Networked Sensing Systems (INSS), Kassel, Germany, 15–18 June 2010; pp. 233–240.
12. Deng, J.; Dong, W.; Socher, R.; Li, L.J.; Li, K.; Fei-Fei, L. Imagenet: A large-scale hierarchical image database. In Proceedings of the 2009 IEEE Conference on Computer Vision and Pattern Recognition, Miami, FL, USA, 20–25 June 2009; pp. 248–255.
13. Smaira, L.; Carreira, J.; Noland, E.; Clancy, E.; Wu, A.; Zisserman, A. A Short Note on the Kinetics-700-2020 Human Action Dataset. *arXiv* **2020**, arXiv:cs.CV/2010.10864.
14. Rey, V.F.; Hevesi, P.; Kovalenko, O.; Lukowicz, P. Let There Be IMU Data: Generating Training Data for Wearable, Motion Sensor Based Activity Recognition from Monocular RGB Videos. In *Adjunct Proceedings of the 2019 ACM International Joint Conference on Pervasive and Ubiquitous Computing and Proceedings of the 2019 ACM International Symposium on Wearable Computers*; Association for Computing Machinery: New York, NY, USA, 2019; pp. 699–708. [CrossRef]

15. Asare, P.; Dickerson, R.F.; Wu, X.; Lach, J.; Stankovic, J.A. BodySim: A multi-domain modeling and simulation framework for body sensor networks research and design. In Proceedings of the 11th ACM Conference on Embedded Networked Sensor Systems, Roma, Italy, 5–11 November 2013; pp. 1–2.
16. Ascher, C.; Kessler, C.; Maier, A.; Crocoll, P.; Trommer, G. New pedestrian trajectory simulator to study innovative yaw angle constraints. In Proceedings of the 23rd International Technical Meeting of The Satellite Division of the Institute of Navigation (ION GNSS 2010), 21–24 September 2010; pp. 504–510.
17. Young, A.D.; Ling, M.J.; Arvind, D.K. IMUSim: A simulation environment for inertial sensing algorithm design and evaluation. In Proceedings of the 10th ACM/IEEE International Conference on Information Processing in Sensor Networks, Chicago, IL, USA, 12–14 April 2011; pp. 199–210.
18. Zampella, F.J.; Jiménez, A.R.; Seco, F.; Prieto, J.C.; Guevara, J.I. Simulation of foot-mounted IMU signals for the evaluation of PDR algorithms. In Proceedings of the 2011 International Conference on Indoor Positioning and Indoor Navigation, Portland, OR, USA, 21–24 September 2011; pp. 1–7.
19. Smith, M.; Moore, T.; Hill, C.; Noakes, C.; Hide, C. Simulation of GNSS/IMU measurements. In Proceedings of the ISPRS International Workshop. Working Group I/5: Theory, Technology and Realities of Inertial/GPS Sensor Orientation, Castelldefels, Spain, 22–23 September 2003; pp. 22–23.
20. Parés, M.; Rosales, J.; Colomina, I. Yet another IMU simulator: Validation and applications. In Proceedings of the Eurocow, Castelldefels, Spain, 30 January–1 February 2008.
21. Cao, Z.; Hidalgo, G.; Simon, T.; Wei, S.E.; Sheikh, Y. OpenPose: Realtime multi-person 2D pose estimation using Part Affinity Fields. *arXiv* **2018**, arXiv:1812.08008.
22. Banos, O.; Calatroni, A.; Damas, M.; Pomares, H.; Rojas, I.; Sagha, H.; del R. Mill'n, J.; Troster, G.; Chavarriaga, R.; Roggen, D. Kinect=IMU? Learning MIMO Signal Mappings to Automatically Translate Activity Recognition Systems across Sensor Modalities. In Proceedings of the 2012 16th International Symposium on Wearable Computers, Newcastle, UK, 18–22 June 2012; pp. 92–99.
23. Kanazawa, A.; Black, M.J.; Jacobs, D.W.; Malik, J. End-to-end recovery of human shape and pose. In Proceedings of the IEEE Conference on Computer Vision and Pattern Recognition, Salt Lake City, UT, USA, 18–22 June 2018; pp. 7122–7131.
24. Elhayek, A.; Kovalenko, O.; Murthy, P.; Malik, J.; Stricker, D. Fully Automatic Multi-person Human Motion Capture for VR Applications. In Proceedings of the International Conference on Virtual Reality and Augmented Reality—EuroVR, London, UK, 22–23 October 2018; pp. 28–47.
25. Mehta, D.; Sridhar, S.; Sotnychenko, O.; Rhodin, H.; Shafiei, M.; Seidel, H.P.; Xu, W.; Casas, D.; Theobalt, C. Vnect: Real-time 3D human pose estimation with a single rgb camera. *ACM Trans. Gr.* **2017**, *36*, 44. [CrossRef]
26. Rogez, G.; Weinzaepfel, P.; Schmid, C. Lcr-net++: Multi-person 2d and 3d pose detection in natural images. *IEEE Trans. Pattern Anal. Mach. Intell.* **2019**, *42*, 1146–1161. [CrossRef] [PubMed]
27. Murthy, P.; Kovalenko, O.; Elhayek, A.; Gava, C.C.; Stricker, D. 3D Human Pose Tracking inside Car using Single RGB Spherical Camera. In *Proceedings of the ACM Chapters Computer Science in Cars Symposium (CSCS)*; ACM: New York, NY, USA, 2017.
28. Omran, M.; Lassner, C.; Pons-Moll, G.; Gehler, P.; Schiele, B. Neural body fitting: Unifying deep learning and model based human pose and shape estimation. In Proceedings of the 2018 international conference on 3D vision (3DV), Verona, Italy, 5–8 September 2018; pp. 484–494.
29. Bogo, F.; Kanazawa, A.; Lassner, C.; Gehler, P.; Romero, J.; Black, M.J. Keep it SMPL: Automatic estimation of 3D human pose and shape from a single image. In *European Conference on Computer Vision*; Springer: Berlin/Heidelberg, Germany, 2016; pp. 561–578.
30. Yao, S.; Zhao, Y.; Shao, H.; Zhang, C.; Zhang, A.; Hu, S.; Liu, D.; Liu, S.; Su, L.; Abdelzaher, T. Sensegan: Enabling deep learning for internet of things with a semi-supervised framework. *Proc. ACM Interact. Mob. Wearable Ubiquitous Technol.* **2018**, *2*, 144. [CrossRef]
31. Li, X.; Luo, J.; Younes, R. ActivityGAN: Generative adversarial networks for data augmentation in sensor-based human activity recognition. In *Proceedings of the Adjunct Proceedings of the 2020 ACM International Joint Conference on Pervasive and Ubiquitous Computing and Proceedings of the 2020 ACM International Symposium on Wearable Computers*; ACM: New York, NY, USA, 2020; pp. 249–254.
32. Radhakrishnan, S. Domain Adaptation of IMU Sensors Using Generative Adversarial Networks. 2020. Available online: https://www.diva-portal.org/smash/record.jsf?pid=diva2%3A1505604&dswid=5801 (accessed on 1 January 2021).
33. Qian, X.; Fu, Y.; Xiang, T.; Wang, W.; Qiu, J.; Wu, Y.; Jiang, Y.G.; Xue, X. Pose-normalized image generation for person re-identification. In Proceedings of the European conference on computer vision (ECCV), Munich, Germany, 8–14 September 2018; pp. 650–667.
34. Sirignano, J.; Spiliopoulos, K. DGM: A deep learning algorithm for solving partial differential equations. *J. Comput. Phys.* **2018**, *375*, 1339–1364. [CrossRef]
35. Raissi, M.; Perdikaris, P.; Karniadakis, G.E. Physics informed deep learning (part i): Data-driven solutions of nonlinear partial differential equations. *arXiv* **2017**, arXiv:1711.10561.
36. Takeda, S.; Okita, T.; Lago, P.; Inoue, S. A Multi-Sensor Setting Activity Recognition Simulation Tool. In *Proceedings of the 2018 ACM International Joint Conference and 2018 International Symposium on Pervasive and Ubiquitous Computing and Wearable Computers*; ACM: New York, NY, USA, 2018; pp. 1444–1448. [CrossRef]

37. Lago, P.; Takeda, S.; Okita, T.; Inoue, S. MEASURed: Evaluating Sensor-Based Activity Recognition Scenarios by Simulating Accelerometer Measures from Motion Capture. In *Human Activity Sensing*; Springer: Berlin, Germany, 2019; pp. 135–149.
38. Kwon, H.; Tong, C.; Haresamudram, H.; Gao, Y.; Abowd, G.D.; Lane, N.D.; Ploetz, T. IMUTube: Automatic extraction of virtual on-body accelerometry from video for human activity recognition. *arXiv* **2020**, arXiv:2006.05675.
39. Radu, V.; Henne, M. Vision2Sensor: Knowledge Transfer Across Sensing Modalities for Human Activity Recognition. *Proc. ACM Interact. Mob. Wearable Ubiquitous Technol.* **2019**, *3*. [CrossRef]
40. Video that was Followed to Produce the Drill Dataset. Available online: https://www.youtube.com/watch?v=R0mMyV5OtcM (accessed on 6 July 2020).
41. Cao, Z.; Simon, T.; Wei, S.E.; Sheikh, Y. Realtime multi-person 2d pose estimation using part affinity fields. In Proceedings of the IEEE Conference on Computer Vision and Pattern Recognition, Honolulu, HI, USA, 21–26 July 2017; pp. 7291–7299.
42. Bai, S.; Kolter, J.Z.; Koltun, V. An empirical evaluation of generic convolutional and recurrent networks for sequence modeling. *arXiv* **2018**, arXiv:1803.01271.
43. He, K.; Zhang, X.; Ren, S.; Sun, J. Deep residual learning for image recognition. In Proceedings of the IEEE Conference on Computer Vision and Pattern Recognition, Las Vegas, NV, USA, 27–30 June 2016; pp. 770–778.
44. Kingma, D.P.; Ba, J. Adam: A Method for Stochastic Optimization. *arXiv* **2014**, arXiv:1412.6980.
45. Raissi, M.; Perdikaris, P.; Karniadakis, G.E. Physics-informed neural networks: A deep learning framework for solving forward and inverse problems involving nonlinear partial differential equations. *J. Comput. Phys.* **2019**, *378*, 686–707. [CrossRef]
46. Video that was Followed to Produce our Seed Motions. Available online: https://www.youtube.com/watch?v=14Cyw7VDsw0 (accessed on 24 March 2021).

MDPI
St. Alban-Anlage 66
4052 Basel
Switzerland
Tel. +41 61 683 77 34
Fax +41 61 302 89 18
www.mdpi.com

Applied Sciences Editorial Office
E-mail: applsci@mdpi.com
www.mdpi.com/journal/applsci

www.ingramcontent.com/pod-product-compliance
Lightning Source LLC
Chambersburg PA
CBHW040825250926
51850CB00007BA/1260